COASTAL HAZARD
MANAGEMENT

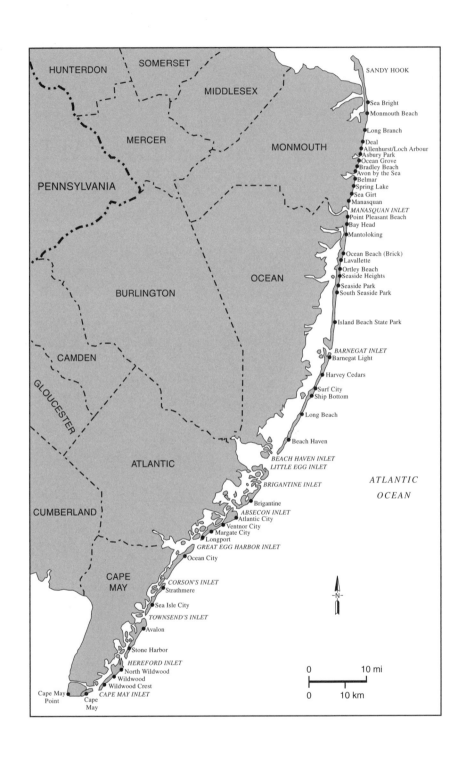

HUNTERDON

SOMERSET

MIDDLESEX

MERCER

MONMOUTH

PENNSYLVANIA

OCEAN

BURLINGTON

CAMDEN

GLOUCESTER

ATLANTIC

CUMBERLAND

CAPE
MAY

SANDY HOOK

Sea Bright
Monmouth Beach
Long Branch
Deal
Allenhurst/Loch Arbour
Asbury Park
Ocean Grove
Bradley Beach
Avon by the Sea
Belmar
Spring Lake
Sea Girt
Manasquan
MANASQUAN INLET
Point Pleasant Beach
Bay Head
Mantoloking

Ocean Beach (Brick)
Lavallette
Ortley Beach
Seaside Heights
Seaside Park
South Seaside Park

Island Beach State Park

BARNEGAT INLET
Barnegat Light

Harvey Cedars

Surf City
Ship Bottom

Long Beach

Beach Haven

BEACH HAVEN INLET
LITTLE EGG INLET

BRIGANTINE INLET

ATLANTIC

OCEAN

Brigantine
ABSECON INLET
Atlantic City
Ventnor City
Margate City
Longport
GREAT EGG HARBOR INLET

Ocean City

CORSON'S INLET
Strathmere

Sea Isle City

TOWNSEND'S INLET

Avalon

Stone Harbor

HEREFORD INLET
North Wildwood
Wildwood
Wildwood Crest
Cape May   *CAPE MAY INLET*
Point
Cape
May

-N-

| 0 | 10 mi |
| 0 | 10 km |

COASTAL HAZARD MANAGEMENT

Lessons and Future Directions from New Jersey

NORBERT P. PSUTY
DOUGLAS D. OFIARA

RUTGERS UNIVERSITY PRESS
New Brunswick, New Jersey

LIBRARY OF CONGRESS CATALOGING-IN-PUBLICATION DATA

Psuty, Norbert P.
    Coastal hazard management : lessons and future directions from New Jersey / Norbert P.
Psuty, Douglas D. Ofiara.
        p.   cm.
Includes bibliographical references (p.   ).
ISBN 0-8135-3150-0 (cloth : alk. paper)
    1. Shore protection—New Jersey.   2. Coastal zone management—New Jersey.   I. Ofiara,
Douglas D.   II. Title.

TC224.N5 P78 2002
333.91'706'09749—dc21   2002020160

British Cataloging-in-Publication information is available from the British Library.

Manufactured in the United States of America

To Syl, my companion in coastal adventures
NPP

To Robin, Jonathan, and Laura
DDO

∼ Contents

~ Preface

During the summer of 1996, researchers at the Rutgers University Institute of Marine and Coastal Sciences completed a significant undertaking: an assessment and update of the primary data and information base applicable to the functioning and management of coastal New Jersey (Psuty et al. 1996). The effort encompassed the disciplines of coastal sciences, coastal engineering, economics, and public policy analysis and resulted in an enormous amount of accumulated and synthesized information. In this book we seek to make this material available and accessible to scientists, politicians, and all other interested parties, including the public.

Central to the evolution of decision making on coastal management is the establishment of the best scientific information. The issues of resource stewardship and allocation of the finite space in the region are multidimensional and present new challenges to economists, coastal scientists, and public officials. The coast's extreme dynamic processes occur irregularly and often partially neutralize or reverse existing shore management efforts. This creates a frustrating management condition that requires improved understanding of the dynamics of the natural and cultural processes that are competing in a restricted coastal zone space. This conflict provokes attempts to shepherd and improve the extant resources, on the one hand, and attempts to build out to the limits of the current planning scheme, on the other. The issues of public involvement, community planning, population caps, zoning, economic development and growth, natural coastal processes, and hazard mitigation are all central to coastal management.

The issues presented here encompass new information on coastal processes, coastal stabilization, relevant economic theory, and hazard management. From this information are derived the fundamental principles required to address the unique challenges facing coastal management. This book seeks to serve not only researchers, individuals, and public decision makers involved in the provision of coastal stabilization, but anyone concerned with coastal New Jersey or similar situations. Concerned citizens can use it to further their understanding of coastal management; practitioners can use it as a handbook to further their understanding of coastal processes, coastal landforms, shoreline stabilization, sea-level rise, economic principles and issues applicable to coastal values, and hazard management implications. The information presented covers local, regional, state, and national perspectives to gain an understanding of the issues and policies involved.

The book contains a number of significant features:

- Its systematic examination of management issues of the coastline combines the areas of coastal sciences and engineering with economics and public policy to achieve a more holistic treatment of public policy design.
- It blends the principles of coastal processes and management with examples from New Jersey to demonstrate the real-world application of techniques and to help improve future approaches to coastal management and public policy.
- It seeks to elevate the discipline of coastal sciences in order to address the new problems that coastal management will face (for example, sea-level rise, changing dynamics of coastal processes, human alterations of coastal resources). Long-range planning can be combined with short-term emergency tools to achieve a more consistent public policy design—that is, research and policy capable of addressing management within a dynamic system.
- The treatment of new information associated with the management of coastal environments and the reduction (or mitigation) of exposure of coastal communities challenges current thinking and causes the coastal public and managers to readdress their views on a variety of dynamic coastal issues.
- It is written for as broad an audience as possible in order to begin a dialogue between the various professions, policy makers, and the public; to bridge the gap in knowledge and communication between coastal scientists, economists, practitioners, and the public; and to foster much-needed collaborative research.

⌒ Acknowledgments

Many people have assisted in the various steps of writing this book. From our early days of gathering information, we express appreciation to the New Jersey Department of Environmental Protection (NJDEP) for its support in our search for new data appropriate to coastal management. Steven Whitney (NJDEP) oversaw the activity and provided helpful guidance. Throughout the data-gathering phase, Mark Mauriello (NJDEP) cheerfully shared his knowledge of the issues and pointed us to sources that were of value to the process.

During the two years of data gathering, public meetings, presentations, and preparation that went into the *Coastal Hazard Management Report* (1996), a small cadre of people worked diligently as a coherent team to meld the many disparate pieces together. Susane Pata was an ardent member who worked tirelessly and creatively on the project; she deserves special recognition for her ability to function under pressure and maintain her pleasant demeanor. Michael DeLuca and Janice McDonnell contributed both to the conduct of our task and to the interpersonal relationships that maintained the harmony of the team and soothed the participants during potentially contentious gatherings. Erica Rohr and Michele Grace were excellent in pulling information together and preparing parts of the report to NJDEP. Also contributing greatly to the final product were Dan Collins, Helen Mattioni, Mike Padula, Michael Siegel, Nam Soo Suk, Marianne Schaffer, George Klein, Jason Tsai, and Greg Martinelli. Michael Bruno of the Stevens Institute of Technology assisted with the information on coastal stabilization, Stewart Farrell of Richard Stockton College of New Jersey shared the recently acquired beach surveys from the New Jersey Beach Profile Network, and Chris Miller of the

U.S. Natural Resource and Conservation Service contributed information on coastal dune plants.

Work on this book started in 1998 as an effort to expand on the great amount of information that had been collected and to share the fruits of the project with a wider audience. At that time, we received support from the New Jersey Sea Grant College Program, which provided assistance for this transformation. As we wrote up our findings for a different audience, the book became more than a repetition of the pieces of information that were submitted to the NJDEP. The intervening period was an opportunity to reflect on the data and their application. It was a time to exchange information with colleagues and with students. It was a time to gain some perspective on the process as well as on the substance of our efforts. We wish to extend a note of appreciation to all the students and faculty who interacted with us as we arranged our thoughts and considered our efforts.

More recently the book took on its own identity, and another cast of participants entered into the mix. Two students at Rutgers University, Jeffrey Pace and Brian Lettini, helped assemble new data and searched through newly found Web sources, bringing a new energy to the project. They worked arduously and without complaint as the final pieces came together. Liliana Costa assisted with the extraction and compilation of data. Their efforts in this production are sincerely appreciated.

Special recognition is extended to Michael Siegel for producing most of the figures in this book. His special skills in cartography are immediately apparent in the quality of the maps and diagrams. Jeffrey Pace and Brian Lettini also contributed to some of the final figures.

We express our sincere appreciation to the members of the Rutgers University Press who worked very closely with us and offered guidance and encouragement throughout the gestation of this book. Their combined experience was very helpful in achieving this final product. We especially acknowledge Elizabeth Gilbert, our copyeditor, for her skill in wordsmithing and massaging our prose.

COASTAL HAZARD
MANAGEMENT

CHAPTER I

Dynamic Coastal Systems:
Illustrations from New Jersey

> A principal goal must be to foster truly long-term mitigation and loss reduction and to avoid burdening future generations with unnecessary hazards.
>
> D. S. Mileti, *Disasters by Design* (1999)

Coastal managers and the citizens of New Jersey share a tremendous concern about the character and quality of the New Jersey shore. Most of the state's coastal zone is developed and classified for either residential or commercial-service land use except for the park and wildlife areas. Few open spaces are left, and the New Jersey coastal zone receives an extremely high level of use by tourists and recreational visitors. In some locations along the shore, the summer population expands by a factor of five to ten or more compared with permanent winter residents. Although the high population density and use during the tourist season contributes to the economic well-being of the coastal zone, the increasing concentration of use and its subsequent impacts have become a source of environmental degradation and have decreased the quality of life in many locations. As a result, many feel that the coast is overdeveloped and that the millions of visitors to the New Jersey shore are exhausting the remnants of the natural character and quality that were once so prevalent. Groins, jetties, rip-rap revetments, and bulkheads (defined in chapter 6) exist nearly everywhere and are another feature of human interaction and interference with shore processes. Beach closings during the summer season have occurred sporadically along the New Jersey coast because of degraded water quality, washups of marine debris, and oil spills. The beach closings in 1987 and 1988 were major news events accompanied by significant declines in beach attendance and accompanying economic losses (Ofiara and Brown 1999).

Long-term interest in the quality of the New Jersey shore and its economic, cultural, and natural resources is evidenced by the creation of the nation's first state commission on coastal erosion, the Engineering Advisory Board on Coastal Erosion of the New Jersey Bureau of Commerce and Navigation, back in 1922. Its first report, entitled *The Erosion and Protection of New Jersey Beaches,* called attention to the problems of narrowing beaches and damaged infrastructure in seaside communities (NJBCN 1922). Likewise, in 1930, the report of the New Jersey Board of Commerce and Navigation emphasized the loss of beach width and recurring damage to the buildings and infrastructure at the shore, and it detailed an array of "hard structures" such as groins and walls that were employed to control coastal erosion (NJBCN 1930). Along most portions of the New Jersey shoreline the same theme of eroding beachfront and damage to property structures can be restated today, despite decades of attempts to "stop the erosion" and "protect the shoreline."

Similar to most shorelines of the world, the New Jersey shore is being slowly eroded through time, and the products of economic development, physical structures, and resources are being threatened. As the earlier reports indicate, this is not a recent revelation; it was also noted in a U.S. Army Corps of Engineers (U.S. ACOE) study (1971). This report classified 82% of the 204 km (127 mile) shoreline of New Jersey as an area of critical shoreline erosion, another 9% as noncritical shoreline erosion, and only 9% as noneroding or stable.

The vitality and resilience of the people in the coastal zone can be seen in their decision to return and to rebuild following destructive events. Investments in housing, commercial ventures, and local infrastructure continue to stimulate development on the spatially constrained buildable areas on the coastal barrier islands. Over time, development has expanded into the beach zone, up to the water's edge, and onto filled land where water or marsh previously existed. Not surprisingly, the quest to move to the water's edge, to transform the barrier islands, and to expand into flood zones (low-lying areas) has been accompanied by an increasing exposure to the natural hazards of the coastal zone—erosion, flooding, and wind and wave damage—and a concomitant interest in securing public protection from these hazards as well as disaster assistance when needed.

Over time, most communities have attempted to "defend a line" at the shore. It may be a building line, a bulkhead line, a dune line, or a shoreline. Whatever it is, there is an outlay of public funds to erect barriers, replace sand, or emplace some sort of structure in defense of the line. However, the costs of maintaining the line can become too much for communities to bear, and they look to the state or federal governments to fund a significant portion of the cost.

Concerns

Serious concerns about the coastal area have been expressed by the general public, coastal residents and visitors, public officials, and researchers. Issues include erosion, coastal dunes, sea-level rise, public safety, beach nourishment, and human interference with coastal processes.

The presence of a wide beach and the existence of some sort of dune system are regarded as "assets" of beach communities that help draw tourists and the spending that contributes to further economic development. Yet there is a general unease among citizens who reside along the shore that the conditions of the natural system are degrading and the beaches and dunes will disappear if left alone. In many areas, there is a grudging acknowledgment that the coastal zone is dynamic, that many of the conditions are hazardous, that it may be overdeveloped, and that it will not be possible to maintain the entire system into the future at the present level.

Consider the following hypothetical scenario regarding coastal processes in New Jersey. Suppose that the shoreline were in some sort of equilibrium, where in some years it eroded and in other years it built back (that is, was depositional), returning to an original position every few years. If so, the concern would be to protect against the bad years (the erosional years), assuming the good years would build back the shoreline. However, the natural conditions along the coastline of New Jersey and many others of the world are characterized by the fact that the shoreline is losing and will continue to lose sediment and recede inland, because the accumulations of the good years do not balance the losses produced during the bad years of severe erosion (Uptegrove et al. 1995 contains evidence of this in recent years for New Jersey). Further, all tidal gauges and monitoring devices in New Jersey show that sea level is rising and drowning the coast, causing the shoreline to be displaced inland even in those years when there is no erosion of the beach sediment.

Dunes are a feature of the coastal zone highly regarded by all. Local ordinances and fines that support dune development and maintenance are fairly elaborate and deserve praise. Coastal dunes offer a certain amount

of protection against natural hazards because of their physical characteristics of mass and elevation, but their presence does not negate erosion.
Dunes have been and can be lost from the effects of erosion, significant
coastal storms, and sea-level rise. Dunes can be more effective barriers to
coastal natural hazards when they are allowed to migrate into a buffer
zone at their inland margin as they are eroded on their seaside margin.
Unfortunately, such a buffer area exists in very few places at the New Jersey shore, and thus dunes have to be considered as short-term to medium-
term approaches to coastal hazard management.

Concern about the safety of those in the coastal zone also exists.
Because of the level of development and the concentration of people in an
area exposed to repeated storms and flooding, public officials and coastal
residents often raise the matter of public safety. Most of the coastal zone
in New Jersey is very low, either naturally or as a result of construction
techniques in developing the land (average height above sea level is about
2 m, or 6–7 ft, on the barrier islands, for example). As a result, the
amount and frequency of damage associated with flooding is increasing.
A continuing rise in sea level exacerbates the situation. Problems also exist
with respect to evacuation across low-lying exit routes and with traffic
congestion. Recognized high-hazard areas frequently suffer damage or
flooding with almost every storm. In some locations the severe erosion
brought by storms repeatedly threatens to undermine dwellings. These
issues raise questions about the exposure of residents and visitors to risk
at the New Jersey shore.

The commitment to occupy coastal space with little or no regard for
or understanding of the dynamic coastal processes in turn creates the need
for shoreline stabilization. Many of the shore communities have initially
relied on engineered structures to solve their erosion problems, with the
exception of the massive beachfill project along most of the coastline of
New Jersey following the famous March 1962 Ash Wednesday northeaster.
Since 1981, however, there has been a shift away from hard structures and
toward the use of "beach nourishment" to counter the effects of chronic
erosion. This placement of new sand on the beach is often eagerly
awaited, reported in local papers, and probably ensures the elections of
local officials. But it is acknowledged that it is a temporary solution available only at a high price. It is expected that the largest share of the cost
will continue to be borne by the state and federal governments as it has
been in the past (historically there has been a 65%–35% split between federal and nonfederal sources, and the state has usually paid about 75% of

the nonfederal share with local communities providing the balance, but a decrease in the federal share in the future is under discussion). The latter half of the 1990s witnessed a major effort to pump large amounts of sand from offshore onto the beaches as a means to counter chronic erosion. Further, the nonfederal share of the cost of shoreline stabilization has been publicly provided in New Jersey, but this allocation has in turn raised the question of whether this is the best use of public funds. With few exceptions, the communities involved would not be able to raise local funds to support the nonfederal share of the cost of beach nourishment programs in their locale.

There is ample evidence of manipulating the coastal system at the local level. In early construction, seawalls, groins, jetties, and revetments were defensive devices to protect a "line" along the shore within coastal communities. There was neither interest in nor concern about the potential effects on neighboring communities. Now that has changed. We have learned that it is not possible to do anything along one portion of the shore without causing effects that cascade downdrift, that is, down the coast. Economists refer to such classic problems as externalities, whereby the actions of an individual (here a community) have subsequent effects on others but are ignored by the individual. The "local approach" has become no longer effective or equitable when externalities are considered. Sediment transport and sediment exchange are occurring in broader regions, often referred to as reaches, or cells, beyond the boundaries of communities. Hence a regional approach should be the basis of coastal and shoreline management. Regional planning could establish targets with appropriate incentives and strategies to achieve specific goals. This could involve specific land-use practices, such as setbacks and buffers, as well as caps on land-use densities. As is all too often found in public policy development, the assistance of state and federal funds is necessary to implement these goals.

Outlook

Two major national policy developments are vital to New Jersey's future role in coastal planning and management. First, the federal role in funding for shoreline stabilization via the Army Corps of Engineers changed in the mid-1990s, when President Bill Clinton did not approve direct funds for shoreline stabilization and redirected the Corps' efforts to more traditional areas for several years. Federal assistance became possible only through congressional mandate. Current discussion on this matter involves changing the cost-sharing formula to reduce or eliminate the federal share in the 50-year maintenance portion of nourishment projects. Consequently the state may have to assume more financial responsibility for any future long-term efforts at shoreline stabilization. Second, a new federal initiative involves the formulation of a National Mitigation Strategy to reduce both the threat and the losses from natural hazards such as coastal erosion and flooding (FEMA 1995a). This national initiative supports using funds to relocate people and structures out of hazardous areas during pre- and post-storm periods. Some of this initiative can be traced back to the Upton-Jones amendments of 1987 to the National Flood Insurance Program (NRC 1990), which sought to encourage the buy-out of property in high-hazard areas.

Hazard mitigation (or "hazard management," in the broader context of this book) is an expensive solution because of the extent of development and the high real estate values at the shore. It will require a dedicated pool of public funds, a consistent public policy, and long-term postdisaster plans to accomplish a reduction in exposure of people and structures. To maintain the existing development and infrastructure at the shore will also require significant, periodic expenditures. The piecemeal approach that the state and individual communities have been using is not adequate to respond to the issues of long-term erosion and sea-level rise. To address these and other matters, decision making must function at the regional level and encompass long-range plans and objectives. The process of developing regional and long-term plans can provide local communities with information and assistance from state and federal levels via appropriate programs that endorse such goals. Researchers recognize that the opportunity to exercise decisions is in the immediate post-storm period, but that plans for those decisions must be in place before that time, or else the opportunity is lost (Olshansky and Kartez 1998; Mileti 1999). In sum, state and federal polices and the availability of shoreline-stabilization funds will influence the future of the New Jersey shore. Furthermore, there is an evolving public policy that is calling for a reduction of exposure in hazardous areas and a reduction in the costs of postdisaster recovery.

What practical, long-range options for coastal management efforts exist for coastal states such as New Jersey? As the effects of a diminishing sediment supply are magnified by rising sea levels, the present coastline of New Jersey is expected to continually erode and drown, allowing the sea to encroach upon coastal communities. In the absence of large subsidies from the federal government or the state to rebuild and defend the present shoreline position, coastal planning should shift toward "managing coastal hazards" rather than strictly "coastal stabilization." More effort should be directed toward increasing public safety on the coast with an eye toward identifying areas of high risk to natural hazards. Hazard management programs can incorporate short-term approaches to the effects of minor storms as well as long-term relocation from high-risk areas.

For much of the coast, the short-term approach associated with coastal dune development and small beach nourishment projects will provide adequate protection against storms of relatively small magnitude. The effects of a slowly diminishing sediment supply and sea-level rise will be masked by the manipulation of the observable shoreline. However, large storms will produce larger displacements of the shoreline that are beyond masking and will require changes in land use or major investments in renourishment to maintain the position of the shoreline. The need for significant investments in shoreline stabilization will escalate in the future because of the greater quantities of sediment required to maintain the shoreline, greater concentrations of people and infrastructure, and higher levels of technology used in development. Such costs are expected to increase in a nonlinear manner as evidenced by the recent trend in damages and losses from catastrophic events (H. J. Heinz Center 2000b; Kunreuther 1998; Ofiara and Psuty 2001).

As the demand for use of the shoreline continues to grow, better information and more creative management strategies are needed to support continued resource use and stewardship. An integrated, coordinated management approach has been used by other coastal states to address shoreline processes that occur on a regional scale and that are more effectively managed at this level. Partnerships that transcend jurisdictional boundaries are desirable and necessary to achieve this aim. A single administrative entity could be developed and charged with the sole responsibility of managing the New Jersey coast and its hazards. It could establish well-defined objectives that would be coordinated through a single office and function in close cooperation with the public and with county and local planning agencies.

Long-term objectives that strive toward increasing the public's safety need to be developed. These objectives involve hazard management strategies and should be determined at the state level, with input from local communities, and then implemented on a regional basis. They should be made consistent with the state's coastal management goals for some future point in time, the year 2050, for example, incorporating expected sea-level rise and a modified coastal zone.

State development of long-term approaches, such as targeting high-risk areas for post-storm changes in land use, is of critical importance because the post-storm period is the only window of opportunity to implement such changes and alter coastal development (Olshansky and Kartez 1998; Mileti 1999). Of special importance are those low-lying areas severely affected by minor storms. Some of these approaches may require new policies, such as enacting zoning ordinances that limit the density in high-hazard areas. It is also imperative to continue to support collection of technical data so that local resource managers have access to accurate, up-to-date information in their decision making. Additionally, there is a need to foster public awareness of the potential dangers associated with coastal settings. Alerting citizens to the risks at the shore promotes the concepts of public safety and recognizes the fiscal limitations on attempts to respond to the effects of very large storms. Findings based on our reassessment (Psuty et al. 1996) emphasize that policies be developed that allow the public to be informed and aware of coastal hazard issues, stress safety, and provide for various incentives to achieve a reduction in the loss from coastal hazards. Hazard mitigation management techniques and certain innovative land-use practices are potentially useful in this respect and, if properly applied, can achieve a reduction in damage. All the stakeholders can potentially gain from such policies, because they foster a return to the functioning of natural coastal systems in dissipating the effects of coastal storms.

CHAPTER 2

New Jersey's Shoreline: Coastal Features and Geomorphological History

The U.S. coastline exhibits a great diversity of shore types, and these variations must be considered when establishing an erosion management program. Differences in the level of development, use, and engineering structures at the shore complicate this natural diversity. Sediment sources and sinks, which are highly susceptible to human activities at the shore and in adjoining rivers and waterways, are also a major concern. . . . As a result of these multiple factors, it is necessary to consider both local conditions and broad regional issues.

National Research Council, *Managing Coastal Erosion* (1990)

New Jersey is a laboratory for the extreme diversity of dynamic coastal processes and geomorphological features found within a coastal zone. Associated with the 203 km (127 miles) of shoreline are barrier islands and spits of varying shapes and sizes, low cliffs, park landscapes with extensive coastal dunes, inlets and adjoining bays and estuaries, small islands, and a broad array of wetlands. Likewise, it is a laboratory for the large human population that occupies the coastal zone and attempts to control the natural variation through the use of hardened structures, sand redistribution, land-use regulation, and mitigation measures. Problems, issues, approaches, and conflicts are ever present. Opportunities exist to reduce the exposure of the public and the associated infrastructure to highly uncertain future events, but all of the current options are both difficult and costly. Given this uncertain, dynamic environment, the best available knowledge and information will be required to pursue appropriate, region-specific short-term, and long-term strategies for reducing future losses.

The multitude of coastal processes, features, and issues are not unique to New Jersey, but are characteristic of many coastal zones with barrier islands, communities dependent on recreation and tourism, and concerns for the management of diminishing shorelines. The factors that influence the shoreline (including waves, tides, water levels, and beach profile) and the new threats that face coastal areas (including diminishing sediment budget, sea-level rise, shepherding of biotic resources) are basic interactions of the natural and cultural processes of concern to all parties interested in improved stewardship of this zone. The situation in New Jersey is further exacerbated by a high population density in the state and in neighboring areas. Yet the characteristics of the state's coastal zone provide examples of problems, concerns, and opportunities common to coastlines elsewhere. It is from this perspective that a description of the geological and geomorphological features of the coastal zone of New Jersey is presented.

The coastal geomorphological history of the New Jersey shoreline is discussed in terms of specific partitions that are referred to as "reaches" by coastal engineers, practitioners, and students of natural coastal divisions. Such an approach can be considered a microlevel or smallest functioning natural system for management as opposed to a macrolevel or broader regional approach. A microlevel approach can identify unique features of the shoreline that would be brushed over in a broader treatment, and it reemphasizes the point that management scenarios and policy approaches should mirror the diversity of the coastline and its inherent dynamic processes. A description of the physical features of the shoreline and their importance in the coastal environment permit an understanding of how and why the shoreline changes over time. Management of the coastal zone then becomes management of the exposure of human occupants and their infrastructure to the dynamic conditions that continually modify natural hazards of the system. Understanding the basis for the changes and their longer-term direction can provide a foundation for the application of management tools and policies to enhance public safety and to encourage appropriate stewardship of the natural and cultural resources of the coast.

Background

The coastal zone may be thought of as that area where the interaction of waves, currents, and winds upon sediments creates and changes features such as barrier islands, spits, inlets, estuaries, beaches, and dunes. People are also situated on this coastal landscape and use these resources. The coast undergoes constant change, either accumulating or losing sand both within the system (as measured by a "sediment budget" methodology) and across the boundaries of the system. Portions of the coast are migrating, are being modified by cultural processes, and are being subjected to both conservation and exploitation. As we gather more information about the coast, we begin to understand the conditions that occur today and that have occurred in the past. The future, in contrast, is not so clear. Trends can be interpreted from past conditions and forecasts of events can be derived from the historical record, but predictions of the future will continue to have a measure of uncertainty about them. We can list the past storms of the century, apply probabilities to their future occurrence, and be reasonably certain that there will be significant storms in the future, but it is nearly impossible to predict the "big storm." Yet it will happen. We can also determine that the sea level rose in the past century, and we can measure its rate of change. In fact, nearly all scientists predict a higher rate of rise in the twenty-first century, although estimates vary about the specific rate. Whether the major influence determining the changing environmental conditions is coastal storms or sea-level rise, it is not so critical to assign the relative impact as it is to understand that the conditions are changing and will continue to change into the future. In bringing together the available information on the conditions that exist at the shore, we seek to understand its dimensions, impacts, and dynamics so that informed decisions can be made both now and in the years to come.

Development of Information

A number of characteristics of the coastal zone are routinely measured. Information gathered includes wave and tidal data that make it possible to chart the principal processes that drive changes at the coast. Other data sets pertain to beaches and dunes as well as engineering structures and other aspects of human development. Agencies and organizations have been assembling these data and creating large data banks of information that together form the basis of a regional geography.

Wave Gauges: Measurement of Wave Action

Although there are, and have been, several wave gauges in operation along the coast of New Jersey, the information is scattered both temporally and spatially. More data (both spatial and time series) are needed to determine the conditions that occur during storms of varying direction, duration, and intensity. Some information is available in the form of a 20-year record of simulated wave data as part of a regional wave information study (WIS) produced by the U.S. Army Corps of Engineers (Jensen 1983). Other wave measurements are available through the National Oceanic Atmospheric Administration's National Ocean Data Center, which had a station at Ambrose Light (no. ALSN6) that collected nondirectional actual wave data from December 1989 to 1997. Additional nondirectional and directional wave data have been collected since April 1991 from a moored buoy (no. F291) located approximately 64 km (40 miles) off Long Branch, New Jersey. A more time-limited wave data set has been collected at the Long-Term Environmental Observation Site (LEO-15) located about 5 km (3 miles) from Little Egg Inlet and operated by Rutgers University. Directional wave measurements at this site have been recorded sporadically from October 1991 to the present. The time periods of record vary from weeks to months.

Wave Information Studies (WIS)

The Army Corps of Engineers wave information studies (WIS) data set (Jensen 1983) was produced by simulating the weather systems over a 20-year period and allowing the wind to blow over the water to generate waves. These waves in turn travel onshore, and associated information on wave heights, lengths, periods, and direction is determined by the physical relationship of the winds moving across the water surface and the transfer of energy to the waves. This methodology of using wind information to back-calculate the dimensions of the waves associated with a storm or wind is known as wave "hindcasting." The model produces what is referred to as the "significant wave," the average dimension of the largest one-third of the generated waves at a site.

In general, waves that are created out at sea and then move out of a storm area to traverse the ocean conserve most of their energy as they pass through the water. They are referred to as *deep-water waves,* because their progress is not impeded by the ocean bottom. As they approach land masses, however, they begin to enter shallower water and the orbital motion of the water particles in waves begins to interact with the bottom, causing most of the wave dimensions and characteristics to change as waves come onshore. Specifically, wave crests begin to bend or refract, resulting in waves that adjust their directional approach to the coast. The WIS data set begins by generating deep-water, or Phase I, waves. As the waves come into shallower water and begin to bend or refract, they

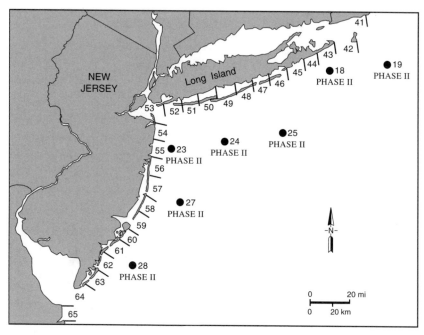

Figure 2.1. Distribution of wave information stations along coastal New Jersey and adjacent areas. Phase II stations are deep-water locations in the computer simulation program. The inshore locations are identified as Phase III stations. (Adapted from Jensen 1983.)

become Phase II waves. In still shallower water, waves enter depths where they interact with the bottom to mobilize available sediment and continue to refract, becoming Phase III waves in about 10 m (33 ft) of water. Inland from this depth, the orientation of the wave crest becomes more parallel to the shoreline and is increasingly affected by the bottom topography. The WIS simulation seeks to develop a wave climatology for segments of the coast based upon a 20-year record of weather systems and their resulting waves. Given the simulations, the nearshore wave climatology consists of a distribution of incoming waves that have both direction and magnitude averaged for the 20-year period. Wave magnitude is important, because it determines the quantity of sediment that can be moved about. Wave direction is important, because it determines whether the flows and transfers will be to the north or to the south and what the net transport may be.

The simulated Phase III wave data sets are associated with specific areas or "compartments" along the shore approximately 16–20 km (10–12 miles) in length (figure 2.1). Each section contains information on the exposure of that section of the coast to wave buildup from offshore direc-

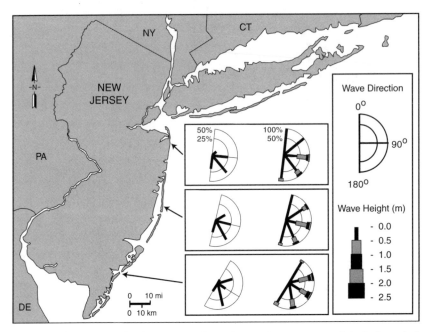

Figure 2.2. Wave roses for three wave information stations: 54, 57, and 61. Left symbols depict the quadrants of oncoming waves, by percent. Right symbols depict the percentage of waves of differing heights. (Jensen 1983.)

tions. An important component of the WIS data sets is the direction of the incoming waves relative to the shoreline. Differing wave climatologies are shown in figure 2.2, in which three wave roses reveal the contrasts of conditions along the relatively short distance of the New Jersey shoreline. (Each rose is composed of a fan-like pattern of measurements of the relative distribution of frequencies and magnitudes of approaching waves at a designated site.) These data depict the shielding effect of Long Island and New England in restricting the direct approach of large waves from the northeast. Waves approaching from storm centers off to the northeast must bend or refract around New England so that their general approach will be from the east or east-southeast by the time they reach the New Jersey coast. As a result, nearly all waves reaching the Atlantic coast in Monmouth County arrive from the east or southeast and produce beach sediment transport to the north (figure 2.2). In Ocean County, by contrast, the protective effect of New England is reduced and more waves arrive from the northeast (figure 2.2). Although the calculations for the Ocean County compartments show that most of the waves are from the southeast, the larger waves are from the east-northeast direction. An

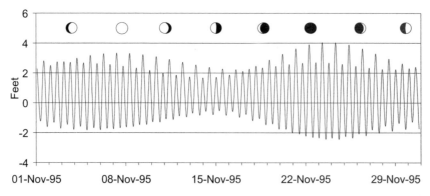

Figure 2.3. Variation in tidal water level through the monthly moon phases, November 1995, Atlantic City, N.J. New moon indicated by black symbol; full moon by white symbol.

application of the wave dimensions produced by the WIS program shows that a southerly drift becomes more pronounced toward the southern margin of Ocean County. For the northern half of Ocean County, however, a considerable portion of waves are directly out of the east, indicating that nearshore and longshore drift may be determined primarily by the local topography within the inshore zone, or that a large proportion of the exchange of sediment is onshore/offshore (perpendicular to the coast) rather than alongshore (parallel to the coast). The protective effect of New England is absent for Atlantic and Cape May counties, and the presence of larger waves out of the east-northeast causes net sediment transport to the south (figure 2.2).

Although the WIS data are excellent for generating the regional variations in wave climatology and potential drift directions, they are not a detailed description of actual wave data. In addition, the data set represents average conditions over a long time period and does not include the 1962 storm conditions; thus it smoothes the individual variation present in storm and wave data.

Tidal Gauges: Measurements of Water Levels

The tides along the coast of New Jersey are semidiurnal, that is, two high tides and two low tides occur each day. In the example of the tidal variation of the water level through one month at Atlantic City, seen in figure 2.3, the trace of the twice-daily oscillation is very apparent. There is also an association with the phases of the moon—actually, it is an association among the earth, moon, and sun. During new moon and full moon, when the earth, sun, and moon are in alignment, the combined gravitational pull maximizes high- and low-tide water levels, increasing the tidal range. Between new and full moon, the planetary arrangement minimizes the water levels and the range is reduced. Thus there is a general variation of two maximum tidal ranges and two maximum high tides per month. These are referred to as spring tides. The reduced tidal ranges are referred to as neap tides. Because tides are based on the gravitational attraction of the sun and moon on the earth's oceans as they pass through their established orbits in a regular manner, it is possible to predict the water levels that will be caused by the interaction of these planetary bodies. The National Ocean Service (NOS) of the National Oceanic and Atmospheric Administration (NOAA) produces annual tide tables for the entire United States based on these celestial relationships. In northern New Jersey, tides at Sandy Hook average 1.42 m (4.66 ft), with a range of over 2.13 m (7.0 ft) during maximum spring tide and only about 0.91 m (3.0 ft) during minimum neap tide. In the southern part of the state, Atlantic City has an average tidal range of 1.25 m (4.1 ft), increasing to about 1.98 m (6.5 ft) during spring tides.

Actual water levels may vary from the predicted level because of wind or storms, which may cause water to accumulate at or be removed from a site. Recently the NOS has begun to make available near real-time water levels recorded at its stations on its Web site *(http://tidesonline.nos.noaa. gov/)*. With the combination of the predicted and actual water-level data, it is possible to determine the effects of winds and storms on predicted water levels and to derive a measure of the accompanying storm's surge (figure 2.4). The trace of the water level at Atlantic City during August 6–8, 1985, is affected by the presence of Hurricane Gloria, which elevated the water at the coast through a combination of storm waves and the low atmospheric pressure in the vicinity of its center. A comparison of the recorded and predicted water level will give a measure of the difference between the two. That difference is the product of the storm, or the storm surge. Figure 2.4 amply demonstrates the rapid rise and fall of the water level, or storm surge, as Hurricane Gloria passed through the region. Storms are part of the dynamics of the coast. Their records are present in the wave data as well as in the tidal data.

Beaches and Coasts

Beaches and coasts are the landforms that are created as the ocean waves and currents mold the margins of the continents. Beaches

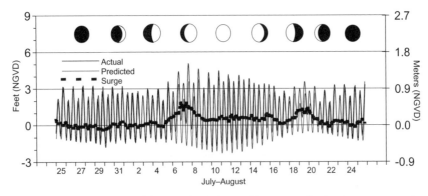

Figure 2.4. Variation in tidal water level during Hurricane Gloria, August 6–8, 1985. Moon phases are depicted (see figure 2.3). The storm surge is the difference between the predicted and the actual water levels. (National Ocean Service website: http://tidesonline.nos.noaa.gov/.)

differ from coasts primarily in terms of scale. The beach is a mass of sand that exists above and below sea level (or near the waterline) in constant interaction with the waves, currents, and wind that move sediment around. The beach is often referred to as a sand-sharing system that extends from the coastal dune out to the offshore bar. This means that sand moves between the beach, dunes, and offshore zones, usually building up one aspect at the expense of another. A coast may be thought of as a broader feature than the shoreline. It is the zone that extends along the shoreline, and it is also considered to be farther inland as well as farther seaward. For example, one often refers to the coastal counties of New Jersey or to the southern coast of the state. The coast includes the beach zone (beach, dune, and offshore bars), but it also includes the barrier islands, the inlets, the wetlands, the estuary, the inland margin of the bay, the bluffs, and anything else that is near the ocean boundary of the particular land mass.

Coastal Processes and the Beach Profile

In areas where adequate sediment is available, the classical beach-dune profile can develop, in harmony with the processes that mobilize sand and shift the sand from one portion of the profile to another (figure 2.5). The complete beach-dune profile extends from the position of the offshore bar through the beach and into the coastal foredune. Waves, currents, and wind interact with the components of the profile to mobilize, transport, and deposit sediments. If the profile receives more sedi-

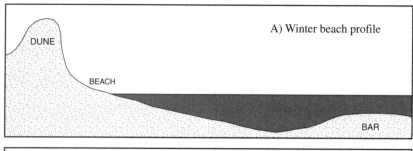

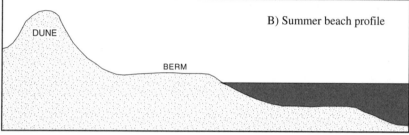

Figure 2.5. Sand-sharing system of dune/beach/offshore bar. (A) Winter profile of narrow, steep beach and well-developed offshore bar. (B) Summer profile of a broad flat berm, flatter offshore.

ment than is lost, the result is accumulation, referred to as accretion, and the profile is displaced seaward. However, if more material is removed from the total profile than is received, the result is erosion and the profile is displaced inland. Because the profile is a product of the processes that are operating on it, the features and forms of the profile are usually retained as the profile either advances seaward or retreats inland. Erosion or accretion is seen as a seaward or landward shift of the profile. An understanding of this basic process-response association can help in locating commercial and residential development in a way that reduces the risk of damage from future coastal storms.

As waves approach the coast, they begin to interact with the beach profile in the offshore zone, because they extend vertically downward through the water column and introduce motion where the water is in contact with the sediment. Generally the initiation of sediment motion on the profile caused by surface waves is considered to occur at a depth of about 10 m (in the range of 30–35 ft). Inland from that depth, waves have a capacity to interact with the bottom and set sediment in motion. The amount of sediment mobilized increases as the waves enter shallower water. As waves come onshore, they may break on an offshore bar or may break directly on the beach if no bar is present. If waves break on an off-shore bar, or on multiple bars, a new and smaller wave will be generated in shallower water, and eventually a wave will break on the beach. The breaking waves cause loose, sandy sediment to be quickly mixed and set in motion, which in turn may settle quickly or may be further mobilized in nearshore currents and transported. The magnitude of this action is related to the energy expended by the breaking wave, and to the size and weight of the loose sediment.

Low, long waves tend to have subdued breakers, and the level of agitation on the bottom is relatively low. In this case, sand tends to move up the beach profile and accumulate on the beach, both building it up and extending it. Much of this sediment is derived from sand that originated from the offshore bar. With larger, steeper waves, there are usually areas of sediment agitation in the vicinities of the bars and at the beach face. Much of this sediment is transferred from the beach to the bars, adding to bar development. Sediment also moves from the beach and bars to even greater offshore depths, however, and if it is transferred to depths greater than 10 m (30–35 ft), it is usually lost to the system, because the calm-water surface waves are unable to remobilize the sands from these depths and return them to the beach. Transfers of sand to the deep-water zone represent a net loss to the beach-dune system and consequently an erosional condition.

Following periods of high-energy waves, the lower wave conditions again produce onshore transport. Manifestations of this process can be seen at an eroded beach in the post-storm recovery period. In the days following a storm, sand will begin to accumulate very low in the beach profile, at the water's edge. In a few more days, the sand will fill the low portion of the profile with a broad low surface and slowly build higher and higher. This is referred to as the sand-sharing system between the beach and the offshore portion of the beach profile. The rate and amount of recovery is related to the loss of sediment to deep water, to downdrift beaches, and to input from updrift beaches. If the total exchange is balanced, the beach profile returns to an original configuration. If the total exchange results in a negative balance, there will be some loss in the profile, which will be represented as a net inland displacement of the profile and the shoreline. Transfers of sediment that stay within the sand-sharing system are not lost. They are temporarily rearranged but can be continually redistributed to rebuild the profile.

Dunes and offshore bars are important components of the beach profile and the sand-sharing system. They represent sand in storage: dunes store sand above water, and offshore bars store sand below water. Where there is a shortage of sand, it is likely that dunes will be relatively small if they are present at all, because there is not much sand in storage above the beach. In areas of sand shortage there will also be a lack of offshore bars. This absence is noted in front of structures such as seawalls, in which the offshore slope continues to deepen seaward without any evidence of sand storage in the form of offshore bars. This means that no sand is available to return and rebuild the beaches after storms. The same condition applies to sand placed in front of seawalls. The slope remains steep, no offshore bars develop, and hence no sand returns after storm events.

Coastal Processes and the
Coastline of New Jersey

The coastal zone is a highly diverse component of the state. It retains aspects of the natural landscape in parts of Sandy Hook, Island Beach State Park, Long Beach Island, Brigantine, Sea Isle City, Avalon, Cape May Meadows, and other locations, but also contains the extensive development of Atlantic City, with its gaming casinos and other high-rise buildings. The zone further displays the relatively wide sand beaches of the Wildwoods in contrast to the much narrower and, at times, diminished beaches in most other portions of the state. Variety and diversity are important characteristics of the coast and provide a wide range of experiences for the citizens of New Jersey and visitors from other states.

If the shoreline were not eroding, the concerns for the shore would relate chiefly to the types of land use and the opportunities available to meet the needs of the citizens. If storms did not threaten the safety of the citizens and cause damage to the infrastructure, community attention would be directed toward managing the resources within the political unit. Long-term erosion, however, is displacing the shoreline and shifting the beaches and dunes inland. In addition, the short-term events such as storms, storm surges, flooding, and wave attack provide dramatic and immediate modification of the attributes of the zone. These long-term and short-term natural processes are responsible for the dynamic nature of the system. Because the zone is replete with many static cultural phenomena, its variable natural processes pose a constant hazard to the lives of coastal residents, to investments in the form of houses, buildings, and general infrastructure, and to many forms of livelihood.

Coastal Geomorphological History

The coastal features of the New Jersey shore are the products of events that began several millennia ago. What is observed now are only the latest forms that have been developing as the sea level has risen, inundating the ancestral coastal zone and as sediment has been redistributed. Conditions have been altered as barrier islands and estuarine habitats have waxed and waned as a result of the wind, waves, tides, and currents that have transported and rearranged sediments along the coast.

The level of the sea was on the order of 135 m (450 ft) lower than today about 20,000 years ago, during the last major stage of glacial ice accumulation in the Pleistocene period. The world's sea level was lower because a great quantity of water was locked up on the continents in the form of large glacial ice masses. At this time, the shoreline of New Jersey was about 160 km (100 miles) seaward from its present position. As the world's temperature rose and the glacial ice began to melt, water returned to the ocean and the sea level began to rise and submerge the margin of the continent and encroach upon the exposed continental shelf. The rate of sea-level rise was not uniform, and probably there were times when the level dropped as small glacial re-advances occurred. By about 7500 years ago, however, the sea was roughly 13–15 m (45–50 ft) below its present level and the shoreline may have been only a few kilometers offshore of its present-day position. During this recent period the rate of sea-level rise was slowing, in comparison with the large change in sea level in the earlier period; the remaining 10% of the rise has occurred in the last 7500 years. An excellent discussion of the characteristics of the continental shelf is found in Teal and Teal (1969) who explore the mobility of sediments during the drowning of the shelf off of New England and the creation of several of the offshore productive fishing grounds such as Georges Bank and Stellwagen Bank.

The sea level continued to rise at a diminishing rate until about 2500 years ago, when it slowed markedly (figure 2.6). At this time, sea level was about 2 m (6–7 ft) lower than today and the shoreline was in the general vicinity of its present location. In the northern portion of the state it may have been somewhat seaward, whereas in the southern part of the state it may have been landward. It is likely that the barrier island shoreline of 2500 years ago was a modest, low, narrow sand ribbon that was frequently overwashed and was extremely mobile. During the rapid sea-level rise of the previous millennia, beaches and dunes were probably poorly devel-

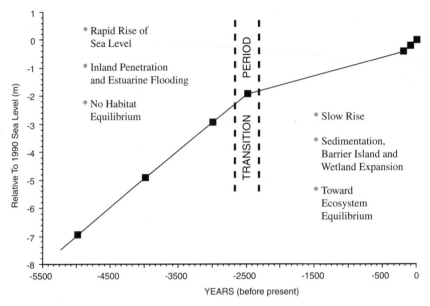

Figure 2.6. Generalized sea-level rise curve at coastal New Jersey, incorporating a major rate change. Most of the coastal features developed after the slowing of sea-level rise that occurred about 2500 years ago. (After Psuty 1992.)

oped and resembled the condition presently found on the northern parts of Brigantine Island (figure 2.7). As the encroaching sea continued to inundate the fringe of the continental shelf, the beach continually retreated inland (figure 2.8, panel A).

With the slowing and cessation of sea-level rise about 2500 years ago, the rate of accumulation of sand at the shoreline was sufficient so that beaches began to build out and the barrier islands began to increase in width and height (figure 2.8, panel B). It is likely that washover became less frequent and the barrier islands either ceased migrating inland or reduced their rate of inland migration. The sand that accumulated and led to the enlargement of the barrier islands originated from the offshore zone, incorporating sediment that had been previously inundated as the sea was quickly rising in the past. Now, because of a relatively stable sea-level position, there existed an opportunity for waves and currents to transfer the sediment to the shore and augment the islands. In addition, sediment began to fill the bays at this time and change them from an open-water habitat to one that was initially composed of a marsh fringe and later of large tidal flats and marsh expanses that extended from the margins of the bays into the diminishing open-water habitat. It is impor-

Figure 2.7. Narrow beach, low dune, breached by overwash: barrier is migrating inland. Old marsh exposed in beach as sand shifts inland. Northern Brigantine Island, N.J.

tant to note that the factors responsible for shoreline growth and wetland development were the relative reduction in sea-level rise and the availability of sediment. The slow rate of rise was now accompanied by a rate of sediment delivery that was greater than that of erosion, and as a result the barrier islands began storing sand and building seaward. The bays

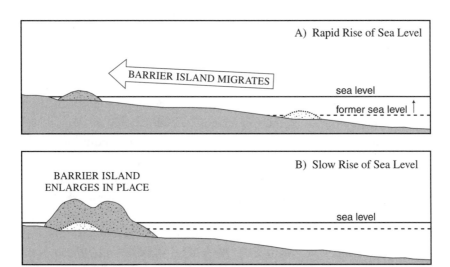

Figure 2.8. Stages of barrier island development associated with sea-level rise. (A) During rapid rise of sea level, barrier was low and narrow. (B) During recent slow rise of sea level, barrier island widened and became higher, and coastal foredune system enlarged.

Figure 2.9. Infilling of the bay between the barrier island and the mainland. Strathmere, N.J.

were also storing sediment and building extensive wetlands where open water had existed previously (figure 2.9).

However, the sand supply in the immediate offshore zone, that area previously inundated by the rising sea, was finite. Once the sediment that was available in this shallow zone was exhausted, the process of accumulation ended and a transition from stability to that of slow loss began. This transition took place at varying rates and depended on whether other sources of sediment replaced that from the offshore zone. Some of the barrier islands had accumulated large masses of sand, and it is likely that sediment from one group of islands helped to nourish some downdrift islands. In sum, the primary source of sediment responsible for the height and width of the islands was exhausted and the situation had changed to one of slow loss of sediment as conditions now transferred sediment away from the New Jersey shoreline in both alongshore and offshore directions.

This general reversal probably began about 500 years ago, although in some areas it was later. At present, there is an overall reduction in available sediment in the coastal zone. As a result, the barrier islands are slowly losing sediment and becoming smaller. In some cases, this loss is observed as erosion of the beaches and dunes and as an inland displacement of the beach profile. In other cases, we see the remains of a low, narrow sand sheet that is displaced inland as washover events are repeated. Washover is a common event in the very early stages of barrier island development and also in the very late stages of barrier island attenuation (figure 2.8). Barrier islands migrate inland during these two stages, when sediment supply is limited and the islands have little height or width. Other examples of the loss are evident in the disappearance both of small islands near the ends of the barrier islands and of the locations of "named" streets and properties on old maps that depict land seaward of today's shoreline. It remains true that many of the New Jersey barrier islands still have great quantities of sediment, washover is uncommon, and island displacement does not occur. But some locations where washovers are frequent, such as at the southern margin of the barrier island at Holgate, Whale Beach in Sea Isle City, and northern Brigantine Island, are symptomatic of barrier island attenuation and the process by which sediment is transferred from the oceanside to the bayside of the island. It is the same process that was dominant when low, narrow islands migrated inland during the time of rapid sea-level rise (20,000 years to 2500 years ago). It is the same process that occurs during major storms, such as the March 1962 Ash Wednesday storm when storm surge spilled across the developed and lowered portions of the barrier islands.

The sequence of New Jersey's barrier islands and their configuration is similar to the coastal barrier island configuration repeated several times along the eastern coast of the United States (Fisher 1967). It is an association of a series of barrier islands that extend from the mainland a short distance in one direction and a great distance in the opposite direction (frontispiece). The sequence for New Jersey begins with the central section of Monmouth County that is currently devoid of barrier islands. In one direction there is a short spit, Sandy Hook, extending alongshore from the mainland. In the other direction, there are several long narrow barrier islands (including Island Beach spit, Long Beach Island, and a relatively open Barnegat Bay), leading to a series of shorter, drumstick-shaped barrier islands (such as the southern section of the state's barrier islands). The numerous active inlets in the south are accompanied by relatively extensive wetlands in the back bays. The end product of long-term natural development is a shoreline that incorporates considerable spatial variability. The coastal processes that are responsible for shoreline development, however, have not stopped. Waves, currents, and wind act upon the existing shoreline and continue to modify its remaining features.

There is also the imprint of cultural development in this zone, and it too is modifying the processes and forms that are created. Through time, both the natural and the human forms will evolve and change.

New Jersey's Coastal Characteristics and Delineation

Although there are few in-depth studies of the wide range of geomorphologies present in coastal New Jersey, a number of general reports and publications assist in understanding the broader systems of barrier island development and historical evolution of the coast, such as those by Nordstrom et al. (1977, 1986). There are also a series of comprehensive studies completed by the U.S. Army Corps of Engineers on a wide range of data based on a micro-approach to the shoreline that provide detailed descriptions of each of the segments of the coast. The *Limited Reconnaissance Report* (see U.S. ACOE 1990 for an introduction to this effort) provides an overview. Other studies have been completed or are in progress for various segments, including Sandy Hook to Manasquan, Manasquan Inlet to Barnegat Inlet, Barnegat Inlet to Little Egg, Brigantine Inlet to Great Egg Harbor Inlet, Great Egg Harbor Inlet to Townsend's Inlet, Townsend's Inlet to Cape May Inlet, and Lower Cape May Meadows to Cape May Point (U.S. ACOE 1996). Each of these projects has produced a *Reconnaissance Report* that includes a bibliography and a tabulation of information.

The state of New Jersey has also generated significant data regarding shoreline change. The shoreline mapping project is part of a N.J. Department of Environmental Protection (NJDEP) initiative to develop and produce a Geographic Information System (GIS) for the collection, storage, retrieval, and analysis of spatial coastal data. It is now possible to determine the past trend of shoreline migration over a 150-year period (NJDEP 1997). In another project, begun in 1986, beach profiles have been annually surveyed at approximately one-mile intervals along the coast as part of the New Jersey Beach Profile Network (NJBPN) (see Farrell et al. 1995). The study area extends from the dune to 15 to 20 feet below the water. The accumulation of these profile data is now beginning to allow more general trends to be distinguished from year-to-year variations. The New Jersey Geological Survey has extended the information derived from these profiles and has produced a comprehensive report on the volumetric changes of sand along the coast (Uptegrove et al. 1995).

Geomorphological Regions

Coastal New Jersey is composed of five broad geomorphological units (Nordstrom et al. 1977). They are the product of coastal processes operating on the continental margin to produce a distinct set of coastal landforms and associated habitats that incorporate the quantity and direction of sediment transport. These areas constitute the fundamental divisions of coastal types from Sandy Hook to Cape May Point:

- Northern Barrier Spit—16 km (10 miles) in length
- Northern Headlands—30 km (19 miles) in length
- Northern Barrier Islands Complex—67 km (42 miles) in length
- Southern Barrier Islands Complex—80 km (50 miles) in length
- Southern Headlands—5 km (3 miles) in length

Within these geomorphological areas exist 13 spatial oceanfront compartments, or reaches (figure 2.10). Reaches are defined as coastal regions where a particular set of coastal processes acting upon available sediment produces the natural physical characteristics of the area. The combination of processes and resulting forms describes each reach. In many cases, inlets define the boundaries and divide the coast into distinct sediment transport cells (littoral cells).

Reaches contain an assortment of geomorphological associations and definitive erosion and deposition patterns. The *New Jersey Shore Protection Master Plan (NJSPMP)* included maps of erosion rates by category, ranging from I through IV to describe a predominant erosion or accretion trend, for particular areas within reaches (NJDEP 1981):

- Category I—Critical erosion
- Category II—Significant erosion
- Category III—Moderate erosion
- Category IV—Non-eroding

The boundaries of the reaches are not entirely closed systems that isolate portions of the coast from one another. For example, inlets are not absolute boundaries; sediments can pass across inlets. Inlets are dynamic portions of the coast that interact with the reaches on either side. Inlets with jetties have both positive and negative effects on sand budgets updrift and downdrift. However, inlets without jetties also store and release sediments and interact with the adjacent shores. Inlets, then, are convenient boundaries and represent a change of processes at a micro-

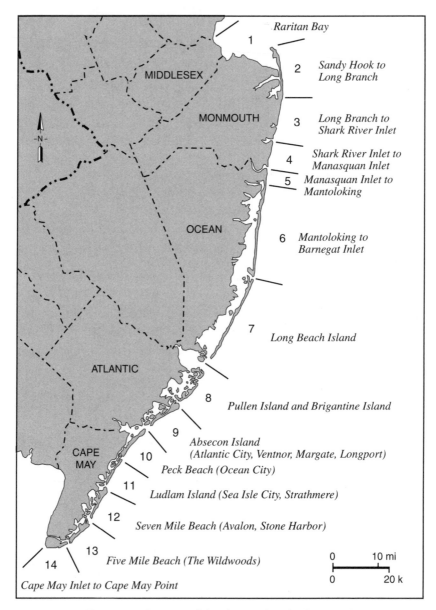

1   Raritan Bay

2   Sandy Hook to
    Long Branch

3   Long Branch to
    Shark River Inlet

4   Shark River Inlet to
    Manasquan Inlet

5   Manasquan Inlet to
    Mantoloking

6   Mantoloking to
    Barnegat Inlet

7   Long Beach Island

8   Pullen Island and Brigantine Island

9   Absecon Island
    (Atlantic City, Ventnor, Margate, Longport)

10 Peck Beach (Ocean City)

11 Ludlam Island (Sea Isle City, Strathmere)

12 Seven Mile Beach (Avalon, Stone Harbor)

13 Five Mile Beach (The Wildwoods)

14 Cape May Inlet to Cape May Point

MIDDLESEX

MONMOUTH

OCEAN

ATLANTIC

CAPE
MAY

-N-

0   10 mi

0   20 k

Figure 2.10. Location of shoreline reaches. (NJDEP 1981.)

level, whereby tidal flows are introduced as a mechanism for sediment transfers in addition to the waves and wave-induced currents.

The following list of New Jersey reaches is derived from the *NJSPMP* (NJDEP 1981). Reaches 1 (Raritan Bay to Sandy Hook), 15 (Cape May Point to Penns Landing), and 16 (Penns Landing to Trenton) are identified as bay and river reaches in the *NJSPMP* and are not included in the Atlantic shoreline reassessment (Psuty et al. 1996); they are also omitted here. Thus 13 reaches are discussed:

2. Sandy Hook to Long Branch
3. Long Branch to Shark River Inlet
4. Shark River Inlet to Manasquan Inlet
5. Manasquan Inlet to the Borough of Mantoloking
6. Mantoloking Borough to Barnegat Inlet
7. Barnegat Inlet to Little Egg Inlet (Long Beach Island)
8. Little Egg Inlet to Absecon Inlet (Pullen Island & Brigantine Island)
9. Absecon Inlet to Great Egg Harbor Inlet (Absecon Island)
10. Great Egg Harbor Inlet to Corsons Inlet (Peck Beach)
11. Corsons Inlet to Townsends Inlet (Ludlam Island)
12. Townsends Inlet to Hereford Inlet (Seven Mile Beach)
13. Hereford Inlet to Cape May Inlet (Five Mile Beach)
14. Cape May Inlet to Cape May Point

The Reach Concept in Management

In the context of coastal zone management, the reach concept is a means to organize and integrate the physical processes, the sediment transport systems, and the geomorphology. It remains relatively simplistic, and conditions develop that blur the boundaries of the reaches and create more or fewer reaches. Although the absolute number of reaches is not of great importance, it is meaningful to approach the coastal zone as a series of segments that respond to stimuli in some holistic manner instead of as a sequence of political boundaries that artificially interrupt sediment-transport cells. Because particular magnitudes and spatial associations of coastal processes are specific to reaches and not jurisdictional regions, management options for New Jersey's coastal region would benefit if based on a "reach" approach. The reach concept further reduces the potential of any shoreline erosion control program to cause adverse effects in adjacent shore areas (for example, downdrift effects). Such adverse effects can be treated as externalities, so that the effect of the actions of one individual (here a locality) on those of others is not taken into account, resulting in added social costs to other individuals or localities. Instead, it is necessary to use knowledge of the range of coastal processes, of sediment delivery, of beach-dune interaction, and of human intervention to formulate strategies that function effectively at a regional level.

The descriptions that follow apply to the conditions that were present up to the 1995–96 period. In some instances since then beach nourishment has widened the beaches and covered many engineering structures. Where beach nourishment has been applied, it has been incorporated into the current analysis as one of the approaches to shoreline management. Yet even with beach nourishment, the conditions that give rise to a deficit supply of sediment have not changed, and it is worthwhile to draw attention to the longer-term trend of shoreline characteristics as well as to the shorter-term management response. Both are instructive.

Northern Headlands

The anchor for the geomorphological development of the coastal landforms in New Jersey is the presence of the mainland in the northern section of the coast. A large portion of the Atlantic coast of Monmouth County consists of the margin of the continental mainland directly exposed to the ocean without any intervening barrier island. This section of the coast is an erosional bedrock promontory and forms a cliff or bluff of varying elevations up to 6–7 m (18–22 ft) that projects farther eastward into the ocean than any other part of the state. This relatively low projection forms the Northern Headlands of coastal New Jersey and is a distinctive component of the coastal geomorphology. It includes Reaches 3 and 4.

REACH 3: LONG BRANCH
TO SHARK RIVER INLET

Reach 3 occupies the northern area of the Northern Headlands. It extends from the communities of Long Branch to Shark River Inlet, 11.5 km (7.2 miles). It consists of Long Branch City, Deal Borough, Allenhurst Borough, Loch Arbour Village, Asbury Park City, Ocean Grove (Neptune Township), Bradley Beach Borough, and Avon-by-the-Sea Borough. Reach 3 is characterized as a cliffed coastal zone with a predominant littoral drift to the north. The elevation of the cliff is the

Figure 2.11. Many groins and other structures at seaward margin of upland. Lake is formed in drowned stream valley. Allenhurst to Asbury Park, N.J.

greatest between Long Branch and Deal, about 7 m (22 ft), and gradually decreases both to the north and to the south.

The crest of the cliff, or bluff, is interrupted in various places by shoreline-perpendicular stream "valleys," created by streams that cut through the cliff edge and extend inland from the shoreline. Some of these valleys are deeper than modern sea level and are occupied by freshwater lakes. These valleys have been blocked at the beachfront by a sand ridge (baymouth barrier) preventing direct water flow to the sea (figure 2.11). Deal Lake, Wreck Pond, and some other valleys have been temporarily open to the sea after storm events, but have been rapidly closed by post-storm wave activity because the stream flow was not sufficient to keep the inlets open. Furthermore, except for Shark River, none of these small valleys have any significant estuary associated with them to result in tidal exchanges.

Reach 3 has been losing sediment offshore and alongshore over an extended period of time. This circumstance explains why there is no natural wide beach in this area and why a cliff exists here. In the mid- to late 1990s, however, the state of New Jersey program of nourishing beaches pumped millions of cubic yards of sand from offshore areas to create a beach at the base of the cliff. In 1996 through 1999, two major beach nourishment projects were completed to construct a beach 30 m (100 ft) wide and 3 m (10 ft) high (above mean low water). The Long Branch area received 3.8 million cubic yards of sand, and from Asbury Park to Shark River Inlet received 2.6 million cubic yards.[1] As a result, the conditions that were generally present in the past were considerably modified because of the input of sediment into a sediment-starved system. The erosional trend present in this area will continue, and the new sand will be continually eroded. However, this will take time to occur and during this period the cliffed area of the Northern Headlands will have an artificially wider beach compared with its pre-1996 situation. Prior to this massive beach nourishment, the beach width was greater in the south and decreased in a northerly direction. Beaches in Long Branch were very narrow, for example, and did not normally exist between the West End area and Allenhurst. They gradually widened in Ocean Grove and narrowed again near the north jetty of Shark River Inlet. According to the New Jersey Beach Profile Network, which measures beaches annually (Farrell et al. 1994, 1995, 1997, 1999; Uptegrove et al. 1995), these beaches did not show rapid changes in accretion or erosion, nor did they reveal trends of chronic accretion or erosion. In other words, the gains and losses did not occur in any particular sequence, because there was too little loose sand to shift about (a characteristic of a cliffed shoreline).

Shoreline-stabilization approaches and management in Reach 3 are varied. They include groins built perpendicular to the shoreline, as well as walls, revetments, and bulkheads that line the face of the bluff, and jetties that flank the navigable inlets (a detailed description of structures is presented in chapter 6). The seawall that exists in Reach 2 continues south into Reach 3, ending just north of 404 Ocean Avenue in Long Branch, near Seven Presidents State Park (figure 2.12). Although there are no shoreline-stabilization structures at the park, a rock revetment and vertical steel bulkhead begin at the north end of Ocean Avenue in Long Branch and extend south continuously to the West End. Both the communities of Deal and Allenhurst have shoreline-stabilization structures such as groins, shoreline-parallel walls (that is, bulkheads), and revetments for most of their length of oceanfront except for the Darlington Avenue site in Deal. From Asbury Park south, shoreline-parallel structures are mostly wooden bulkheads, a few with rocks at their base. Little beach nourishment activity occurred in the period from 1986 to 1992, with only modest beachfill in Allenhurst in 1989 (*NJBPN* 1993). Between Allenhurst

Figure 2.12. Upland margin at Long Branch, N.J., 1994. Many groins and revetments line the contact. Longshore transport of water and sediment is to the north.

and Deal there is a very long groin; it acts to restrict the littoral transport of sediment to the north, adding to the scarping or steepening of the cliff face in Deal. Because of the very limited sediment that was available, beaches were largely restricted to a few hundred feet on the south side of a groin with no dry beach on the north side of the same groin. Small coves in the bluff edge created localized "pocket" beach segments that have been used intensively, as in Long Branch's Seven Presidents State Park.

Recent development in the Reach 3 area indicates a trend toward townhouse and condominium units. The area inland of the seawall in Long Branch is extensively developed, with many stately homes lining the shorefront. This is also true at the southern margin of Long Branch, atop the low cliff. Land use in Deal is characterized by single-family residences on large lots perched on the "cliff top." The coastal area has also experienced an increase in year-round housing. There are very few open-space sections of the shoreline in this reach; the Seven Presidents State Park is the only public-use open space, along with a municipal beach club and private beach clubs.

Long Branch City. The oceanfront in Long Branch is 6.8 km (4.3 miles) in length. In 1981 the area from Lake Takanasee south was classified as Category I, critically eroding, and Long Branch from Lake Takanasee north as Category II, significantly eroding *(NJSPMP)*. The city has approximately 34 groins and 5 T-groins with lengths of 60 m to 150 m

Figure 2.13. Rip-rap revetment protecting erosional margin of Northern Headlands. Narrow beaches and re-entrants in coastal bluff. Deal, N.J., 1994.

(200–500 ft) in front of a stone-and-timber seawall-bulkhead that extends for 3,180 m (10,450 ft) along the shorefront (U.S. ACOE 1990). The groins are composed of several materials: 13 are of stone-timber, 1 timber, 23 stone, and 2 stone-steel (U.S. ACOE 1990).

Deal Borough. Deal's oceanfront area is 2.5 km (1.6 miles) long and is characterized by narrow pocket beaches in between stone groins, backed by a cliff-bluff that varies from 3 m to 6 m (10–20 ft) in height (figure 2.13). This bluff erodes during storms, and in most cases there are no dunes at the back of the narrow beach. Deal was classified in 1981 as Category I, critically eroding *(NJSPMP)*. Short-term stabilization efforts involving concrete "rip-rap" (interlocking concrete-stone blocks) and bulkheads have been constructed at the base of the bluff by Deal residents. The borough has 10 groins extending from 60 m to 180 m (200–585 ft): 6 are made of stone and 4 of stone-timber (U.S. ACOE 1990). Its revetment is made of stone and stone-steel, and is sectioned into three parts measuring 210 m (700 ft), 200 m (650 ft), and 450 m (1500 ft) in length (U.S. ACOE 1990). Deal's steel bulkhead is 7 m (23 ft) high and 150 m (500 ft) long (U.S. ACOE 1990).

Loch Arbour Village and Allenhurst Borough. The communities of Loch Arbour and Allenhurst have a very short shoreline that essentially acts as a single unit located between 2 long groins (figure 2.11). The ocean-front section of Allenhurst is 0.4 km (0.3 miles) in length and that of Loch Arbour is 0.3 km (0.2 miles) long. The beachfronts of Loch Arbour and Allenhurst, however, have different profiles; Loch Arbour has one vegetated dune at the back of the beach and Allenhurst has no dunes. These beaches were classified in 1981 as Category I, critically eroding *(NJSPMP)*. Two groins exist in Allenhurst, 1 of stone and 1 of stone-timber, ranging from 39 m (105 ft) to 170 m (564 ft) in length. In addition, Allenhurst has a large concrete bulkhead about 6.5 m (22 ft) high and 450 m (1500 ft) long (U.S. ACOE 1990).

Asbury Park City. The beach in Asbury Park extends for 1.5 km (0.9 miles). The elevation of the backshore prior to the latest nourishment in 1999 was low (1.0 m) (3–4 ft), and storm waves have frequently washed sand onto the road. Erosion has occurred just downdrift of a relatively long groin, and the shoreline has been displaced inland of the groin. This groin was augmented by a rip-rap wall constructed parallel to the shore, but erosion has undermined it. There are no dune systems on the narrow beaches. A boardwalk runs along the beach with a bulkhead beneath it to prevent storm surge penetration. There are 5 groins in Asbury Park. One of them is in the shape of an "L" and is located in front of the Paramount Convention Hall. Three are constructed of stone and 2 of stone-timber, and they range from 170 m to 185 m (548–603 ft) in length (U.S. ACOE 1990).

Ocean Grove (Neptune Township). Ocean Grove has slightly less than 0.9 km (0.6 mile) of beach, with a modest dune system at the back preserved with dune grass and dune fencing. Ocean Grove has 4 groins made of stone-steel that range from 135 m to 165 m (444–548 ft) in length, and two of these are "notched," that is, there is a gap of 10 m (33 ft) or more in the groins to allow (littoral) water and sediment to pass through (U.S. ACOE 1990).

Bradley Beach Borough. Bradley Beach, with slightly less than 1.6 km (1.0 mile) of ocean frontage, is deprived of sediment in the littoral zone because of a northern littoral drift and the impact of a relatively long jetty at Shark River Inlet. Bradley Beach was classified in 1981 as Category I, a critically eroding zone *(NJSPMP)*. Bradley Beach has 3 long stone groins and numerous dilapidated wooden groins. The stone groins have been reduced in length and notched, and extra stone was placed at the inland margin of the groin on the beach to support the timbers. There are no dunes, but there were intentions to create a dune system once a beach nourishment project scheduled for 1999 was completed. A new boardwalk is made of patio blocks and has been reconstructed 12 m (40 ft) inland of its previous position as a result of extensive storm damage in 1992. This is a good example of selective retreat practiced by a municipality because of repeated damage in the more exposed seaward location.

Avon-by-the-Sea Borough. The oceanfront section of Avon-by-the-Sea is 0.8 km (0.5 mile) long and is immediately updrift of Shark River Inlet. Avon has very narrow beaches and no dune system. There is a jetty at the northern side of Shark River Inlet, and 4 long groins in Avon; the jetty is 188 m (620 ft) long while the groins are between 58 m and 187 m (192–615 ft) in length (U.S. ACOE 1990). Beach nourishment is scheduled seaward of the low bluff (2.0 m) (6–7 ft) during 2000.

REACH 4: SHARK RIVER INLET TO MANASQUAN INLET

Reach 4, encompassing both residential and commercial land use, lies between Shark River Inlet and Manasquan Inlet and is approximately 9.4 km (5.9 miles) long. The reach consists of the municipalities of Belmar Borough, Spring Lake Borough, Sea Girt Borough, and Manasquan Borough and is a continuation of the Northern Headlands portion of the New Jersey coastline. As in Reach 3, the littoral drift in this area is predominantly to the north, and the seacoast is primarily a low cliff (2–5 m) with some sand accumulation at the base of the cliff.

The area south of Shark River Inlet has the widest beach between this point and Sandy Hook to the north. The inlet jetty traps large volumes of sand being transported to the north (figure 2.14). Prior to the late 1990s' beach nourishment project, the beach narrowed in the southern part of the oceanfront of Belmar to a minimum width at the boundary with the community of Spring Lake, and beach widths remained narrow between Manasquan and Sea Girt as well.

Reach 4 contains a modest dune system that extends along portions of the coast. Belmar has a low, narrow, vegetated dune system alongside the boardwalk at the back of the beach. The dune system ends in the vicinity of Lake Como in South Belmar. Then there is a dune system that covers the bluff edge in Spring Lake (figure 2.15) and National Guard Beach (part of the National Guard Armory, a state-owned shorefront between Sea Girt and Manasquan). The original foredune still exists in Manasquan, largely covered with housing. A new, man-made dune system was built in the 1980s seaward of the Manasquan boardwalk.

Figure 2.14. Widening beach caused by jetty at Shark River inlet. Groins line margin of upland. Lakes form in drowned river valleys. Northerly drift, Belmar, N.J.

Figure 2.15. Narrow beach at contact with upland, boardwalk over beach. Dunes cap upland contact. Lake is formed in drowned river valley. Spring Lake, N.J.

A variety of structures occur throughout this reach. A bulkhead extends for nearly the entire length, and there is a jettied inlet at either end. Groins are ubiquitous and extend from the low cliff seaward into the Atlantic Ocean. A beach nourishment project of 5.5 million cubic yards of sand was applied throughout this reach in 1997–1999 to create a sandy beach seaward of the low cliff. An experimental offshore submerged breakwater was installed in the southern portion of Spring Lake in July 1994. There did not seem to be any noticeable beach accretion in its vicinity prior to the beach nourishment project.

Shark River Inlet. Shark River Inlet, the northernmost portion of Reach 4, was stabilized in the early twentieth century with stone jetties to prevent the navigation channel from meandering at its entrance into the ocean. The accumulation of sediment at the beach, as a result of the jetties, is substantial on the updrift side (southern side) of this inlet. However, the downdrift sediment deficit caused by this jetty construction has resulted in accelerated erosion in Reach 3.

Belmar Borough. Belmar's oceanfront is composed of a relatively low bluff (1–2 m, or 5–6 ft) along a length of 2.3 km (1.5 miles). Its widest beach seaward of the low bluff occurs at the northern end near the projecting jetty and narrows to the south. Lake Como is another example of a drowned river valley in the upland that extends to the Atlantic Ocean. In 1981 the area from Shark River Inlet to Lake Como in South Belmar was classified as Category III, moderately eroding, and the area in front of Lake Como as Category II, significantly eroding *(NJSPMP)*. The shore structures at Belmar consist of 4 notched stone groins and 1 stone jetty (U.S. ACOE 1990). The jetty is 270 m (885 ft) long and the groins range from 182 m to 190 m (600–620 ft) in length.

Spring Lake Borough. The municipality of Spring Lake has 3.2 km (2.0 miles) of oceanfront. The northern end had a narrow beach, with an old bulkhead that was no longer effective. There is a boardwalk located in front of a dune system at this beach (the dune masks the low bluff) and parts of this coastal section had an experimental chain of concrete discs (circa 1.0 m, or 3–4 ft, in diameter) located in front of the boardwalk as a form of shore stabilization to retain sand. As with the case of the offshore breakwater, there did not seem to be noticeable accretion in the vicinity of this structure. Spring Lake was classified in 1981 as Category III, moderately eroding *(NJSPMP)*, and has 11 stone groins ranging from 100 m to 195 m (325–640 ft) in length (U.S. ACOE 1990).

Sea Girt Borough. Sea Girt contains 2.3 km (1.4 miles) of oceanfront.

Before the beach nourishment project in 1999, the cliff rose to more than 5 m (15 ft) in the central portion and was slightly lower both to the north and to the south. The cliff face was masked by some sand, especially at the southern end, to resemble a dune system. In the northern portion of Sea Girt, buildings are farther back from the beach. Some community infrastructure and a boardwalk are located seaward of residential buildings. The center of Sea Girt's coastal zone contains a boardwalk over the beach seaward of the low cliff. The southern portion of Sea Girt consists of private residences and has no boardwalk; there is no infrastructure adjacent to the beach. The coastal zone in Sea Girt in 1981 was classified as Category II, significantly eroding *(NJSPMP)*. Sea Girt contains 6 stone groins and 7 timber groins, and the National Guard Armory's beachfront has 3 stone groins and 1 timber groin (U.S. ACOE 1990). All structures range from 36 m to 170 m (120–550 ft) in length.

Manasquan Borough. Manasquan is the southernmost community of Reach 4 and has a beachfront 1.6 km (1.0 mile) long characterized by a narrow beach, a result of the effects of the jetty at Manasquan Inlet. The geomorphological feature that dominates this area is a type of baymouth barrier island that extends across a sizable cove, connecting Sea Girt with Point Pleasant. The cove is drained by the Manasquan River, which has relatively long jetties controlling its navigational channel between the Atlantic Ocean and the canal leading into northern Barnegat Bay. The seaward side of the baymouth barrier has a macadam or asphalt walkway separating the beach-dune area from commercial and residential development (figure 2.16). Because of the effects from the Manasquan jetty and northerly drift at the beach, the portion of the town immediately downdrift (northern side) of the jetty has a steep and narrow beach. Manasquan in 1981 was classified as Category II, significantly eroding *(NJSPMP)*. Manasquan has 12 groins along the coast to try to keep the sand in place and the long jetty at Manasquan Inlet that causes downdrift sediment starvation. Of these structures, 3 are constructed of stone, 1 of stone-timber, and 3 of stone-steel (U.S. ACOE 1990). The groins range from 45 m to 166 m (150–545 ft) and the jetties are 375 m (1230 ft) and 313 m (1030 ft).

Northern Barrier Spit

The Sandy Hook spit is the product of sediment contributed from the erosion of the Northern Headlands and a supply of sediment from offshore that is propelled northward by the persistent northerly current. In addition, the small drainage basins of the Navesink

Figure 2.16. Boardwalk and artificial dunes. Steep beach, with structures. Manasquan, N.J., 1995.

and Shrewsbury rivers previously discharged sediment to the ocean via inlets that existed in the past that cut through the spit. Construction of an extensive seawall initiated at the beginning of the twentieth century has essentially stabilized the continuity of the spit, terminated the contributions of sediment from the local rivers through these inlets, and all but extinguished the presence of a beach seaward of this seawall until recently. In 1994–1996, a major beach nourishment project pumped 7.9 million cubic yards of sand seaward of this seawall and generated a beach for its entire length. The communities of Monmouth Beach and Sea Bright occupy the basal portion of the spit, whereas the National Park Service (NPS) operates the Sandy Hook Unit of Gateway National Recreation Area at its northern end, all comprising Reach 2.

REACH 2: SANDY HOOK TO LONG BRANCH

Reach 2 forms the northernmost portion of the Atlantic coast of New Jersey. It extends 17.9 km (11.2 miles) from the northern boundary of Long Branch, New Jersey, to the northern tip of the Sandy Hook spit. This portion of the coast receives the protective effects of Long Island and New England and thus has waves that approach either from the east or more commonly from the southeast throughout the year. These conditions result in a longshore drift carrying sediment primarily

Figure 2.17. Base of Sandy Hook spit attached to the upland at Long Branch. Sea Bright and Monmouth Beach development. Seawall with groins lining oceanside of spit.

to the north. Over time, the northerly drift has transported great quantities of sediment from the eroded cliffs of the Northern Headlands and from offshore toward Raritan Bay and has created and extended Sandy Hook spit (figure 2.17).

The attachment of Sandy Hook to the Northern Headlands, however, has waxed and waned as variations in sediment availability, storm effects, and drainage from the Navesink River–Shrewsbury River systems have alternately breached and sealed inlets in the spit. At various times Sandy Hook has been an island, has been connected to the Headlands, or has been attached to the Northern Headlands at Long Branch as it is today.

Most of Reach 2 is described as sediment starved. The only portion that has a history of accumulation (that is, accretion) is the northwest section at the end of Sandy Hook. Initiated around the turn of the twentieth century and completed in 1926, a seawall 4.6 m (15 ft) high was constructed to prevent breaching of the spit, to protect a coastal railroad, and to protect the very narrow southern portion of the spit. The seawall eventually lined the entire oceanfront of Monmouth Beach, a length of 2.6 km (1.6 miles), and Sea Bright, which has 5.9 km (3.7 miles) of oceanfront. The seawall, along with its attendant characteristics and impacts, is primarily responsible for the present-day landforms and situations in this reach. The human manipulation of the shoreline of this section of New Jersey is known by coastal geomorphologists throughout the world for the long-term record of shoreline change and the many attempts to cope with the issues of stabilization. The beach nourishment seaward of the seawall is the latest of many projects proposed to stabilize the shoreline. Its progress is being carefully monitored.

In this portion of Reach 2, hardened structures and a "hard" engineering approach have dominated. The seawall has been defending the shoreline position for nearly 100 years. It has undergone repairs and renovation countless times. Numerous groins of varying dimensions and shapes extend outward into the ocean from the seawall in efforts to catch and hold minor amounts of sand being transported north. The national park at Sandy Hook has 5 timber groins and 6 stone groins that extend from 30 m to 182 m (100–600 ft) in length; the community of Sea Bright has 25 groins: 6 of stone, 12 of timber, 4 of stone and timber, and 3 of stone and steel, with lengths varying from 48 m to 182 m (160–600 ft); and the community of Monmouth Beach has only 4 groins: 1 of stone and steel, 2 of stone, and 1 of stone and timber (U.S. ACOE 1990). At times small pocket beaches have been created in the corners of the attached groins. For long stretches, waves would break directly on the seawall with no beach to intercept them or dissipate their energy. Once the seawall was constructed, erosion continued to remove sediment from the submarine portion of the spit seaward of the wall. As a result, the offshore zone has been deepened, beaches have a steepened angle at the water's edge, and the seawall has been undermined by the loss of supporting sediment. Storm conditions have caused portions of the wall to collapse.

Prior to 1926 and construction of the seawall, the portion of the spit presently situated at the downdrift terminus of the seawall, now part of the NPS Gateway National Recreation Area, was very wide at that location. Severe erosion at that site has been common in recent decades (figure 2.18). By 1978 erosion had progressed so far that the hook was in danger of being breached. In 1978 the first of several overwashes occurred, and eventually a small breach destroyed the road and threatened to sever the connection to the northern portion of the spit. Three major beach nourishment projects (2.95 million cubic yards in 1982–1984, 2.5 million cubic yards in 1989–90, and 287,000 cubic yards in 1997–98) temporarily filled the critical zone at the end of the seawall, but shoreline erosion continues and a similar situation will most likely recur. This portion of Sandy Hook was classified in 1981 as Category I, critically eroding, in the

Figure 2.18. Narrow Sandy Hook spit. Seawall at oceanside ends in middleground, causing critical erosion area. Sandy Hook Unit, Gateway National Recreation Area.

NJSPMP. The area to the north of the critical zone was relatively stable during the twentieth century, an indication that there was a sediment balance. The extreme northwest portion of Sandy Hook is accumulating material and extending into Sandy Hook Channel and Raritan Bay, and occasionally the shipping channel has to be dredged.

The shoreline of the communities of Sea Bright–Monmouth Beach was directly at the base of the seawall for many decades. In 1994–1996, however, a major beach nourishment project designed to create a beach 30 m (100 ft) wide, at an elevation of 3 m (10 ft) above mean low water, was conducted (figure 2.19). In 1981 both areas had been classified as Category I, critically eroding *(NJSPMP).*

There are locations of well-developed dunes in the national park, north of the critical zone. Some of the foredune areas are natural; others have been constructed by the managers of the park or by the previous occupants of this section of Sandy Hook (including the U.S. Coast Guard and the Department of Defense). There are several ridges of large natural dunes 6–7 m (20–23 ft) curving across the northern portion of the hook. These unique and well-preserved dunes represent the position of former shorelines in the geomorphological developmental history of the extending hook. A very low, vegetated dune ridge (1.0 m, or 3–4 ft) has been constructed at the inland margin of the beachfill throughout most of Sea Bright and in parts of Monmouth Beach.

Figure 2.19. Beach nourishment seaward of seawall, 1994–1996. Sea Bright, N.J.

Sea Bright, adjacent to Sandy Hook Park, is predominantly a recreational community. Residential development with commercial development is concentrated in the center of town. There has been a recent trend toward townhouse and condominium units, and an increasing trend to year-round housing. Elevations are very low (1–3 m, or 3–10 ft) above sea level, excluding the 4.6 m (15 ft) seawall, and flooding is a recurring problem even with the present beach in front of the seawall. Monmouth Beach is a residential community and, as in Sea Bright, a trend toward more townhouse and condominium developments and more year-round housing is evident.

Northern Barrier Islands Complex

As one continues south of the Northern Headlands, the next major geomorphological feature of the New Jersey coast is a series of long narrow barrier islands, referred to as the Northern Barrier Islands Complex. This is the beginning of the barrier island system in coastal New Jersey. Although there is evidence of inlets that have existed in the past, the first two barrier islands are continuous masses of sand where once sizable dune systems existed, but which have been nearly completely altered by residential and tourist development. Attempts at shoreline stabilization are evident in the many groins present in the sandy beach of most communities. The topography of Island Beach State Park demonstrates the quality and character of the natural features of the barrier islands prior to intensive manipulation. This area consists of Reaches 5, 6, and 7.

REACH 5: MANASQUAN INLET
TO MANTOLOKING BOROUGH

Reach 5 begins at Manasquan Inlet and extends south to Mantoloking Borough. It represents the transition from the Northern Headlands to the long barrier islands. The area near Manasquan Inlet is a very eroded portion of the Northern Headlands and consists of deep coves bordered on the seaward side by the initiation of the barrier islands. The orientation of the barrier is almost north-south in this location, whereas the dissected upland tends to the southwest direction. Thus the barrier becomes more completely separated from the upland land mass toward the south, and the intervening water body of Barnegat Bay is formed. This reach is 4.9 km (3.1 miles) in length and consists of the communities of Point Pleasant Beach Borough and Bay Head Borough. The coastal zone in this reach is characterized by moderate-to-narrow beach widths, except updrift of the Manasquan Inlet jetties, and steeply sloped beaches.

The dissected bluff area, which is primarily in Point Pleasant Beach Borough, has an elevation of 0–2 m (0–7 ft), masked by beach and dune sand. Forsythe Avenue in Bay Head marks the southernmost point on the New Jersey coast where the older coastal plain sediments are exposed directly adjacent to the ocean shoreline. The shoreline south of Forsythe Avenue is a combination of spits built from local headland segments, bay-mouth barriers, and barrier islands. All the individual coastal features evolved and merged into the present-day "spit-like" peninsula that ends at Barnegat Inlet. It has been constructed of sand eroded from the upland bluffs and from sand transported westward on the continental shelf by wave processes as the sea level has risen over the past several thousand years.

The beach associated with the Reach 5 northern barrier island segment gradually narrows to the south. The net littoral sand transport direction is northerly, as evidenced by the accumulation of sand on the south side of the Manasquan Inlet jetties. South of Manasquan Inlet, there is an increasing proportion of southerly littoral drift as the shielding influence of New England decreases. Eventually the littoral drift becomes more balanced toward the south, and then is succeeded by a net southerly drift.

Figure 2.20. Widening beach north to Manasquan Inlet, sand against upland margin. Lakes are formed in drowned valleys. Southern Barrier Islands Complex system extends from upland at top of figure. Point Pleasant, N.J.

Point Pleasant Beach Borough. The beachfront of Point Pleasant Beach, 2.9 km (1.8 miles) long, narrows toward the south. In 1981 the northern section was classified as Category IV, noneroding, and the southern section was classified as Category III, moderately eroding *(NJSPMP)*. The beachfront at the northernmost portion of Point Pleasant Beach is at its greatest width and dunes should exist, but extensive recreational activities have eliminated natural dune development (figure 2.20). A small amusement park area and accompanying recreational facilities classify the community as one of recreational land use. A low dune line has been constructed seaward of the boardwalk along portions of the community.

Bay Head Borough. The oceanfront in Bay Head is 1.9 km (1.2 miles) in length. The beach is of moderate width and has a moderate-to-large dune system that is well vegetated and fenced. This dune system is continuous throughout Bay Head; it functions as a buffer and fortifies the gaps at street ends. Walkways through the dunes exist and are perpendicular to the beachfront. In 1981 the beachfront in Bay Head was classified as Category III, moderately eroding *(NJSPMP)*, and contains 10 groins, which are exposed at low tide and covered at high tide. Eight of these groins are made of timber and 2 are made of stone (U.S. ACOE 1990).

Figure 2.21. Exposed rip-rap seawall at inner margin of beach, under artificial dune ridge, after storm. Part of barrier island. Bayhead, N.J., December 1992.

These groins range from 45 m to 76 m (150–250 ft) in length. Bay Head also has a stone seawall that is 1260 m (4150 ft) long (U.S. ACOE 1990), located under the coastal foredune (figure 2.21).

REACH 6: MANTOLOKING BOROUGH
TO BARNEGAT INLET

Reach 6 extends from Mantoloking Borough to Barnegat Inlet, is approximately 32.5 km (20.3 miles) long, and includes the municipalities of Mantoloking Borough, Normandy Beach, Ocean Beach (Brick Township), Lavallette Borough, Ortley Beach (Dover Township), Seaside Heights Borough, Seaside Park Borough, South Seaside Park (Berkeley Township), and Island Beach State Park.

Major storms have occasionally broken through this reach's long peninsula and created new inlets. The Ortley Beach–Seaside Heights boundary was once the site of Cranberry Inlet, which existed immediately seaward of the mouth of the Toms River estuary. This historical inlet site is an example of inlet creation and migration within a barrier island system. The southern 16 km (10.0 miles) of this spit is the site of Island Beach State Park, the longest section of continuous open space on the New Jersey coast and the largest state-owned section of the shoreline.

Longshore transport of sediment is to the south at Barnegat Inlet. Old maps and charts indicate that Barnegat Inlet migrated 1125 m (3700 ft) south between 1866 and 1932, 17 m (56 ft) per year, and then became stabilized when construction of the first rock jetty structure was started (Farrell and Leatherman 1989).

Beach surveys conducted from 1986 to 1995 indicate that sediment loss has occurred for most of this reach, whereas there has been substantial accretion at the southern end in Island Beach State Park adjacent to the jetty (*NJBPN* 1986–99). Reach 6 has beaches of moderate width, and many communities have promoted the development of dunes. Dunes are present throughout most of this reach, except for the community of Seaside Heights and the northern portion of Seaside Park.

The development in Reach 6 varies from a predominantly privately owned shorefront in Mantoloking to extensive public amusements and boardwalk-oriented recreation in Seaside Heights. South of Seaside Heights, the development is mostly single-family homes and small motels. Island Beach State Park occupies the southernmost section of this reach. It is largely undeveloped, incorporating a natural area to the north, a public bathing area with accompanying infrastructure in the center, and a modestly disturbed southern portion.

Mantoloking Borough. The beachfront in Mantoloking is 3.2 km (2.0 miles) long with a moderately wide beach that has sizable dunes from 5 m to 6.5 m (16–21 ft) in elevation. Throughout most of the community, residential lots are large and houses are built well back from the beach (figure 2.22). In the vicinity of South Mantoloking, the dune system serves as a buffer to the overwash and inlet creation processes previously characteristic of Reach 6. In 1981 Mantoloking was classified as Category III, moderately eroding *(NJSPMP)*.

Normandy Beach Borough. Normandy Beach has a beachfront 0.8 km (0.5 mile) in length, and the beach is of moderate width. Normandy Beach was classified in 1981 as Category III, moderately eroding *(NJSPMP)*. Near Normandy Beach, houses are protected by a dune, although it is narrow and discontinuous in places. Earlier houses were often built on top of the dune, and these homes are vulnerable during severe storms because erosion of the dune base can undermine their foundations and result in structural failure and collapse.

Ocean Beach (Brick Township). Ocean Beach extends for 1.2 km (0.8 mile) of moderate-width oceanfront. Portions of the beach have dunes with vegetation and fencing. A large, exposed condominium located on

Figure 2.22. Well-maintained coastal foredune. Mantoloking, N.J.

the beach is the closest structure to the waterline within the area, and it does not have dunes or much beach width in front of it. This beach was classified in 1981 as Category III, moderately eroding *(NJSPMP)*.

Lavallette Borough. The generally narrow oceanfront of Lavallette is 2.1 km (1.3 miles) in length. A boardwalk lines most of the shorefront in this community (figure 2.23). A dune system, which measures 3.6 m to

Figure 2.23. Single-family residences, boardwalk, and artificial foredune. Lavallette, N.J.

4.5 m (12–15 ft) in elevation, is sparsely vegetated and fenced. Lavallette was classified in 1981 as Category III, moderately eroding *(NJSPMP)*. There are 9 stone groins in Lavallette ranging from 90 m to 106 m (300–350 ft) in length (U.S. ACOE 1990), along with a timber bulkhead that is 638 m (2100 ft) long.

Ortley Beach (Dover Township). The beachfront of Ortley Beach is 1.2 km (0.7 ft) long and is of moderate width. There is a boardwalk and a gazebo at each end of a boardwalk within the community and a vegetated and fenced dune system along the back of the beach, seaward of the boardwalk. The beach does not contain any hard shoreline-stabilization structures. Ortley Beach was classified in 1981 as Category III, moderately eroding *(NJSPMP)*.

Seaside Heights Borough. Seaside Heights has 1.2 km (0.8 miles) of oceanfront with moderate beach width. A boardwalk extends throughout Seaside Heights and into part of Seaside Park, with a bulkhead under the boardwalk. The town does not have any dunes or any particular beach management plan other than maintaining a beach of moderate width for the recreation season. Seaside Heights was classified in 1981 as Category III, moderately eroding *(NJSPMP)*, a designation that has remained accurate. Two piers extending seaward are the only structures on the beach in Seaside Heights.

Seaside Park Borough. The community of Seaside Park is 2.7 km (1.7 miles) long and its beachfront, which does not have any hard shoreline-stabilization structures, is of moderate width. This area was classified in 1981 as Category III, moderately eroding *(NJSPMP)*. The town utilizes a dune maintenance program for its artificial dune system, with heights of 2–3 m (7–10 ft), located seaward of its boardwalk. A continuous line of fenced dunes is fertilized regularly and has a significant amount of vegetation. Paths from the boardwalk through the dunes are oriented toward the southeast to slow inland penetration by storm waves from the northeast. Dunes are completely absent in the area of the amusement pier.

South Seaside Park (Berkeley Township). South Seaside Park has approximately 1.1 km (0.7 mile) of oceanfront and a moderately wide beach with no hard stabilization structures. Fenced dunes in this area have developed over the past 15 years into substantial size, 4–6m (12–20 ft), and most of the vegetation is natural. The area was classified in 1981 as Category III, moderately eroding *(NJSPMP)*.

Island Beach State Park. Island Beach State Park has 15.2 km (9.5 miles) of natural beach of moderate width that is sectioned into different

Figure 2.24. Large dissected foredune, northern portion of Island Beach State Park, N.J.

areas, each with separate rules and regulations for maintenance and use. The northern natural area has limited beach use for recreation (figure 2.24), the central area is used for bathing and recreation, and the southern end allows beach buggies onto the beach as a recreational activity. Throughout most of the park, the dunes are relatively natural and presently unaffected by construction or modification, except seaward of the governor's summer house. There are two distinct dune lines present in the northern and central portions of the park, and the inland dunes are thickly vegetated with brush, grasses, and small trees. Some dune breaches that exist are wide enough to permit four-wheel-drive vehicles to pass through, and although the access roads are not straight, they could permit entry of large storm surge waves into the interior parts of the park during significant storm events. The coastal foredune is a largely continuous line ranging from 5 m to 9 m (15–20 ft) in height, with beachgrass vegetation. There are significant blowouts, or gaps where sand is blown inland, in parts of the foredune, but they lead to high ridges immediately inland. This is the process by which the dunes are able to be maintained while transgressing landward. The dunes of Island Beach State Park are primarily the product of natural processes, but there is evidence of some assistance by fence lines and piles of debris. The major structure in Island Beach State Park is a stone jetty located at Barnegat Inlet, which extends for approximately 1500 m (4900 ft). The beach width increases toward the jetty, indicative of a net southward littoral drift and sediment transport.

REACH 7: BARNEGAT INLET TO LITTLE EGG INLET (LONG BEACH ISLAND)

Long Beach Island (LBI) is a barrier island that marks a change in orientation of the shoreline, now extending south-southwest from Barnegat Inlet to Little Egg Inlet. The island is 28.8 km (18 miles) in length and is the longest of any of New Jersey's barrier islands. It is the southernmost component of the Northern Barrier Islands Complex. There are six municipalities within this reach: Barnegat Light Borough, Harvey Cedars Borough, Surf City Borough, Ship Bottom Borough, Beach Haven Borough, and Long Beach Township (consisting of three named places: Loveladies, Brant Beach, and Beach Haven Crest). The entire island is classified for recreational land use and is heavily developed with the exception of the southernmost 2.9 km (1.8 miles), which is devoted to open space as part of the Forsythe National Wildlife Refuge (Holgate Unit). A dense network of groins is present throughout most of the barrier island, which is slowing the alongshore transport of sand and sometimes produces offsets in the shoreline configuration. The island is vulnerable to storm damage because most of its topographic elevation has been lowered and flattened in the process of development, its beaches are steep and narrow, and the threat of breaching is a concern.

The past 50 years of development have greatly changed the oceanfront and the bayside of the island. Before 1950, neither homes nor motels existed within approximately 150 m (500 ft) of the coastal dunes in large sections of the oceanfront. Today very few sites exist where buildings are greater than 60 m (200 ft) from the mean high-tide line. Storm evacuation is a difficult task on LBI because only a few roads run its length, and more important, there is only one bridge to the mainland. Most of the bayside margin has been modified, as wetlands have been filled and housing now exists at very low elevations on fill and along canals that extend into the bay.

The littoral drift is predominantly to the south along much of LBI. Barnegat Inlet has been stabilized by very long jetties that have a downdrift effect on sediment transport, whereas Beach Haven Inlet at the southern terminus has no structures and is actively migrating at the rate of up to 60 m (200 ft) per year. Most of the reach in 1981 was classified as Category III, moderately eroding *(NJSPMP)*. A few areas of significant erosion (Category II) occur in portions of Ship Bottom Borough, Brant Beach, and Beach Haven Borough. With the exception of these locations, there is sufficient setback between the eroding beach and the nearest buildings and roads.

There are large natural dunes present in Barnegat Light, where they occupy a wide buffer immediately downdrift of the jettied Barnegat Inlet. Most of the remainder of LBI has a single built dune ridge about 4–5 m (12–17 ft) high, which is maintained with a myriad of fence lines, controlled walkovers or access points, and programs of fertilization and grass planting. The dunes in the wildlife refuge are lower, 1–2 m (3–7 ft), and are breached in numerous places.

Shoreline-stabilization methods within this reach have featured both "hard" and "soft" engineered approaches in the form of groins and beach nourishment. The entire barrier island received beach nourishment in 1962–63 following the Ash Wednesday northeaster in 1962 (Podufaly 1962). In recent years beach nourishment has mostly been confined to the northern quarter of the island, because the sand supply has been derived from dredging Barnegat Inlet. During the early 1990s in Loveladies there was an episode of beach nourishment in which sand was trucked in, and Harvey Cedars also received a beach nourishment project in 1990. Sand has also been trucked in to provide beachfill intermittently at Brant Beach. Groins present in this reach are characterized as low rubble mound and located 240–365 m (800–1200 ft) apart. Rock groins in Harvey Cedars are an exception and are relatively higher and extend back to the street end, as does a similar rock groin in Beach Haven.

Barnegat Light Borough. Barnegat Light contains about 2.7 km (1.8 miles) of oceanfront that has a wide beach backed by well-vegetated dunes. A 1991 south jetty renovation project at Barnegat Inlet had a major impact on the beach position of the northernmost part of Long Beach Island, when the shoreline advanced seaward by hundreds of feet within a thousand feet of the jetty (figure 2.25). Accretion tapers off near the Barnegat Light–Loveladies boundary. Barnegat Light has 13 groins, mostly buried, and 1 jetty. One of these structures is constructed of stone-timber-core, 3 of timber-stone, 5 of timber, and 5 of stone. The groins range from 50 m to 155 m (165–506 ft) and the jetty is 900 m (2950 ft) long (U.S. ACOE 1990).

Harvey Cedars Borough. Harvey Cedars contains approximately 3 km (1.9 miles) of beachfront, which is very narrow (figure 2.26). A dune system exists at the back of the beach, and the community contains 11 stone

Figure 2.25. Barnegat Inlet jetties, widening of beach adjacent to jetty, and erosional embayment in island downdrift of the inlet.

Figure 2.26. Narrow beach and dune zone, dense residential development. Northern portion of Long Beach Island, N.J.

groins up to 98 m (320 ft) in length (U.S. ACOE 1990). In 1981 the area was classified as Category III, moderately eroding *(NJSPMP)*, but has potential for critical erosion during major storm events because of downdrift effects from the Barnegat Inlet jetties.

Surf City Borough. Surf City contains approximately 2 km (1.3 miles) of oceanfront with moderate beach width. Dunes are maintained at 6 m to 7 m (22–24 ft) in height and are never lower than 4.8 m (16 ft). All dunes are vegetated and walkovers for beach access have been constructed above the dunes. Surf City has 7 timber-stone groins that are about 100 m (335–340 ft) in length (U.S. ACOE 1990).

Ship Bottom Borough. Ship Bottom contains 2 km (1.3 miles) of oceanfront, which is of moderate width. Dunes are fertilized and fenced for stabilization and are maintained at a 4.8 m (16 ft) height at the building line. At some street ends groins are present, along with bulkheads maintained at a 4.25 m (14 ft) height. There are 7 timber-stone groins in Ship Bottom, which are about 100 m (335–340 ft) in length (U.S. ACOE 1990).

Long Beach Township (Loveladies, Brant Beach, Beach Haven Crest). The municipality of Long Beach Township, interspersed among the island's other municipalities, consists of approximately 14.9 km (9.3 miles) of oceanfront with beaches of moderate width. A dune system, 4.25 m to 4.8 m (14–16 ft) high, is maintained with fences and vegetation throughout the township. In total there are 65 groins ranging from 70 m to 130 m (235–420 ft) (U.S. ACOE 1990). Sixty are constructed of timber-stone, and 5 are constructed of stone.

Beach Haven Borough. Beach Haven contains 2.9 km (1.8 miles) of oceanfront with beaches of moderate width. Dune development has been enhanced with a myriad of fences and sizable setbacks. A relatively high groin (considerably above high tide) has caused a downdrift beach offset at Holyoke Avenue of more than 45 m (150 ft) (figure 2.27). Ten groins are present in Beach Haven: 8 of stone, 1 of timber and 1 of timber-stone (U.S. ACOE 1990). These range from 90 to 100 m (300–340 ft) in length.

Edwin B. Forsythe National Wildlife Refuge (Holgate Unit). The wildlife refuge is a natural, undeveloped, federally owned portion of coast that has a moderately wide beach and is 2.7 km (1.7 miles) in length. There are several lines of low coastal dunes that curve around the end of the island. The well-vegetated primary dune has been breached in several locations by overwash, and several large sand fans have spread across the inland into the bay (evidence of overwash).

Figure 2.27. Multiple groins; offsets in island shoreline are associated with structures and sequence of development. Southern portion of Long Beach Island, N.J.

Southern Barrier Islands Complex

In southern New Jersey, the barrier island form changes from a long and narrow configuration to a shorter configuration with a wide updrift portion and a narrow downdrift portion. This shape is referred to as a "drumstick" barrier island. The barrier islands tend to be offset at the inlets, and the wide part of the drumstick usually extends farther seaward than the adjacent narrow portion. The wide end of the barrier island interacts with the sandy shoal of the adjacent inlet, which causes the bulbous end to receive, store, and release sand as part of a downdrift movement of the sediment supply. The result of this sequence is that the updrift inlet and ocean shorelines are very dynamic and pass through considerable shifts in position. Wave refraction at the inlet offshore shoal and the bulbous end tends to foster sand storage and maintain the general shape of the island, and it leads to a sediment deficit immediately downdrift of the seaward projection. The narrow, downdrift ends of the drumstick may be locations of elongation of the barrier and accretion into the next inlet. In conjunction with these processes, the natural dune system produces larger, irregular, and dissected dune forms at the updrift end of the islands, grading downdrift to a more coherent single foredune ridge that tends to decrease in height in the downdrift direction, and finally to an accretionary group of low foredunes at the downdrift terminus.

The barrier islands in the southern complex are closer to the mainland than the group that formed the Northern Barrier Islands Complex, and the adjacent bays are shallow with large areas of wetlands. Five of the seven inlets separating this complex of islands are not completely controlled (that is, do not have pairs of jetties controlling the navigation channel). Groins and bulkheads are common at the beach and a seawall is present in Avalon and in Cape May City. With the exception of portions of the Wildwoods, erosion is a common problem on these sandy barrier islands. Beach nourishment has been used to stabilize the shoreline throughout this complex. This portion of the New Jersey coast consists of Reaches 8, 9, 10, 11, 12, 13, and part of Reach 14.

REACH 8: LITTLE EGG INLET
TO ABSECON INLET

Reach 8 is 9.6 km (6 miles) in length. It is composed of Pullen Island and the Brigantine barrier island, which in turn consists of the Edwin B. Forsythe National Wildlife Refuge (EBFNWR–Brigantine Unit) and the city of Brigantine. The littoral drift is predominantly in a southerly direction.

Pullen Island and Brigantine barrier island have extensive estuarine systems on their western boundaries that separate the barrier islands from the mainland. This estuarine area consists of small uninhabited islands, shallow bays, tidal marshes, creeks, and lagoons. The marshland exists because of the large amount of sediment that has been overwashed and transported through the inlets into the backbay to fill parts of the embayment (figure 2.28). The sand that makes up the beaches and dunes in much of this reach contributes a thin veneer of coarse sediment on top of a thick accumulation of silt and organic peat. This sand unit is slowly migrating inland across the marsh and bay sediments. Following a storm, the beach is usually swept clean of sand and the old marsh is exposed along much of both islands. Both the sand and the entire barrier island system are migrating inland.

Edwin B. Forsythe National Wildlife Refuge (Brigantine Unit). The barrier island portion of the Edwin B. Forsythe National Wildlife Refuge is composed of Pullen Island and a 3.2 km (2 mile) section of the undeveloped northern end of Brigantine barrier island. Pullen Island was breached in 1994, and is presently two small uninhabited islands (figure 2.28). Overwash is commonplace throughout the wildlife refuge. Although Pullen Island has experienced dynamic migration, it has no devel-

Figure 2.28. Undeveloped barrier islands, unique on the East Coast. New inlet breached the island in 1994. Pullen Island, N.J.

opment that can be threatened and therefore the area was not classified with an erosion rate in the *NJSPMP*. The northern section of Brigantine barrier island is the remaining part of the EBFNWR and is characterized by a consistently narrow beach. Brigantine Inlet, also called "Wreck Inlet," lies between Pullen Island and Brigantine barrier island, interrupting the EBFNWR area. There are no jetties at Brigantine Inlet. Pullen Island has a well-developed coastal dune at its northern margin, nearly 6 m high. It is part of an old shoreline that is now nearly 50 m from the shoreline because the inlet margin has shifted. The remainder of the shoreline has a very low coastal foredune, 0.5–1.5 m (2–4 ft), which is frequently breached and overwashed.

Brigantine City. The city of Brigantine occupies the remainder of the Brigantine barrier island, with slightly over 6.4 km (4 miles) of beach-front. There is an offset in the beach at the border of the EBFNWR and the city of Brigantine, because the wildlife refuge is shifting inland naturally under a negative sediment scenario whereas the city is defending its infrastructure with a metal bulkhead, rip-rap, and beach fill (figure 2.7). The northern part of Brigantine City has a very low, intermittent, and narrow dune system in front of a narrow beach. A coherent coastal dune system exists south of the Brigantine Hotel and continues for the length of the island, becoming larger as the beaches widen in a southerly direc-

Figure 2.29. Widening beach and dune zone updrift of Absecon Inlet. Brigantine City in foreground, Atlantic City in background; islands are offset at the inlet.

tion. The southernmost dune zone attains an average width of 55 m to 70 m (190–230 ft) and a height of 2 m to 3 m. The foredune ridge is well vegetated. In 1981 the developed northern portion of Brigantine City was classified as Category I, critically eroding *(NJSPMP)*, because of a low, narrow beach and the poor condition of its groin structures. Existing development in this area is very close to high water, and storm waves pose a considerable threat to the buildings. Although Brigantine barrier island has a wider beach in its central area, it has been classified as Category II, significantly eroding *(NJSPMP)*, because the developed areas are threatened by storm penetration and damage. The southernmost portion of the island is influenced by the north jetty of Absecon Inlet (figure 2.29). Sand accumulates on the updrift side of the jetty, forming a wider protective beach. This area was classified in 1981 as noneroding, Category IV *(NJSPMP)*. Brigantine City has 28 groins along its shoreline from 20 m to 190 m (70–630 ft) in length, and 1 stone jetty that is 1135 m (3730 ft) long bordering Absecon Inlet (U.S. ACOE 1990). The groins vary slightly in construction: 21 groins are made of timber and 7 are made of timber-stone. Brigantine also has 8 bulkheads: 6 are made of timber, 1 is made of timber-stone, and 1 is metal (U.S. ACOE 1990).

REACH 9: ABSECON INLET TO GREAT EGG HARBOR INLET

Reach 9 extends from Absecon Inlet to Great Egg Harbor Inlet and consists of the Absecon barrier island with an oceanfront over

12.8 km (8 miles) in length, and a beach that gradually decreases in width from north to south along with noncontinuous (segmented) coastal dune systems. It is the most intensively developed barrier island in New Jersey, consisting of the four communities of Atlantic City, Ventnor City, Margate City, and Longport Borough. Absecon Island has an extensive estuarine habitat on its western boundary consisting of numerous shallow bays, small uninhabited islands, tidal marshes, creeks, and lagoons. It is part of the Southern Barrier Islands Complex, and the dominant direction of longshore transport is southerly.

Dynamic littoral change has been characteristic of this area. Around the mid-nineteenth century, Absecon Island was configured as two drumstick-shaped barrier islands separated by a small tidal inlet, which is presently occupied by the community of Ventnor. The smaller islands merged to create the modern-day Absecon Island.

Common to many of the drumstick barrier islands, the updrift bulbous end is the site of a very dynamic inlet with extensive sand shoals on both the oceanside and the bayside of the inlet. As the large sand shoals gather and release sediment, the adjacent beaches wax and wane in their development, especially on the downdrift margin and ocean beach. Absecon Inlet is controlled by a pair of jetties that have stabilized the position of this channel, but the downdrift exchanges of sediment continue to produce oscillation of the ocean beaches in northern Atlantic City. Presently this area is eroding and has led to the installation of groins and a relatively high bulkhead under the boardwalk, and episodes of beach nourishment have been applied to stabilize the shoreline.

Another set of inlet processes has affected the southern end of Absecon Island, where a large portion of the narrow tip separated from the rest of the island earlier in the twentieth century (10 blocks of the city were lost). This is part of the process of sediment transfer from one island to the next in the downdrift direction, and similar events have been recorded on the downdrift ends of other New Jersey barrier islands.

Atlantic City. Atlantic City has 5.4 km (3.4 miles) of shoreline. The Absecon Inlet jetty has helped to retain sediment at the inlet's border, but the beach narrows toward the south. The beach is especially narrow in front of an intensive development of hotels and casinos (figure 2.30). Atlantic City also has several piers extending seaward and a boardwalk that extends the length of the city. Geo-tubes, a stabilization measure, are used as a core for an artificial foredune ridge in the northern part of Atlantic City's beach, in front of the casinos. The remainder of the beach

Figure 2.30. Casino development at northern bulbous end of barrier island. Atlantic City boardwalk above small dune remnants after storm, December 1992.

area has artificial dunes without the solid core. Northern Atlantic City was classified in 1981 as Category II, significantly eroding, and to the south as Category III, moderately eroding *(NJSPMP)*. Atlantic City has 19 groins from 50 to 185 m (165–600 ft) in length and 1 stone jetty that is 360 m (1177 ft) long (U.S. ACOE 1990). Six groins are constructed of timber-stone, 7 are made of stone, and 7 are made of timber. In addition, there are 2 timber bulkheads and 1 stone revetment.

Ventnor City. Ventnor City contains 2.7 km (1.7 miles) of narrow beachfront with sporadic areas of small dunes. A boardwalk is located at the high-tide line and has piers extending from it to the street ends (figure 2.31). Ventnor consists of residential development along its coastline, and all oceanfront buildings are located behind bulkheads. To the south, the bulkhead and boardwalk are separated by a distance of approximately 45 m (150 ft), which serves as a type of backbeach area. In 1981 northern Ventnor City was classified as Category III, moderately eroding, and its southern portion as Category II, significantly eroding *(NJSPMP)*. Ventnor City has 13 bulkheads, 5 of which are constructed of timber, 6 of concrete, and 2 of concrete-timber (U.S. ACOE 1990).

Margate City. The beachfront in Margate is 2.6 km (1.6 miles). The beach is narrow, with remnants of dunes in small coves in the bulkhead line. The beach at the southern end of the community almost triples in width. Margate is characterized by a "zig-zag" development line, in which some developed areas extend onto the beach. In 1981 areas of Margate

Figure 2.31. Boardwalk over beach. No dunes on profile, bulkhead is in upper beach. Ventnor, N.J.

City were classified as Category II, significantly eroding, or III, moderately eroding *(NJSPMP)*, depending on the local condition of beaches and shoreline-stabilization structures. There are 14 groins in Margate City, ranging from 38 m to 130 m (125–425 ft) in length: 9 are made of timber, 3 are made of stone, and 2 are made of timber-stone *(NJSPMP)*. Margate also has 19 bulkheads: 2 are made of concrete, 12 of timber, 4 of timber-concrete, and 1 of concrete-block (U.S. ACOE 1990).

Longport Borough. The southernmost community of Reach 9, Longport Borough, contains 2.4 km (1.5 miles) of beachfront and occupies the narrow portion of the drumstick-shaped barrier island. The beach is characterized by a relatively low elevation, narrow width, and no dune system (figure 2.32). The entire length of the beach is lined with either a bulkhead or a seawall, both of which have the potential to be overtopped in storm events. Residential development sits a few feet behind these structures. On the basis the severity of the erosion problem and the potential for danger to private property and the infrastructure, Longport was classified in 1981 as Category I, critically eroding *(NJSPMP)*. Longport has 11 groins that range from 75 m to 155 m (250–507 ft) in length. One groin is made of timber, 8 are made of timber-stone, and 2 are made of stone. In addition, there are 2 concrete seawalls and 1 stone revetment (U.S. ACOE 1990).

Figure 2.32. High-density development on southern end of barrier island, no dunes. Bulkhead is at top of narrow, low beach; terminal groin at inlet. Longport, N.J.

REACH 10: GREAT EGG HARBOR INLET
TO CORSON'S INLET

Reach 10 is approximately 12.6 km (7.8 miles) long and is part of the Southern Barrier Islands Complex. It extends from Great Egg Harbor Inlet to Corson's Inlet and consists of Ocean City and Corson's Inlet State Park, both of which are classified for recreational land use *(NJSPMP)*. Reach 10 is the only barrier island reach that consists of only one municipality, Ocean City. Corson's Inlet State Park occupies an undeveloped portion of the southern end of the barrier island.

The northern end of Ocean City, bound by Great Egg Harbor Inlet, is presently an area of considerable erosion potential and exposure to hazard, similar to other inlet throats. In 1981 the *NJSPMP* classified the area as Category II, significantly eroding. Corson's Inlet, which bounds the southern end of the barrier island, is not stabilized and has not been dredged since 1971, when hopper dredging was suspended. The state presently has no plans to stabilize this inlet, and it has been "officially closed to navigation" since 1984. Corson's Inlet has been narrowing over the years as a result of sediment accumulation within the inlet.

Ocean City. The oceanfront is 11.4 km (7.1 miles) long. The northern portion of the bulbous drumstick has had considerable shoreline fluctuation over the years. The community has been involved in pumping sand from the bay and from the inlet shoals to the beach for nearly a half century. A recent emphasis on dune construction and preservation has

Figure 2.33. Dunes in occupying offset in development. Central portion of Ocean City, N.J.

witnessed dune growth seaward of the boardwalk in the north and in isolated pockets in other portions of the community (figure 2.33); in the south, however, dunes are absent (figure 2.34).

Low timber groins in the northern and central portion of the island have slowed the downdrift transport of sediment (figure 2.35). There is a bulkhead in the central part of this section and a rip-rap seawall in the southern section. In 1981 the *NJSPMP* classified Ocean City's beach as Category I, critically eroding. However, more recent data from the New Jersey Beach Profile Network indicate that the beach erosion problems are not as critical in the central portion of the barrier island as they are to the north (Farrell et al. 1994). It may be that periodic beach nourishment in conjunction with bulkheads and groins has helped reduce the rate of net sediment loss within the area. In the past, Ocean City pumped small quantities of sand from Great Egg Harbor Bay across the island and onto the beaches. In the 1990s, however, a major project was begun that pumped 6.2 million cubic yards of sand to the beaches in the northern part of the island (1992–93), and has placed smaller amounts at short intervals (606,000 cubic yards in 1994, 1.7 million cubic yards in 1995, and 800,000 yards in 1997). Comparison of storm damage caused by ocean waves from the October 1991 and January 1992 events and the more

Figure 2.34. Bulkhead at inland margin of beach, small dune forms, narrow beach. Extensive wetlands. Southern portion of Ocean City, N.J.

Figure 2.35. Groins at northern bulbous end of barrier island. Development completely covers the island. Coastal dunes in segments. Ocean City, N.J.

Figure 2.36. Bulbous end of barrier island. Accretion at inlet, erosion downdrift. Bulkhead and structures in beach in background. Strathmere, N.J.

intense December 1992 storm demonstrate the protective and sacrificial effects of the beach nourishment projects.

Ocean City has 48 groins ranging from 25 m to 325 m (80–1070 ft) in length, and their construction varies: 32 are made of timber, 10 of stone, 2 of timber-steel sheet, 2 of stone-timber core, 1 of stone-timber, and 1 of stone-concrete (U.S. ACOE 1990). Ocean City also has 13 bulkheads with an average height of about 3.5 m (11 ft); their length varies from 40 m to 365 m (130–1200 ft) (U.S. ACOE 1990). Twelve of these bulkheads are constructed of timber and 1 is made of sheet-pile steel. In addition to the groins and bulkheads, Ocean City has 2 groin breakwaters made of timber-stone, 1 revetment made of sandbags, and 3 bulkhead-revetments: 1 is made of timber, 1 of timber-stone-timber, and 1 of timber-stone (U.S. ACOE 1990).

REACH 11: CORSON'S INLET
TO TOWNSEND'S INLET

Reach 11, Ludlam Island, is 13.8 km (8.6 miles) in length and extends from Corson's Inlet to Townsend's Inlet. It consists of two communities, Strathmere (Upper Township) and Sea Isle City, both classified for recreational land use. The bulbous northern end of the barrier island, occupied by Strathmere, is characterized by a narrow beach (figure 2.36), and the beach widens at the southernmost area of Sea Isle City.

Ludlam Island is unusual among the barrier islands of New Jersey because the shorelines near Corson's Inlet as well as Townsend's Inlet have been relatively stable to accretional over the past 50 years, while the midsection has retreated and is presently sand starved. Much of the development, found mostly in Sea Isle City, however, is located close to the mean high-water line and can be damaged during moderate storms. In 1981 the entire reach was classified as Category I, critically eroding *(NJSPMP)*. The northern inlet shoreline of the island was classified as Category II, significantly eroding, and in the 1990s was undergoing strong erosion effects. Furthermore, the shoreline on Corson's Inlet was classified as Category III, moderately eroding *(NJSPMP)*. Corson's Inlet has never been stabilized or managed by dredging. In recent times, the inlet opening has been narrower than at any other time on record (Farrell and Leatherman 1989), because of the growth of the northern tip of Strathmere into the inlet opening.

Ludlam Island has a very limited natural sand supply. Beach nourishment and hard shoreline-stabilization efforts have not been able to sustain beach width, and its beach is presently critically eroding. Ludlam Island has received beach nourishment several times; Strathmere's most recent beach fill was completed in 1984 and that of Sea Isle City in 1999. The storms of the 1990s were particularly devastating in this reach, completely eroding dunes and overwashing the island in places.

Strathmere (Upper Township). Strathmere contains 3.7 km (2.3 miles) of oceanfront with a narrow beach that decreases to no beach in central Strathmere (near Vincent Street, where the ocean laps against a bulkhead). The widest beach is along the northern inlet shoreline. In many locations the beach is too narrow to support dune systems; constructed dunes have been destroyed and rebuilt numerous times as storms eroded the dune form and overwashed some of the sediment westward onto the marsh and onto properties located west of the roadway. Strathmere has 13 groins ranging from 38 m to 152 m (125–500) in length (U.S. ACOE 1990). Five are made of timber and 7 are made of timber-stone. Strathmere also has 1 timber bulkhead that is 966 m (3175 ft) long and 2.75 m (9.0 ft) high and lines of wooden pilings in the beach designed to dissipate waves before they reach the dunes (referred to as wavebreakers) (U.S. ACOE 1990).

Whale Beach is a segment of narrow beach on Ludlam Island in both Strathmere and Sea Isle City and is an area of critical hazard (figure 2.37). Wave refraction on the inlet shoals and a low sand supply produce erosion along this stretch of shoreline that is not reduced by the presence of

Figure 2.37. Overwash across road with January 1992 storm. Rebuilding sand dike seaward of road. Northern portion of Sea Isle City, N.J.

beach-stabilization structures (2 groins) along adjacent shoreline segments. Attempts at dune construction have not been successful because of the low and narrow beach.

Sea Isle City. Sea Isle City has approximately 7.7 km (4.8 miles) of narrow beach with a steep slope. Sea Isle City, an intensely developed section of the barrier island, is much wider at its southern portion, but it is low in elevation. A dune system was developed seaward of an asphalt promenade that extends from 28th Avenue south to 58th Avenue, which provided significant buffering for the community during storm events prior to the 1990s. North of 58th Avenue, much larger dunes, 3–5 m (10–17 ft) high, have protected houses from wave attack. These relatively wide and high, well-vegetated dunes were destroyed in Sea Isle City by the combined efforts of the October 1991 storm and the storm of December 1992 (figure 2.38). The southern shorefront has no walkway, and private homes directly front the beach with only modest buffering provided by low dunes. Within Sea Isle City are 16 groins that range from 90 m to 218 m (300–716 ft) in length (U.S. ACOE 1990). Four are made of timber, 11 of timber-stone, and 1 of stone. A geo-tube was installed seaward of the southernmost development in the community, where the beach is very narrow. The geo-tube functions as a seawall.

Figure 2.38. Groins in beach, absence of dunes in central portion. Dense residential development with boardwalk, bulkhead. Sea Isle City, N.J.

REACH 12: TOWNSEND'S INLET
TO HEREFORD INLET

Reach 12, known as Seven Mile Beach, is 21.6 km (13.5 miles) long, extends from Townsend's Inlet to Hereford Inlet, and contains two recreational communities, Avalon Borough and Stone Harbor Borough. This drumstick-shaped barrier island has had a history of erosion/accretion on its northern, bulbous end and significant erosion at its southern end.

Northern Avalon has lost land to Townsend's Inlet, an inlet that began to migrate in a southwest direction during the early part of the twentieth century and gradually eliminated almost all of five blocks of the area's street plan. In 1978, when Townsend's Inlet was dredged for sand to be used as beach nourishment for Sea Isle City, the main inlet channel began to migrate again and resulted in additional erosion. Beach nourishment projects were implemented as an attempt to counter the erosion, but did not stabilize the area. In 1981 the *NJSPMP* classified this inlet shoreline as Category III, moderately eroding.

The southernmost portion of Reach 12 has become a classic example of shoreline displacement downstream of a terminal groin. Hereford Inlet is now the widest inlet in the state because of the long-term erosion of the

Figure 2.39. Bulbous end of barrier island. Beach nourishment has widened the beach seaward of the seawall. Avalon, N.J.

southernmost two miles of the barrier island, which was once Stone Harbor Point. The Hereford Inlet shoreline, classified in 1981 as Category I, critically eroding *(NJSPMP)*, has no jetties, but is heavily lined with rock revetment and short groins. This portion of the barrier island is undeveloped and is currently protected by the borough of Stone Harbor as a nature sanctuary.

Avalon Borough. Avalon has 12.5 km (7.8 miles) of oceanfront with a beach that is relatively stable and of moderate width. In 1981 it was classified as Category IV, noneroding *(NJSPMP)*. As a result of a beachfill operation in 1995, the seawall at the back of the beach was covered by sand (figure 2.39). A dune system located in a section east of Dune Drive has been preserved, and is one of the most extensive fields of large dunes, greater than 10 m (33 ft) high, found in New Jersey (figure 2.40) outside of those within the state and federal parks. A portion of Avalon is adjacent to Townsend's Inlet and is continuously protected, with bulkheads and revetments along the inlet frontage. There are 6 groins in Avalon that range from 70 m to 243 m (228–800 ft) in length. Four are constructed of stone, 1 is made of stone-filled timber, and 1 is made of timber (U.S. ACOE 1990). Avalon also has a stone seawall that is 395 m (1300 feet) long, a timber-stone bulkhead that is 1215 m (4000 ft) long, and a timber-stone bulkhead-revetment (U.S. ACOE 1990). In addition to these shore

Figure 2.40. Large dune field preserves the topography of the predevelopment conditions on the barrier island. Avalon, N.J.

structures, a submerged geo-tube breakwater was constructed near the entrance to Townsends Inlet in 1993 as an experiment.

Stone Harbor Borough. The municipality of Stone Harbor has 5.8 km (3.6 miles) of oceanfront. The beach is of moderate width at its northern portion and gradually narrows to the south (figure 2.41). A coastal dune system is present, consisting of low and narrow dunes (1–2 m) created between 1986 and 1991. The northernmost area of Stone Harbor was classified in 1981 as Category IV, noneroding *(NJSPMP)*. Stone Harbor's shoreline incorporates a series of widely spaced groins that retain sand on the beaches, but at the expense of the downdrift area. A combination of sediment starvation, refraction, and diffraction around the last groin in Stone Harbor has produced acute downdrift erosion within the southern area. Because of the narrow beach and the proximity of development to the high-water line, the southern end of Stone Harbor was classified as Category I, critically eroding *(NJSPMP)*. Stone Harbor has 11 groins: 5 are made of timber-stone, 3 are made of stone, and 3 are made of timber-cribstone (U.S. ACOE 1990). They range from 106 m to 245 m (350–804 ft) in length. In addition, there are 4 bulkheads, ranging from 135 m to 2020 m (450–6550 ft) in length, and 2 revetments, 88 m and 245 m (290 and 800 ft) in length (U.S. ACOE 1990).

Figure 2.41. Groins in narrow beach, with artificial foredune line. Bulkheaded shoreline in background. Stone Harbor, N.J.

REACH 13: HEREFORD INLET
TO CAPE MAY INLET

Reach 13 extends from Hereford Inlet to Cape May Inlet, a distance of 12.2 km (7.6 miles). It consists of North Wildwood City, Wildwood City, Wildwood Crest Borough, and a U.S. Coast Guard base. Reach 13, made up entirely of recreational communities, contains no Category I (critically eroding) erosion areas, and has significantly wide beaches (figure 2.42). Wildwood has the widest beach, measuring approximately 365 m (1200 ft). The beachfront within this reach, however, has relatively low elevations, around 0.5 m (1–3 feet), and floods during spring tides and storm surges.

Natural accretion occurs at the northeastern portion of the island, although a jetty at Cape May Inlet accumulates sand at the southwestern portion. Accretion began several decades ago when large quantities of sand began to be transferred from the Stone Harbor side of the inlet to the bulbous end of the drumstick at Wildwood and North Wildwood. However, a combination of tidal currents and storm waves have removed much of the sand along the inlet shoreline of North Wildwood so that the beach is very narrow in the inlet throat (or entrance). Beach-stabilization structures such as bulkheads, seawalls, and groins have been built along the inlet shoreline to counter erosion. The westernmost inlet shoreline in

Figure 2.42. Wide bulbous end of barrier island. Accreting beach with vegetated dune zone. North Wildwood and Wildwood, N.J.

North Wildwood was classified in 1981 as Category III, moderately erod-ing, whereas the inlet shoreline on the easternmost side of the northern tip was classified as Category II, significantly eroding *(NJSPMP)*.

North Wildwood City. North Wildwood City contains 4.6 km (2.9 miles) of wide oceanfront with dunes at the back of the beach. This beach has not always been wide. In 1963 the mean high-water line was at North Wildwood's boardwalk. However, sediment transfers from the updrift barrier island have significantly expanded this beach in recent decades. In 1981 the southernmost portion of North Wildwood was classified as Cat-egory IV, noneroding *(NJSPMP)*. North Wildwood has 3 groins con-structed of rubble and concrete that range from 23 m to 57 m (77–187 ft) in length. In addition, there are 4 bulkheads that range from 264 m to 1580 m (933–5200 ft) in length and 3.5 to 3.8 m (11.3–12.5 ft) in height. The bulkheads vary in construction: 2 are made of timber-sheet pile, 1 is made of steel piling, and 1 of a concrete-stone-brick mix. There are also 3 revet-ments, 1 of stone-timber-rubble, 1 of concrete rubble, and 1 of stone and grout (U.S. ACOE 1990).

Wildwood City. There are 2.2 km (1.3 miles) of beachfront in Wild-wood, characterized by the widest beaches in the state and a very modest dune system (figure 2.43). The width of this beach offers more protection

Figure 2.43. Low flat beach, amusement piers stranded over accreting beach. Patches of vegetation, no dune line. Wildwood, N.J.

against wave attack than any other location in southern New Jersey. In 1981 most of Wildwood was classified as Category IV, noneroding, although the southernmost section of Wildwood was Category II, significantly eroding *(NJSPMP)*. Presently the beach is not threatened by any critical erosion. This coastal community has no hard-engineered structures on the beach. Plans are in progress to build a new convention center to extend the season beyond the typical summer months. This structure will be constructed on the accreted beach.

Wildwood Crest Borough. The municipality of Wildwood Crest contains 3 km (1.9 miles) of oceanfront. The beach is low in elevation and narrower than that of Wildwood. Wildwood Crest has natural, well-vegetated dunes about 2 m (6–7 ft) high, and there are multiple dune lines on the beach. The southern part of Wildwood Crest contains dunes that have formed against the bulkhead. In 1981 Wildwood Crest was classified as Category II, significantly eroding, for most of its length, and Category III, moderately eroding, at its southern end *(NJSPMP)*. Presently this beach is fairly wide and is not undergoing any critical or significant erosion. In addition, Wildwood Crest has a bulkhead that is 1580 m (5200 ft) in length and about 3.4 m (11 ft) in height (U.S. ACOE 1990).

Figure 2.44. Accretion updrift of jetty at Cape May Inlet. Groins in beach, offset of shoreline at the inlet. Coast Guard base and Wildwood Crest.

The U.S. Coast Guard base makes up the southernmost portion of this barrier island. Its coastal area is approximately 1.9 km (1.2 miles) in length (figure 2.44). This area has had accretion on the order of thousands of feet since the construction of jetties at Cape May Inlet. The stone jetty at the inlet that affects the Coast Guard beach is 1383 m (4548 ft) long (U.S. ACOE 1990).

REACH 14: CAPE MAY INLET
TO CAPE MAY POINT

Reach 14 extends 10.1 km (6.3 miles) along the shore. It is the southernmost ocean shoreline in New Jersey, extending from Cape May Inlet to Cape May Point and includes a U.S. Coast Guard base, Cape May City, and Cape May Point Borough. The U.S. Coast Guard base and Cape May City occupy the barrier island, part of the Southern Barrier Islands Complex. Cape May Point occupies the southwest corner of the reach and is part of the Southern Headlands region. Cape May Point State Park and a portion of land owned by the Nature Conservancy (South Cape May Meadows, Lower Township) forms the shoreline between Cape May City and Cape May Point.

The principal cause of the lack of sand in this reach and the subsequent erosion is due to the jetty at Cape May Inlet (figure 2.45). Additionally, localized erosion is exacerbated by the groins and seawall that run

Figure 2.45. Trapping of littoral sand updrift of Cape May Inlet jetty causes downdrift erosion of most of adjacent beach.

along this reach's shoreline. The division between Cape May City and the South Cape May Meadows in the Lower Township lies southwest of the Third Avenue groin. The series of groins in Cape May has effectively blocked the limited volume of sand being transported southwest (figure 2.46). As a result, South Cape May Meadows, Cape May Point State Park, and Cape May Point were relatively sand starved. Because of the beachfill projects during the 1990s, sand is currently bypassing the terminal groin at Third Avenue and is building a beach seaward of the meadows. The initial emplacement of pumped sand was along the entire seaward margin of Cape May City, amounting to 1.4 million cubic yards in 1991, 415,000 in 1993, 331,000 in 1995, 366,000 in 1997, and 360,000 in 1999. It is expected that similar amounts will be pumped to the beaches at two-year intervals.

The region from the Cape May Inlet (formerly known as Cold Spring Inlet) to Cape May City at Ocean Avenue was classified in 1981 as Category I, critically eroding *(NJSPMP)*, because of accelerated beach loss and the deteriorated conditions of the seawall that may provide protection from very small storms, but not from severe storms. The beach at the western portion of Cape May City was classified as Category II, significantly eroding *(NJSPMP)*, because of the dependence on the condition at the updrift beaches. This area is characterized by a narrow to moderate beach with stone groins. Despite the erosion of Lower Township and Cape May Point State Park, these areas were classified as Category III, moderately eroding *(NJSPMP)*, because sufficient setback distance exists

Figure 2.46. Groins have further limited sediment transport beyond Cape May City, causing a large offset in the shoreline. Cape May City, N.J.

between erosion forces and the developed areas or infrastructure. Cape May Point was classified as Category II, significantly eroding *(NJSPMP)*, because of previous erosion trends.

The Coast Guard base stretch of shoreline northeast of Cape May does not have many buildings, although it included an airport at one time. This area has had rapid erosion, estimated at 6 m (20 ft) per year, as a result of its location downdrift of the jetties at Cape May Inlet.

Cape May City. Cape May City has 6.6 km (4.1 miles) of shoreline. Prior to a beach nourishment operation in 1990 and subsequent renourishment, there had been either a narrow beach or no beach at high tide in this area. Continued erosion led to the construction of groins and reinforcement of the seawall. Extensive shoreline development and shore management structures have prevented shoreline retreat, and as a result, offshore slopes are steep. However, with the advent of beach nourishment, the groin compartments are full of sand and there is a dune-building program seaward of the seawall. Cape May City has 14 groins and 1 stone jetty (U.S. ACOE 1990). The groins range from 45 m to 240 m (150–786 ft) in length, and the jetty is 1333 m (4385 ft) in length. Five of the groins are made of timber crib and 9 of them are made of stone. Cape May also has 5 stone seawalls, which range from 122 m to 1345 m (400–4426 ft) in length and are about 4.25 m (14 ft) in height, and 1

Figure 2.47. Exposure of the upland margin forms the southwestern point of New Jersey. Sediment starvation results in narrow beaches and scarped dunes. Groins limit alongshore transport. Cape May Point.

timber bulkhead that is 1125 m (3703 ft) long and 4.4 m (14.3 ft) in height (U.S. ACOE 1990).

Cape May Point Borough. Cape May Point contains 1.8 km (1.1 miles) of oceanfront with a narrow beach and an artificial dune system on most of its length (figure 2.47). Cape May Point is an older part of the continent referred to as the Southern Headlands and is characterized by a low-cliffed coast at this southwestern point of New Jersey. Sand from the barrier island system has masked the exposure of the cliff and has provided a narrow buffer seaward of the Cape May Meadows, the wetlands that were formed between the barrier islands and the interior upland of the continent. In 1981 Cape May Point's eastern section was classified as Category III, moderately eroding, and its western portion as Category II, significantly eroding *(NJSPMP)*. Cape May Point has 9 groins, 6 made of timber-stone and 3 made of stone (U.S. ACOE 1990). These groins range from 85 m to 152 m (275–500 ft) in length. Cape May Point also has 1 timber bulkhead, which is 3.6 m (12 ft) in height and 425 m (1400 ft) in length (U.S. ACOE 1990). In addition to these structures, a submerged breakwater was constructed offshore of the eastern part of Cape May Point in May–June 1994.

CHAPTER 3

Shoreline Change: Historical Displacement

Utilize the historical shoreline change method to immediately begin mapping erosion hazard zones. . . . Care should be exercised in extrapolating erosion rate data owing to variable geologies and other factors. . . . Methods should be established by FEMA to provide effective and regular notice to all land owners in E-zones as to the existence and magnitude of erosion.

National Research Council, *Managing Coastal Erosion* (1990)

Shoreline erosion is a fairly easy concept to understand, because it represents the loss of sediment from the coastal system and the spatial displacement of the waterline inland. However, both the causes and the measurement of shoreline erosion offer some complications. The inherent natural variability in the coastal system results in a shifting of the shoreline over several time spans. In the long term, as the sea level rises, the shoreline-waterline will shift inland, primarily as a result of the encroachment of the sea. Within that shift will be variations produced by spatial and temporal differences in sediment supply, temporal differences in storm frequencies, and the general pulsing of sediments through the system. Indeed, there are so many short-term variations that the trend of shoreline change may take decades or longer to decipher. But with longer-term records and information on beach characteristics, dune forms, and offshore profiles, it becomes possible to understand the physical components of shoreline change and to incorporate that understanding into effective management policies. According to several studies conducted by the H. J. Heinz Center (2000a, b), there is inadequate recognition of the costs of coastal erosion. These analysts point to the need to generate better data sets on the measures of shoreline change and the societal costs that are borne by coastal populations in their efforts to defend the shoreline. Basic to these observers' premise is a fuller appreciation of both long-term and short-term shoreline changes and a better understanding of the processes that are causing them.

At present, the two most important natural variables influencing shoreline change are (1) the quantity of sediments available to the system and (2) the rate of sea-level rise. Further, human alterations of the system are another component of change that is superimposed on the natural variation. It is the combination of these factors that is responsible for the shoreline that we witness today.

Sediment Supply

Beaches are traditionally maintained by natural processes of sediment transport from external sources that accumulate at the point of water-land contact (the littoral zone). Generally the most important source of beach sand is from rivers that transport eroded sediment from inland areas of the continent and discharge their loads in the vicinity of the shoreline. The Mississippi River Delta region is a good example of this sediment accumulation. A map of the state of New Jersey, however, indicates that there are no major rivers discharging sediment at the shoreline (figure 3.1). Most of the state's rivers enter bays or estuaries that catch and trap the available sediment, and therefore a riverine source of sediment does not exist to nourish the beaches. New Jersey is not alone in this situation. A glance at a map of the eastern coast of the United States reveals a similar situation along most of the shoreline.

A second source of beach sand stems from erosion of the continental margin by waves and currents, and is found in cliffs and bluffs at the water's edge. In New Jersey, portions of coastal Monmouth County and a small portion of Cape May County are characterized by low, eroding bluffs. Although this is a source of sand that can be transported to supply adjacent beaches, the rate of sediment delivery is relatively slow and the amounts are insufficient to maintain adequate beaches. The supply has been further reduced by the many structures and walls that have been erected to slow the rate of recession of the bluffs. Thus sediment from the erosion of cliffs and bluffs is of minor importance in New Jersey's coastal zone.

The remaining natural source of sand for New Jersey beaches is from offshore. During the rise of the sea level that occurred over the past several thousand years, considerable quantities of sand were submerged as

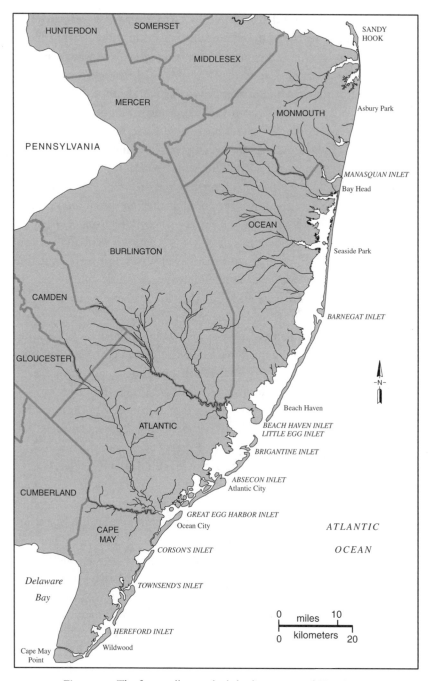

Figure 3.1. The few small watersheds leading to coastal New Jersey.

the ocean rose and inundated the sandy coastal plain. Waves can move sand landward along the bottom and cause accretion at the beach. Indeed, the recovery of beaches following a storm is the process of wave transport of sand from the offshore zone to the profile of the beach. The offshore source is finite, however, and once the available sand from within the depth of the wave disturbance has been mobilized and transported, the supply becomes exhausted. No new sands are generated to replace the materials that have been moved. As a result, over a period of time, the offshore sand supply will have contributed as much as is available. For New Jersey, coastal scientists believe that this limit has been reached. There are sand sources that remain in the offshore zone, but they are simply too deep or too far away to be mobilized by the surface waves and currents. Thus the offshore sources of sediment are essentially exhausted in the context of landward transport by the natural processes of waves and currents within the limiting depth. However, these deep-water sand sources can be reached by offshore dredges, and thus they are a source of sediment for artificial replenishment of eroded beaches.

In lieu of human manipulation of the coastal sediment supply, a fundamental characteristic of the New Jersey shoreline is the slow but continuous loss of sediment over time. There are no new natural sources of sediment entering the beach environment to balance the losses of sand caused by the processes of wave and current transport from the beaches into deeper waters offshore and/or downdrift. These losses are slow but, over an extended period, there is a net decrease of sand in the beach zone, on the barrier islands, and in front of the low bluffs that line the Atlantic coast of New Jersey. The overall result is a slow displacement, recognized as shoreline erosion and represented as an inland shift of the shoreline. This process is a natural one, and the conditions responsible for it are found along most of the world's shorelines.

In the past some shoreline displacement was reduced by the natural sacrifice of the sand that had accumulated on the barrier islands in the form of large dunes. This dune sand contributed sediment to slow the losses from erosion and thus buffered the rate of displacement. However, by now the dunes have been removed along much of the New Jersey Coast and that source of sand is not available in much of the zone. Rates of sand loss are also affected by human intervention. Some of the engineered structures that have been placed to maintain inlets or to protect development interfere with the natural transport processes and, in some instances, the structure directs some of the sand into deeper water at rates faster than would naturally occur and where it is less likely to return to nourish the beaches. Other structures affect the distribution and rate of alongshore transport by producing accumulations in one location while causing greater erosion at an adjacent site. These structures may not diminish the overall sediment supply, but do cause a redistribution that is beneficial to one area and adverse to another.

New sand may be added to the beach-dune system by physically transferring it by truck or by pipeline. Additional sediment, both dredged from sites more than a mile offshore at depths greater than 20–25 m (65–82 ft) and brought by trucks from the mainland, has made possible a series of additions to the existing beaches. Moving sand by these means is costly, and each program that adds sand to the beaches must be carefully evaluated. Any quantity of sand added to the system modifies the local sediment budget but does not reverse the causes of sediment removal and longer-term losses; eventually this new sand will be removed by natural processes. Sand management is an important issue in the coastal area, involving decisions on the retention of the sand that is being lost, the identification of new potential sand sources, and the opportunity to bring new sand into the system.

Sea-Level Rise

As the sea level rises, the shoreline is displaced inland, except in those areas where sufficient sediment is accumulating to build the shoreline seaward. In coastal locations where a local shortage of sediment is accompanied by sea-level rise, the problem is compounded and the result is an increased rate of shoreline displacement. In areas of gentle slopes, such as those associated with barrier islands, the amount of horizontal shift produced by sea-level rise alone is thought to be on the order of 100 times the amount of vertical rise. At coastal bluffs, however, the horizontal shift is much less. It is likely that the response of the shoreline position to sea-level rise has a time lag associated with it. The inland displacement may not occur incrementally each year, but it will become noticeable when a significant storm mobilizes large quantities of sediment and the returning sand does not build the beach back to its former position, because the profile is now assuming a sediment distribution in equilibrium with the higher sea level. The failure of the beach to return to its original position will add a stepwise or periodic response to the effect of sea-level rise along with changes in the sediment budget.

Changes in Shoreline Position

In addition to sediment supply and sea-level rise, a variety of specific coastal situations either increase or decrease the rate of displacement, such as differences in exposure, persistence in alongshore drift, local sediment delivery, and the occurrence of beach-stabilization structures. They all contribute to the variation in the rates of shoreline erosion and displacement. Major differences are evident in the erosion-displacement responses of the barrier islands versus those of the cliffed coasts. Variations also occur across and within the barrier islands, and there is further variation in response to the effects of groins, jetties, and seawalls.

An analysis of the historical changes of the shoreline reveals considerable variation in both a spatial and a temporal sense. Many coastal areas show periods of net erosion interspersed with periods of net accumulation. This variation was demonstrated in a comparison of shoreline positions by Nordstrom et al. (1977) that applied rates of positive and negative displacement over a number of time periods (figure 3.2). These investigators consulted old coastal surveys and existing aerial photographs to derive displacement for a number of points along the shoreline. Rather than comparing the position of the shoreline between the oldest source and some modern spatial equivalent, they separated the nineteenth-century data from that of the twentieth century. The rationale was that in many places the shoreline was natural in the earlier period and more subject to human manipulation in the later period. Thus two numbers are presented. An example of this data set is figure 3.2, which contains the results of shoreline point comparisons for Cape May County. The time periods do not cover the entire century and may vary for particular points. However, it is instructive to note the presence of agreement of rates of change in some locations and disagreement in others. This is an element of the natural variation as well as of the human manipulation of the shoreline.

Information on historical shoreline change is available on a New Jersey Department of Environmental Protection CD-ROM (1995) that shows shoreline positions at different times and allows for empirical evaluations across any combination of shoreline positions as far back as the 1830s relative to a base shoreline position of 1986. This approach depicts the entire shoreline at a particular point in time and provides excellent information about the trends of shoreline change, the interplay between the updrift and downdrift ends of islands, the effects of structures, and

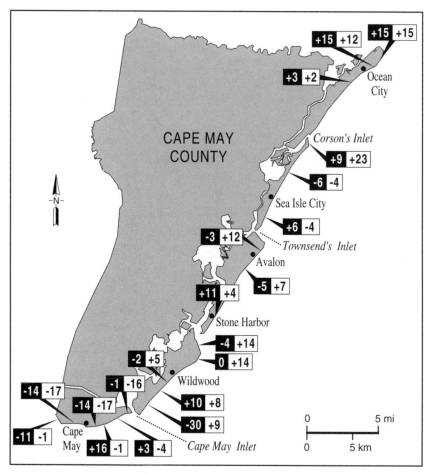

Figure 3.2. An example of variations in shoreline displacement rates in feet per year in Cape May County, covering more than one century of record (Nordstrom et al. 1977). Measurements from the nineteenth century appear in the white boxes and from the twentieth century in the black boxes. The latter period incorporates the construction of structures at the shoreline and at the inlets.

episodes of beach nourishment. The data have been entered into a GIS program and put on the CD. An example (figure 3.3) of this data source shows the position of the shoreline of Brigantine Island at various times from 1836 to 1986. The northeastern terminus of the island (updrift end) is shown to be very dynamic, shifting about 1.0 km (0.6 mile) over the period of record. In addition it is apparent that the Absecon Inlet end of the island was being displaced inland until the construction of the jetty in 1940, and then the shoreline built seaward to its present position. Further,

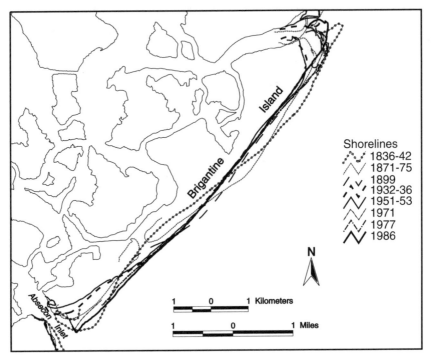

Figure 3.3. A comparison of shoreline positions on Brigantine Island, 1836–1842 to 1986. (NJDEP 1995.)

one of the earliest sources represented in the data set (1871–1875) depicts an inlet in the island about 3 km (1.8 miles) updrift from the current Absecon Inlet. This is very valuable information, because old inlet sites are locations of weak links in the barrier islands which are vulnerable to storm breaching.

A general pattern of shoreline displacement is complicated by human attempts to stop or reverse the trend of erosion. The use of shore-parallel structures such as seawalls creates an artificial shoreline position while permitting continued erosion of the sand immediately seaward of the structure (in the subaqueous zone) and causing increasingly steep offshore slopes in these locations. Shore-perpendicular structures such as groins and jetties interfere with the alongshore transport of sand in the beach zone and cause offsets in the shoreline or beaches of unequal widths. Sand usually accumulates on the updrift side of these shore-perpendicular structures but is withheld from the downdrift portion, causing an accelerated inland displacement downdrift of these structures. Beach nourishment projects produce a positive shoreline response, both rebuilding the beach and adding sand to the sediment budget, temporarily displacing the shoreline seaward.

All of these approaches to stabilization manipulate the shoreline position and affect the general analysis of trends of displacement. It is therefore necessary to qualify any identification of change or rate of change by taking account of the natural as well as the cultural processes that have affected the outcome of the shoreline position over time and considering the changes to the total profile from the dunes, through the beach, to offshore.

Despite the temporal variations, nearly all of the New Jersey shoreline was experiencing a net erosion and was being displaced inland prior to the advent of the massive, spatially limited beach nourishment projects of the late 1990s (U.S. ACOE 1971, 1990; Nordstrom et al. 1977, The downdrift ends of barrier islands or tips of barrier spits were exceptions to this erosional trend, but these accumulations were primarily elongations of the islands or spits and were downdrift shifts rather than seaward displacements. The updrift, bulbous ends of barrier islands were accumulating and losing sand on a lengthy cycle that can be seen in charts and aerial photos from the twentieth century. Furthermore, as the updrift, bulbous ends of the barrier islands release sand, this tends to buffer the shoreline displacements of downdrift portions of the islands. Changes in the position of the shoreline have responded to both long-term and short-term processes of sediment exchange. General descriptions of the shoreline position in New Jersey implicitly include the effects of natural cycles and of cultural attempts at stabilization. When we disaggregate these two factors in a description of the temporal trend of shoreline displacement, their effects must be associated with a specific period of time, and extensions beyond that period are cautioned. Analyses over longer time periods tend to smooth the naturally and culturally produced variations and identify the general net situation. Therefore two levels of analysis are useful: one that examines short-term changes at a local-regional level and one that examines longer-term changes over a broader area.

New Erosion Data: New Jersey Beach Profile Network

A relatively new effort and data set, supported by the state of New Jersey, has been developed to describe shoreline change. Beginning in 1986, beach profiles have been surveyed by the New Jersey Beach Profile Network at approximately one-mile intervals for most of the New

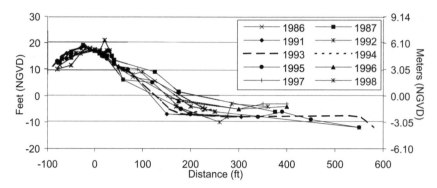

Figure 3.4. Variations in topography at Profile 151, Normandy Beach. New Jersey Beach Profile Network, 1986–1998. (Farrell et al. 1988, 1994, 1995, 1997, 1999.)

Jersey shoreline. Some public coastal landholdings (such as parks and wildlife areas) were initially excluded from the survey program but were later added in the 1990s. Such profiles provide complementary information to spatially contiguous photo comparisons. The profiles include the dune zone, beach zone, and the immediate offshore zone.

Each survey portrays the condition at a single line (figure 3.4), and there are now more than 15 years on record. Originally the profiles in the offshore zone extended to a depth of 1–2 m (3–7 ft), but they now extend to an average depth of 6 m (20 ft) below sea level. Because of this length of record, there is now a basis for comparing the beach condition through an extended period of time and determining both the trends and the amount of variation. This significant accomplishment gives coastal researchers and planners alike a powerful data set with which to unambiguously examine spatial and temporal details of shoreline change. The data derived from the profile surveys represent changes in the cross-section of the total profile. If an equal amount of sand shifts from dune to offshore or vice versa, for example, the net change in the profile measurement is zero. Therefore any net positive or negative change is a respective gain or loss to the cross-section of the total profile and not just a shift of sand from one position to another position within the profile. This characterization is especially important because these profiles are surveyed only once per year (in the fall) and there is the possibility that more spatially restricted profiles (to shallow wading depth, for example) would miss and misrepresent the cross-shore (onshore-offshore) transfers of sediment associated with storms and post-storm recovery at the beach. The more recent, deeper surveys are not so affected by the balanced exchanges associated with cross-shore transfers that do not leave the general profile area. This restricted variation is amply depicted in figure 3.4 through the large variation in the beach zone versus the more limited variation in the dune location as well as far offshore.

Furthermore, the profiles capture localized manipulations of the beach profile such as dune-building or beach nourishment projects that deposit volumes of sand on the beach within the profile bounds. Nourishment projects create a large positive gain on the profile and will significantly skew any average condition at the site. The annual surveys record the initial change and then track the subsequent modifications to the site and its interaction with adjacent sites.

Sediment Budget

Each of the profiles records the configuration of the topography for each site in cross-section terms, a snapshot of the dune-beach-offshore conditions at one specific point in time. A comparison of successive cross-sections over time then provides a measure of two-dimensional area change relative to the profile line across survey periods. It is a usual convention to extend the cross-sectional information derived from the comparison of profiles to create a volume measure of the amount of sand gained or lost. For example, an expression such as a change of 10 cubic yards of sand per foot of beach (the original data are in feet and yards only) would indicate that the cross-sections of the profiles differed by 270 square feet (each 27 square feet of change between two profiles equal one cubic yard of sand per foot of alongshore beach length). Extending this information to a stretch of beach between two adjacent profile lines spaced one mile apart, the total volume for this segment can be calculated from the change in the cross-section area between successive surveys for each of the two profiles, summing them and dividing by two. This calculation produces an average change in volume in cubic yards per foot of beach between the two profile lines for the intervening period, usually one year. The total change in volume may be derived by multiplying its average value by the length of the beach between the relevant profile lines, and by 5280 if the intervening distance is one mile. Subsequently, the annual change in volume for each compartment may be tracked as mean change for the entire eleven-year survey period (table 3.1).

Net profile changes in cubic yards of sand per foot along beach lengths between adjacent profiles over the 1986–1997 period provide information on the scale of the changes for the eleven years and their

Table 3.1 ∼ AVERAGE ANNUAL VOLUME CHANGES
OF PROFILE COMPARTMENTS, 1986–1997
(CUBIC YARDS PER FOOT OF BEACH LENGTH)

Stations	Mean for area	Beachfill
285–284	−7.87*	$
284–184	−8.82*	
Gateway entrance		
184–183	23.47	$
183–282	65.90*	$
282–182	66.39*	$
182–181	23.42	$
181–180	17.44	$
180–179	10.94	$
179–178	12.23	$
Long Branch city limits		
178–177	13.14	$
177–176	1.26	
176–175	−.008	
175–174	1.36	
174–173	0.72	
173–172	−.012**	
172–171	−0.39**	
171–170	−2.55	
170–169	−1.55	
169–168	−.037	
168–267	−1.31*	
267–167	−1.7*	
167–166	−1.51	
166–165	−1.05	
165–164	−1.42	
Shark River Inlet		
163–162	3.75	$
162–161	9.21	$
161–160	7.65	$
160–159	11.27	$
159–158	18.42	$
158–157	17.15	$
157–256	22.92*	$
Manasquan Inlet		
156–155	−0.92	
155–154	−2.7	
Metadeconk River		
154–153	−0.81	
153–152	−0.74	
152–151	0.18	
151–150	0.07	
150–149	0.1	

(*continued*)

Table 3.1 ∼ (continued)

Stations	Mean for area	Beachfill
149–148	0.82	
148–147	2.01	
147–247	−1.61*	
247–246	−2.7*	
246–146	0.61*	
Barnegat Inlet		
245–145	16.85*	$
145–144	−5.84	$
144–143	−9.05	
143–142	−2.68	
142–241	0.88*	
241–141	2.5*	
141–140	1.72	
140–139	0.20	
139–138	−2.48	
138–137	2.52	
137–136	2.29	
136–135	−1.08	
135–234	1.91*	
Little Egg Inlet		
134–133	−1.68	
133–132	2.6	$
132–131	6.37	
Abescon Inlet		
130–129	2.05	$
129–128	1.67	
128–127	−0.99	
127–126	−0.81	$
Great Egg Harbor Inlet		
225–125	−15.39*	$
125–124	12.17	$
124–123	21.24	$
123–122	6.68	
Corson's Inlet		
121–120	−10.52	
120–119	−2.61	
119–118	−0.54	$
118–117	−0.90	$
Townsend's Inlet		
216–116	25.28*	$
116–115	5.85	$
115–114	3.59	
114–113	−1.37	
113–212	11.73*	

(continued)

Table 3.1 ~ *(continued)*

Stations	Mean for area	Beachfill
Hereford Inlet		
111–110	14.42	
110–109	9.51	
109–208	7.04*	
Cape May Inlet		
108–107	19.66	$
107–106	5.13	$
106–105	5.6	$
105–104	6.36	$

SOURCES: Uptegrove et al. (1995); Farrell et al. (1994, 1995, 1997, 1999); *http://www. usace.army.mil/.*

*1994–1997.

**1987–1992 only.

spatial distribution (figures 3.5, 3.6). These average values per beach seg-ment represent part of the longer natural trend as well as the human manipulation of sand on the beaches. Many of the positive averages cor-respond to sites of beach nourishment, such as at Sea Bright and Mon-mouth Beach, lines 184-178; Shark River Inlet to Manasquan Inlet, lines 163-256; Ocean City, lines 125-122; and Cape May, lines 108-104. Other positive effects observed are due to the presence of structures, such as the areas updrift of the Barnegat Inlet jetty, the Absecon Inlet jetty, and the Cape May Inlet jetty. There are also downdrift negative average effects, but many have been countered by beach nourishment projects. With the exception of the results for locations known to be affected by beach nour-ishment or structures, the amounts of change are small over the eleven-year period. Few profile lines have been consistently positive or negative throughout the duration of the surveys. Furthermore, there are no reaches that are consistently positive or negative. Although a reach may have a net positive or negative budget, the pattern seems to be one of variation or oscillation within the reach, perhaps indicative of episodic alongshore transfers of sediment. As more current information from the New Jersey Beach Profile Network becomes available, it will no doubt provide addi-tional insight into the effects of current beach nourishment efforts.

Shoreline Displacement

Another view of shoreline change from the NJBPN data set is the average annual variation of the location of the 1929 NGVD

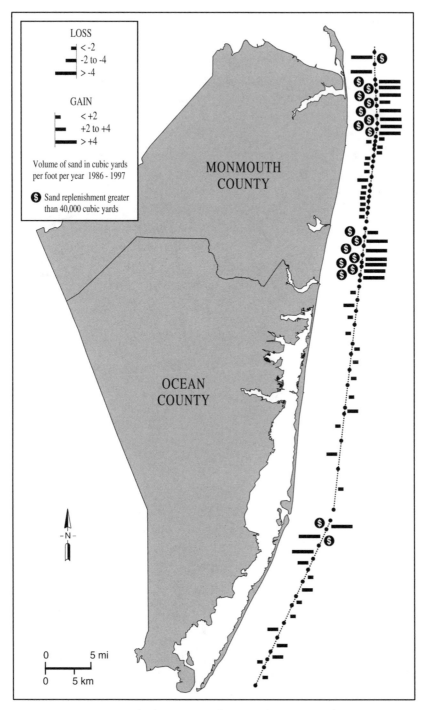

Figure 3.5. Average annual changes in sediment volume on profile compartments, 1986–1997, Monmouth County and Ocean County. (Uptegrove et al. 1995; Farrell et al. 1994, 1995, 1997, 1999.)

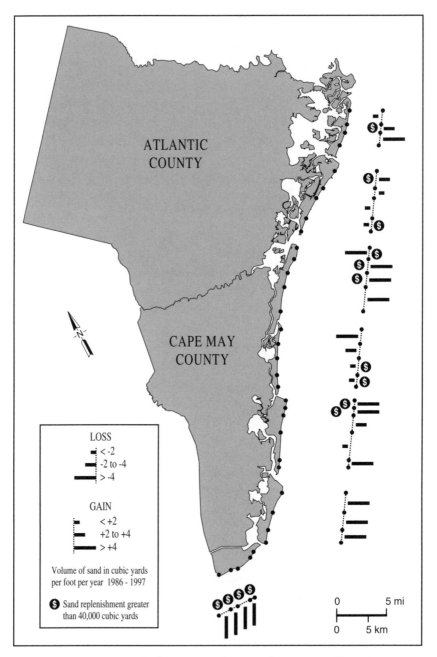

Figure 3.6. Average annual changes in sediment volume on profile compartments, 1986–1997, Atlantic County and Cape May County. (Uptegrove et al. 1995; Farrell et al. 1994, 1995, 1997, 1999.)

(National Geodetic Vertical Datum) shoreline for each profile over time (figures 3.7 and 3.8 and table 3.2). NGVD, a reference benchmark established in 1929, is a constant elevation that was about 0.16 m (0.59 ft) lower than mean sea level calculated for the 1960–1978 tidal epoch, and about 0.27 m (0.9 ft) lower than mean sea level in the year 2000. Although using the NGVD reference plane to establish a shoreline position is common practice, comparing surveys one year apart is problematic because of short-term variation or "noise" in the NGVD shoreline position due to transfers of sand on the profile and from alongshore variation in the beach configuration. Therefore the data set presented in figures 3.7 and 3.8 displays mean annual change, and the data in table 3.2 include mean annual change and net change, for the survey period. On one scale, the year-to-year variation is especially instructive in describing the effects of beach nourishment and subsequent adjustments, because they happen quickly. Furthermore, the episodes of beach nourishment and effects of structures will influence the measurements and add to the variation and should be accounted for. The individual annual changes in the NGVD shoreline incorporate both the natural and the beach nourishment variations. However, given a long period of beach profile data at these sites, the short-term variations, or noise, will become less significant and a general trend will emerge. As examples, Monmouth Beach (lines 180–178) and most of Ocean City (lines 125–123) show large net seaward displacements that coincide with short-term episodes of beachfill. On the other hand, most of the changes of shoreline position in other areas over the eleven-year period represent minor changes in beach widths and could be the result of interannual variation. The presence of bulkheads, seawalls, artificial dune lines, and other cultural manipulations tends to further limit the inland penetration of the shoreline and partially constrains the displacement of the total profile. Despite the inherent noise and problems associated with the temporally varying waterline position measurements, such data can be used in combination with other measures of change to determine general trends of shoreline displacement.

Identification of High-Hazard Zones

In order to manage and plan for natural coastal hazards so as to reduce the damage from their occurrence, it is necessary to identify the coastal areas that are most exposed and at highest risk. FEMA now requires communities to identify areas of high hazard as a condition for

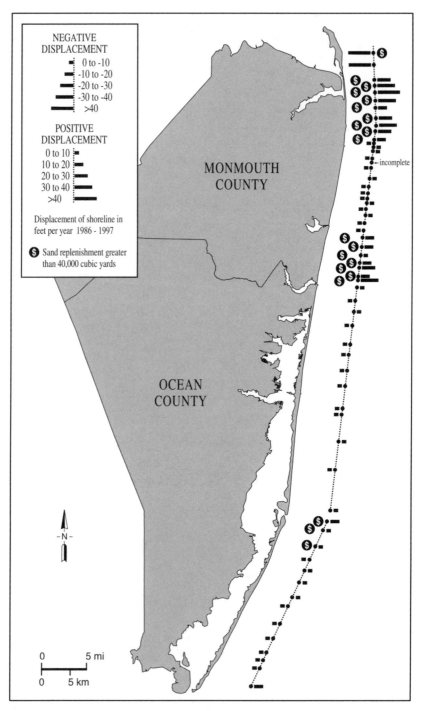

Figure 3.7. Average annual shoreline displacement of NGVD intercept on profile, in feet, 1986–1997, Monmouth County and Ocean County. (Uptegrove et al. 1995; Farrell et al. 1994, 1995, 1997, 1999.)

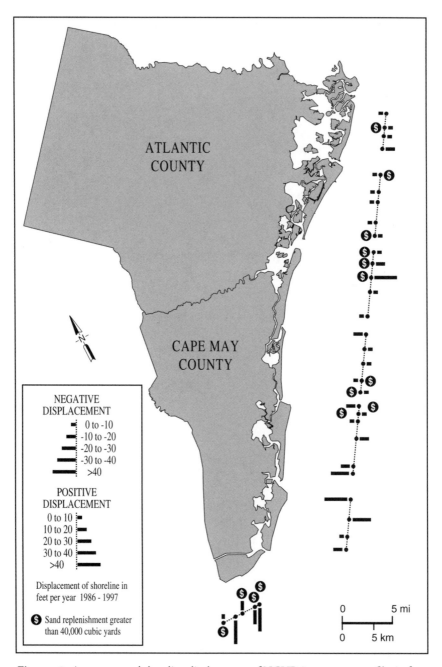

Figure 3.8. Average annual shoreline displacement of NGVD intercept on profile, in feet, 1986–1997, Atlantic County and Cape May County. (Uptegrove et al. 1995; Farrell et al. 1994, 1995, 1997, 1999.)

Table 3.2 ～ ANNUAL DISPLACEMENT OF SHORELINE, 1986–1997 (IN FEET)

Stations	Net change	Mean change	Beachfill
285*	−255.1	−101.8	$
284*	−171.1	−52.4	
Gateway entrance			
184	+234	+21.3	$
183	+380	+34.5	$
282*	+360	+128.6	$
182	+405	+36.8	$
181	+215	+19.6	$
180	+238	+21.6	$
179	+435	+39.5	$
178	+301	+27.4	$
Long Branch city limit			
177	+211	+19.2	$
176	−1	−0.1	
175	+36	+3.3	
174	+34	+3.1	
173	−43	−3.9	
172		Incomplete	
171	−20	−1.8	
170	+41	+3.7	
169	−35	−3.2	
168	−10	−0.9	
267	−45	−4.1	
167	+8	+0.7	
166	−51	−4.6	
165	+1	+0.1	
164	−31	−2.8	
Shark River Inlet			
163	−9	−0.1	
162	+117	+10.2	$
161	+121	+11.0	$
160	+83	+7.6	$
159	+210	+19.1	$
158	+279	+25.4	$
157	+180	+16.4	$
256*	+110	+39.3	$
Manasquan Inlet			
156	+63	+5.7	
155	−47	−4.3	
154	+2	+0.2	
Metedeconk River			
153	−25	−2.3	
152	−15	−1.4	
151	−41	−3.9	
150	−48	−4.4	

(continued)

Table 3.2 ～ (*continued*)

Stations	Net change	Mean change	Beachfill
149	−48	−4.4	
148	−22	−2.0	
147	−40	−3.6	
247	+95	+8.6	
246	−31	−3.8	
146	+92	+9.2	
Barnegat Inlet			
245*	+50	+17.8	$
145	+82	+7.5	$
144	+28	+2.5	$
143	−29	−2.6	
142	+12	+1.1	
241**	+15	+5.0	
141	+15	+1.4	
140	−36	−3.3	
139	−8	−0.7	
138	−32	−2.9	
137	−36	−3.3	
136	−75	−6.8	
135	+57	+5.2	
234**	+60	+21.4	
Little Egg Inlet			
134	−39	−2.3	
133	+44	+4.0	$
132	+44	+4.0	
131	+157	+14.3	
Absecon Inlet			
130	−30	−2.3	$
129	−11	−1.0	
128	−45	−4.1	
127	−28	−2.5	
126	+34	+3.1	$
Great Egg Harbor Inlet			
225**	+15	+5.0	$
125	+220	+10.0	$
124	+475	+43.3	$
123	+102	+9.3	
122	−79	−7.2	
Corson's Inlet			
121	−182	−16.5	
120	+6	+0.5	
119	+37	+3.4	
118	−36	−3.4	$
117	+15	+1.4	$

(*continued*)

Table 3.2 ⁓ (*continued*)

Stations	Net change	Mean change	Beachfill
Townsend's Inlet			
216*	+50	+17.8	$
116	+72	+6.6	
115	−22	−2.0	
114	+117	+10.6	
113	−120	−11.8	
212**	+90	+30	
Hereford Inlet			
111	−441	−40.1	
110	+346	+31.6	
109	−40	−3.6	
208**	−50	−16.9	
Cape May Inlet			
108	+511	+46.6	$
107	+312	+28.4	$
106	−184	−16.7	$
105	+609	+55.4	
104	−70	−6.4	$

SOURCES: Uptegrove et al. (1995); Farrell et al. (1994, 1995, 1997, 1999); *http://www.usace.army.mil/*.
*January 1995–October 1997.
**October 1995–October 1997.

inclusion in the National Flood Insurance Program. This effort will help decision makers and planners in developing policies that address those areas in relative need. Hazard management tools can be used in such areas to help reduce public and private exposure to coastal hazards. The NJBPN coastal profile data can be quite useful in this regard as a means to identify an area's vulnerability. One of the earliest references to hazardous areas is found in a National Shoreline Study conducted by the Army Corps of Engineers (1971), which describes and classifies about 82% of the New Jersey coast as critically eroding (figure 3.9), the most severe category. The 1981 *New Jersey Shore Protection Master Plan* classified areas of critical erosion on the basis of development at risk (figure 3.9), and determined that almost 32% of the New Jersey shore was in that category (NJDEP 1981). Later, Nordstrom et al. (1986) also described areas of critical erosion in the coastal area of New Jersey (figure 3.9) and classified about 20% of the coast as critically eroding. There is some agreement among the three sources regarding the areas of erosion, but their objectives were different and thus the conclusions are not geographically coincident.

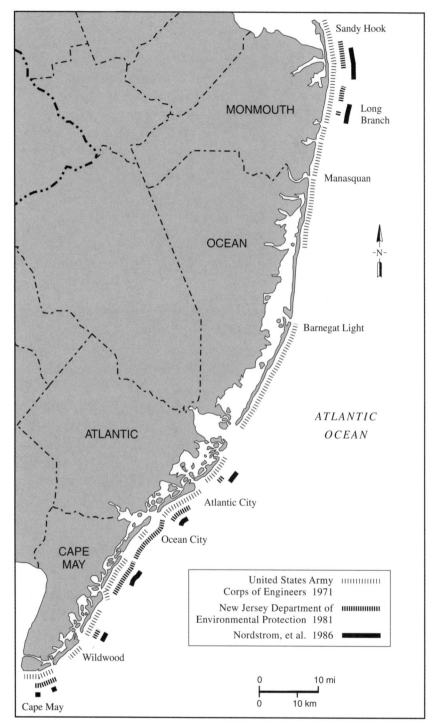

Figure 3.9. Distribution of critically eroding shoreline areas. (U.S. ACOE 1971; Nordstrom et al. 1986; NJDEP 1981.)

These sources used different criteria in their analyses regarding what areas of the New Jersey coast were experiencing critical erosion. From their experience it is clear that any system of classifying the shore in terms of erosion categories will first have to develop a set of criteria that can be used in determining levels of risk and identifying those areas at high risk at the shore. This approach will be most valuable if based on a procedural identification of the factors of exposure to hazards and their magnitudes, which will then lead to an identification of exposed locations. Five criteria have been used previously:

- Erosion rate—shoreline movement
- Inlet proximity—position
- Barrier island breaches
- Overwash areas
- Elevation

Erosion Rate—Shoreline Movement

Hazard areas along the coast based on erosion data can be identified by using factors such as the (relative) beach width and height, presence of dunes, sediment budget, and density of development (NJDEP 1981). Property situated in high-erosion areas is generally expected to be exposed to the effects of coastal storms at higher intervals, resulting in repetitive damage. Exceptions are expected in areas where structural approaches and beach nourishment projects have occurred. Hence other variables are needed to determine shoreline stability versus movement over different temporal spans.

Inlet Proximity

By their nature, inlets are highly dynamic, and therefore adjacent areas are subject to frequent episodes of erosion and accretion (NJDEP 1985). The downdrift ends of barrier islands are often low and especially exposed to overwash. The updrift ends are frequently the sites of better dune development but may be characterized by severe erosion. Physical structures (such as jetties) further affect mobility at inlets.

Island Breaches

Areas of former inlet breaches and locations of former inlets have been identified through historical records (NJDEP 1985). Many of these areas remain relatively vulnerable during severe storms, because they are low in elevation and because underlying sediments in such areas are not as cohesive as sediments in adjacent locations. During severe storms these sites are more likely to experience serious erosion or breaching. As a result, development located in these areas will be at higher risk to flood and storm-related damage. The community of Harvey Cedars provides a good example. During the 1962 Ash Wednesday storm, this town lost almost 50% of its property tax base (that is, ratables) as a result of the storm and the breaching that occurred at the location of an old inlet (Savadove and Buchholz 1993).

Overwash Areas

Previous overwash areas can also be identified from storm damage records and aerial photographs (NJDEP 1985). It is expected that these areas are susceptible to repetitive flooding, and therefore structures situated near them are also likely to be prone to repetitive flood-related damage.

Elevation

Low-lying areas are more easily flooded and are often in exposed locations. Frequently marshes that have been filled for development are only slightly above the high-tide level. FEMA uses elevation as its main criteria to determine risk of flood damage.

Application to Management

Given the multitude of physical structures and barriers to sediment transport and the many episodes of sediment manipulation within New Jersey's coastal zone, it should come as no surprise that at present many portions of the shoreline in New Jersey do not freely evolve. Because there have been so many efforts to manipulate the beaches, snapshots of them at different points in time may be inadequate to fully document and describe the conditions that characterize the shoreline. Shoreline position data sets are most helpful when they are accompanied by a narrative of the human efforts to alter the shoreline associated with data on shore-stabilization efforts. This is especially true when shore-parallel structures or exceptionally long shore-perpendicular structures have been constructed that interfere with the transport of sand.

Shoreline management can be thought of as sand management. There is a net long-term loss of sand in New Jersey, as reported in the *NJSPMP*

(NJDEP 1981) and by the Army Corps of Engineers in its *Limited Reconnaissance Report* (1990). Some portions of the state's shoreline have been classified as areas of critical erosion; other portions are eroding at a moderate rate or are fairly stable. Management options, to be most effective, must be specific to each reach or region, and thus will vary in accordance with the history of sediment loss, the presence of structures, and the objectives for continued use of that portion of the shore. From the shoreline data sets, a number of insights can be generalized:

- Shoreline change and sediment management must be approached on a regional scale.
- Three-dimensional data and volume of sediment as a measure of change are preferable to two-dimensional descriptions of the shoreline position.
- Objectives should be established that are attainable within a reach (region), recognizing the natural variation in shoreline change, especially at inlets.
- Much of the Northern Headlands shoreline has been relatively stable over time because it is a cliffed coast that is sediment poor, and the bluffs are not as easily eroded as the barrier island sediments.
- Structures, especially at inlets, cause updrift and downdrift shoreline displacements.
- The exchange of sediment has many short-term variations; longer-term records are needed to determine trends.
- Beach nourishment can cause short-term positive displacements in local areas.
- Inland displacement of the shoreline is expected in the medium- to long-term time period because of sediment deficits and the effects of sea-level rise.
- The identification of high-risk areas is a step toward the application of hazard management measures to reduce loss and damage. Criteria and levels of risk should be developed and applied to identify the exposed at-risk locations.

Conclusion

Displacement of the shoreline is a product of the human and natural processes that both shield and expose the coast to the major driving forces of waves and currents acting upon an increasing sea level.

A natural loss of sand from the shoreline takes place on both sides of the barrier islands, and an inundation of the land margins on the barrier islands and on the mainland is occurring at a slow but measurable rate. Transfers of sand from offshore or other local sources will help to reduce some of the displacement. A systematic identification of hazardous areas should be established to assist in applying the principles of coastal hazard management.

CHAPTER 4

Coastal Storms: Their Importance to Coastal Systems and Management

> The potentially devastating effects of the storm surge are further illustrated if one considers that a cubic yard of seawater weighs nearly three-fourths of a ton[,] which pretty much guarantees destruction of anything in its path.
>
> J. M. Williams and I. W. Duedall, *Florida Hurricanes and Tropical Storms* (1997)

> Hurricane Andrew radically changed the perceptions regarding the vulnerability of coastal areas and the nation to the financial consequences of severe Atlantic hurricanes. Before 1992, most insurers, natural disaster agencies, and financial analysts believed that only a catastrophic California earthquake would have the capability of disabling national insurance and financial markets. Hurricane Andrew demonstrated that a severe storm . . . could have similar consequences.
>
> Insurance Research Council and the Insurance Institute for Property Loss Reduction (1995)

Storms are the major natural driving force causing permanent changes in the coastal zone. They mobilize and transfer sediment across the beaches, the dunes, and through inlets. They reshape the barrier islands, erode the low bluffs, flood and nourish the wetlands, and inundate the low-lying areas. There is a certain fascination with coastal storms, because they overwhelm many of the features with which we are familiar. The beach disappears, the dunes are cut back and leveled, boardwalks tumble into the ocean in response to the storm surge and intense wave action. The management response to these storm events is critical, as decisions are exercised at an early stage regarding evacuation and other safety measures. Other sorts of management decisions are required following the storm, when it is necessary to assess damage and to determine post-storm recovery options that can run the gamut from replacing sand on the beach, rebuilding dunes, repairing infrastructure, and filing insurance claims to condemning properties, rezoning, and changing land use.

The challenge is to learn from the dynamics of past storms, in order to make decisions that do not increase the risk of future damage. A review of earlier storm conditions is enlightening, because it establishes a scale of coastal change. But bringing together information on coastal storms is more than providing a litany of events; it is an opportunity to raise issues regarding risk and exposure. It is an opportunity to pose questions about coastal management, damage reduction, public safety, and the mitigation of storm effects in the highly developed coastal zone.

Post-1980 Stormy Weather

There is both a scientific interest in coastal storms, as demonstrated in an excellent review of weather systems in New Jersey by Ludlam (1983), and a popular interest, as shown in a 1993 highly illustrated summary of the effects of major storms on the New Jersey coast and its development (Savadove and Buchholz 1993). This latter theme is repeated in a number of illustrated histories of portions of the coast that show the effects of particular major storms (such as Methot 1988 and Lloyd 1994). A discussion of post-1980 storms helps us to understand, in a tangible way, the relative impact such storms can have upon coastlines.

Northeasters and hurricane weather systems are major contributors in shaping, eroding, and redefining New Jersey's coastal zone. Since 1980 the state has experienced an increased frequency and intensity in storm activity compared with previous decades. From 1980 through 1998, twelve storms were considered "major" northeasters, and their damage to aspects of the infrastructure (figures 4.1 and 4.2) or to the beach system was clearly documented (figure 4.3). These include the storms of March 1984, January 1987, October 1991 (the Halloween or "All Hallow's Eve" storm), January 1992, December 1992, March 1993, March 1994, the blizzard of January 1996, March 1996, and November 1997. The period of December 1997 through February 1998 was especially stormy and marked by a number of strong events in close succession. A few mid-Atlantic hurricanes in the post-1980 period did influence the conditions at the shore, notably Gloria in 1985. Hurricane Floyd, though responsible for major damage in interior New Jersey because of excessive rainfall, did not produce commensurate changes at the shore. The data collected regarding these storms

Figure 4.1. Damaged boardwalk and bulkhead, overwash onto road. Sea Girt, N.J. December 1992 storm.

Figure 4.2. Sand dune eroded and boardwalk damaged. Manasquan, N.J. January 1992 storm.

Figure 4.3. Coastal dune eroded, old fence lines exposed in beach, rip-rap at bulkhead. Ocean City, N.J.

indicates the hazards that continue to visit the New Jersey coast. The information presented in this chapter is derived from a number of different sources, including the National Weather Service, NJDEP, the U.S. ACOE, the National Oceanic and Atmospheric Administration, the National Ocean Service, and local newspapers within the New Jersey coastal zone. Data on storm water-level elevation were taken from the NOAA website at *http://co-ops.nos.noaa.gov* and from reports produced by the U.S. ACOE and the NOS.[1]

To compare storms' relative strengths and to rate storm events, classification systems have evolved that establish a hierarchy of severity, such as the Saffir-Simpson scale for hurricanes and the Dolan-Davis scale for northeasters (Dolan and Davis 1992). Another method developed to classify storms is the use of probability of occurrence, which draws on historical data to statistically describe the probability of the repetition of specific water levels as well as to rate storm events, such as a 100-year storm flood (a probability of 1 in 100 years). Through an application of the scales or the probability values, it becomes possible to consider both the natural and the management responses to storms of varying magnitudes. Scenarios can be developed based on the severity of past storms to estimate the effects of future high-magnitude storms and to propose management strategies to mitigate their potential disruptions.

Identification of Northeasters and Hurricanes

Northeasters

Northeasters, or nor'easters, as they are sometimes called, continue to be a relatively frequent storm occurrence along the New Jersey coastline. When the center of a northeaster affects or hits the coast, damage to the natural system and to the built environment can equal and sometimes surpass that produced by hurricanes. Northeasters are technically known as cold-core cyclones, a frontal system that is produced over cooler regions than hurricanes. Northeasters are also influenced by the circulation of high-altitude atmospheric winds. The northeaster storm season occurs when the jet stream is in a southerly pattern, usually between the months of October and April. Northeasters are known for their duration: they can last for several days and thus for several tidal cycles. The winds, which typically range from 32 to 80 km/h (20–50 mph), can exhibit a wind fetch (the distance wind is in contact with the water surface) that ranges from hundreds to thousands of kilometers. For example, the Halloween storm of 1991 was associated with a wind fetch of over 3125 km (1800 miles). The combination of wind fetch and a long duration can produce extremely large wave conditions and accompanying high storm surges along eastern coastal areas. Northeasters' characteristic high water and high wave energy are responsible for the majority of the property damage, loss of life, and beach erosion in New Jersey. Dolan and Davis's (1992) classification system of northeasters places these storms into five categories (table 4.1); it was originally developed by comparing the wave height and duration characteristics of 1347 storms that occurred at Cape Hatteras, North Carolina. This quantitative treatment produced a distribution of relative wave-power measurements that was subsequently separated into the five classes, with Class One having the most frequent occurrence (about 50% of the total number analyzed) of weak storms, grading to Class Five with the rarest occurrence (1%) of very severe storms. The classification was extended to a description of the effects of the classes on beaches, dunes, and community development. The system it is very useful for researchers as well as for coastal planners and emergency management personnel, both in rating the severity of these storms and in analyzing characteristics of exposure and hazard.

Table 4.1 ~ DOLAN-DAVIS CLASSIFICATION OF NORTHEASTERS

Storm class	One: Weak	Two: Moderate	Three: Significant	Four: Severe	Five: Extensive
Beach erosion	Minor	Modest	Across beach	Erosion and recession	Extensive
Dune erosion	None	Minor	Significant	Severe on low-profile beaches	Destruction
Overwash	None	None	None	Communitywide	Massive in shoals and channels
Property damage	None	Modest	Local	Communitywide	Regionwide
Average peak wave height	6.6 ft	8.2 ft	10.8 ft	16.4 ft	23 ft
Average duratioin	8 hrs	18 hrs	34 hrs	63 hrs	96 hrs
Relative frequency	49.7%	25.2%	22.1%	2.4%	1%

SOURCE: Dolan and Davis (1992).

Hurricanes

Hurricanes, known formally as warm-core cyclones, are areas of low pressure around which winds flow in a counterclockwise direction in the Northern Hemisphere. They are associated with violent weather and strong winds that can bring devastation to any area the storm passes near or over. An Atlantic hurricane is a nonfrontal system that develops and reaches maturity in tropical or subtropical waters, usually in the Gulf of Mexico, the Caribbean sea, or the waters surrounding the southern tip of Florida. Hurricanes are believed to originate from low-pressure systems that are spawned near the west coast of Africa (Doehring, Duedall, and Williams 1994). Their size and intensity can vary greatly depending on several factors, including wind fields, rain fields, and the duration of time spent over warm water, which strengthens the storms. Because warm water is required for hurricane production, hurricanes tend to be seasonal, starting in early June and continuing through November with their peak in mid-September. They are powerful storms and usually develop quickly, move quickly, and reach a point of landfall quickly. Compared with northeasters, the area at which a hurricane makes landfall is typically small, and damage from wind and storm surges tends to be more localized (Dolan and Davis 1994). The Saffir-Simpson scale (table 4.2) was devised to classify hurricanes on the basis of wind speed, but also incorporates measures of atmospheric pressure, storm surge, and damage

Table 4.2 ～ SAFFIR-SIMPSON HURRICANE SCALE

Category	Definition and Effects
One	Winds 74–95 mph or storm surge 4–5 ft (1.22–1.52 m) above normal: No real damage to building structures. Damage primarily to unanchored mobile homes, shrubbery, and trees. Some coastal road flooding and minor pier damage.
Two	Winds 96–110 mph or storm surge 6–8 ft (1.83–2.44 m) above normal: Some roofing material, door, and window damage to buildings. Considerable damage to vegetation, mobile homes, and piers. Coastal and low-lying escape routes flood 2–4 hours before arrival of center. Small craft in unprotected anchorages break moorings.
Three	Winds 111–130 mph or storm surge 9–12 ft (2.74–3.66 m) above normal: Some structural damage to small residences and utility buildings with a minor amount of curtainwall failures. Mobile homes are destroyed. Flooding near the coast destroys smaller structures, with larger structures damaged by floating debris. Terrain continuously lower than 5 ft (1.52 m) above sea level may be flooded inland as far as 6 mi (9.6 km).
Four	Winds 131–155 mph or storm surge 13–18 ft (3.96–5.49 m) above normal: More extensive curtainwall failures with some complete roof structural failure on small residences. Major erosion of beach areas. Major damage to lower floors of structures near the shore. Terrain continuously lower than 10 ft (3.05 m) above sea level may be flooded, requiring massive evacuation of residential areas inland as far as 6 mi (9.6 km).
Five	Winds greater than 155 mph or storm surge greater than 18 ft (5.49 m) above normal: Complete roof failure on many residences and industrial buildings. Some complete building failures with small utility buildings blown over or away. Major damage to lower floors of all structures located less than 15 ft (4.57 m) above sea level and within 500 yards (455 m) of the shoreline. Massive evacuation of residential areas on low ground within 5–10 mi (8–16 km) of the shoreline may be required.

SOURCE: *www.nhc.noaa.gov.*

NOTE: Actual storm surge value will vary considerably depending on coastal configurations and other factors.

to communities (*http://www.nhc.noaa.gov*). This scale has become one of the tools used to characterize the level of hazard in the coastal communities, resulting in "hazard maps." Although hurricanes rarely make landfall in New Jersey, the high water levels and waves associated with hurricanes do reach the coastline and cause erosion, flooding, and damage.

Significant Coastal Storms

Significant Northeasters (since 1980)

From 1980 through 1998, the New Jersey coast was affected by more than 25 northeasters. Twelve of these systems were considered

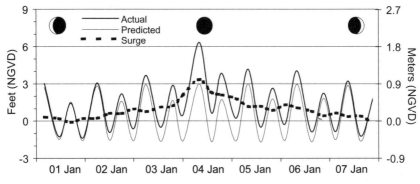

Figure 4.4. January 1992 storm. *(http://co-ops.nos.noaa.gov/)*

"major" storms, because of their unusually long period of duration, high water levels, extensive damage, or some combination of these factors that caused the storm water elevation to exceed the probability of occurring once in five years. A brief description of some of these significant storms follows, which serves to characterize the conditions attained along the New Jersey coast. Water-level elevations are based on the National Geodetic Vertical Datum (NGVD) of 1929. As noted in chapter 3, the NGVD is a national reference plane established for the year 1929, and it represents a benchmark surface for measuring water-level elevations in the coastal zone. In this presentation, the NGVD plane is 0.16 m (0.59 feet) below mean sea level at Atlantic City for the tidal epoch 1960–1978. Because the NGVD plane refers to a constant surface, it does not have to be adjusted to take account of the rise of sea level or the subsidence of land, and thus it is possible to compare the surge and inundation of storms over extended periods of time.

The signature of a particular storm is available in the water-level elevations recorded at Atlantic City, and it is possible to determine the onset, the duration and magnitude, and the end of the storm effects by comparing the actual water-level elevations to the elevations predicted by normal tidal influences. Consider the effects of a storm in early January 1992 and its water-level record (figure 4.4). Three values are shown. The thin solid line is the predicted tidal-water elevation, which shows its characteristic two highs and two lows per day. One of the daily highs is higher than the other, and that pattern is repeated each day. One of the daily lows is lower than the other. January 4 is the time of new moon, the condition described as a spring tide, when the tidal range is increased. Superimposed on the predicted tide is the dark line of the recorded water elevation at the pier in the Atlantic Ocean. It is apparent that for several days, several tidal cycles, the actual water level was higher than the predicted water level because of the effects of the storm. The timing of the maximum recorded water level coincides with the time of the maximum predicted water level, suggesting that the effects of the storm were fairly coincident with the timing of the spring tide. The third line is dashed, and it shows the difference between the predicted and observed water levels. The difference is maximized on January 4. The record of the storm is the storm surge, the difference between the predicted and observed water levels. The storm effects began on the morning of January 2 and continued until the morning of January 7 (five days). However, the duration that the water was elevated more than 30 cm (1 ft) above the predicted level is much less (three days), the duration the water was above the predicted spring-tide high-water level was still a bit less (two days), and the major surge of about 0.9 m (3 ft) was concentrated in one day. Calculated on the basis of the surge water level attained, this was a storm with a recurrence interval of slightly greater than once in 10 years. Of additional importance in causing change to the area's natural and developed systems is the five high-water periods that were greater than the spring high-tide level. These were the times when the upper beach and the dunes were eroded. This was when water penetrated into the streets and flooded low-lying properties. This was when the infrastructure was exposed to damaging conditions. This was when the wetlands and the bayside developments were completely flooded. The storm signature depicted in figure 4.4 displays the conditions associated with a moderate-level meteorological storm combined with a spring tide and a repetition of spring-tide flooding through several tidal cycles. It deserves to be classified as a major coastal storm, because of the elevation and penetration of water along the coast. Other major storm events during this post-1980 period have their own impressive characteristics. The descriptions presented here are derived from their water-level signatures, similar to that presented in figure 4.4, and also from their impacts on the coast.

MARCH 28–31, 1984

The northeaster that peaked during March 28–29, 1984, was the first major northeaster to strike the New Jersey shoreline since the now infamous Ash Wednesday storm of 1962.[2] Fortunately, this storm passed quickly; the center of the storm was in the coastal zone for only two high tides when the surge reached above usual high-tide levels.

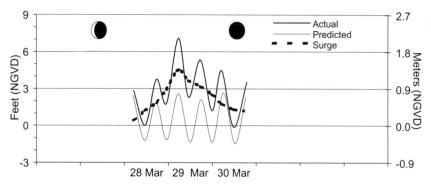

Figure 4.5. March 1984 storm. *(http://co-ops.nos.noaa.gov/)*

Water-level elevations recorded at the NOS tidal gauge in Atlantic City indicated that the storm water level was up to 2.07 m (6.8 ft) above NGVD on March 29 (figure 4.5). The highest open ocean waves, 3.6–6.4 m (10–20 ft), were also recorded on March 29 (U.S. ACOE 1985). Water levels throughout New Jersey reached more than 2.13 m (6.5 ft) above NGVD during the storm. Tidal surges in the back bay areas were relatively lower than those along the coast, measured at 0.61 (2 ft) above the predicted high water. The most severe part of the storm occurred on the morning of March 29, lasting approximately 12 hours. Flooding occurred throughout the coast, damaging houses and commercial buildings and closing several roadways, along with considerable beach erosion.

JANUARY 1–2, 1987

The effects of a northeaster that peaked on January 2, 1987, were intensified by an unusual astronomical alignment that occurs about once every decade, the combined effects of the planetary relationships known as perigee and syzygy. *Perigee* occurs when the moon is closest to the earth, causing a greater gravitational pull on the oceans and resulting in unusually high tides. *Syzygy* is an astronomical event that occurs when the sun, moon, and earth are in a line to produce the maximum gravitational pull on the oceans, creating spring tides. When the two planetary relationships occur simultaneously, tides and tidal ranges throughout the world are abnormally high. When coupled with a major storm system, water-level elevations can become extreme (figure 4.6). For New Jersey, lunar tides were predicted to range 1.98 m (6.5 ft). Average tidal ranges typically are 1.35 m to 1.7 m (4.5–5.5 ft). However, the combination of perigee, syzygy, and a moderate storm caused the water levels to range

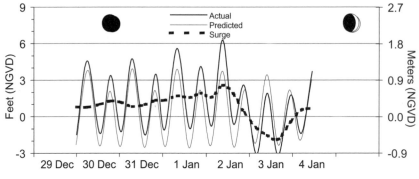

Figure 4.6. January 1987 storm. *(http://co-ops.nos.noaa.gov/)*

more than 2 m and reach about 1.92 m (6.3 ft) above NGVD, with a surge of about 0.76 m (2.5 ft). A surge greater than 0.52 m (1.7 ft) occurred for three successive high tides, reaching into the upper beach and low areas. This storm produced minor to moderate flooding and beach erosion along New Jersey's coastline, with the southern part of the state reporting the most damage (*Philadelphia Inquirer* 1987).

OCTOBER 28–NOVEMBER 3, 1991

This northeaster, named the "Halloween storm" or "All Hallow's Eve storm" of 1991, struck the New Jersey coast with its full force on "Mischief Night," October 30, 1991. It was a very unusual storm, generating power from three separate weather systems (NOAA 1992). Steady winds generated seas in the open ocean in excess of 12 m (40 feet) in height and created a "dam-like" structure that restricted outward tidal flows and tidal river discharge. Consequently, inland areas were flooded over 160 km (100 miles) away from the coast (NOAA 1992). Along the New Jersey coast, storm surges reached elevations of 2.23 m (7.3 ft) above NGVD (figure 4.7). Under bright sunny skies, high waves pounded the entire shore (figure 4.8). Water levels reached 4.43 feet above the predicted tidal elevations and remained high for seven successive high tides. The storm persisted for 114 hours (Ludlum 1991), and high water levels were still present 96 hours after the peak of the storm. The storm surges resulted in major flooding to the barrier islands as well to the backbay areas and inland communities. Dunes were damaged in some locations and completely breached in others, and some roads were (completely) closed because of excessive flooding. In addition, boardwalks were badly damaged by the extended period of high water and wave action (NOAA 1992).

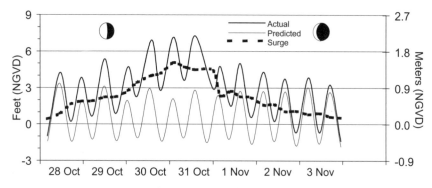

Figure 4.7. October 1991 storm. *(http://co-ops.nos.noaa.gov/)*

JANUARY 3–7, 1992

The characteristics of the January 1992 northeaster differed greatly from that of a typical northeaster. This storm was rather small, developed very quickly, and was a fast-moving system that swept through the New Jersey coastal area in approximately 24 hours (Shelby 1992; U.S. ACOE 1993a). However, as in January 1987, the storm coincided with a spring tide and the planetary relationship known as perigee, which amplified the normal tidal range. Although the influence of the storm was felt

Figure 4.8. Storm waves crashing over seawall, Sea Bright, N.J. Halloween storm, October 1991.

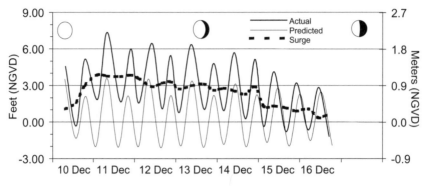

Figure 4.9. December 1992 storm. *(http://co-ops.nos.noaa.gov/)*

for several days, there was a single pronounced peak in storm surge (figure 4.4). The maximum water elevation recorded at Atlantic City by the NOS was slightly over 1.92 m (6.3 ft) above NGVD on January 4, incorporating a storm surge of 3.34 feet. This storm caused damage along the entire coast of New Jersey. Many beaches were severely eroded, with small cliffs (scarps) up to 1.5 m (4.9 ft) high in some areas, such as Ocean City, whereas other beaches were completely breached, causing sand to pile up along roadways (U.S. ACOE 1993a). Road damage occurred as a result of flooding, and dune fencing was lost in several locations.

DECEMBER 10–17, 1992

A northeaster, classified as *the* "storm of the century," struck the New Jersey coast in mid-December 1992 (*Asbury Park Press* 1992) (figure 4.9). A long wind fetch, approximately 1600 km (1000 miles), and unusually high waves accompanied this storm into the Delaware to New York region (U.S. ACOE 1993b). Rainfall of 5–10 cm (2–4 inches) occurred during the period of December 10–13 and added to the flooding. During the night of the 11th, the strongest period of the storm coincided with a spring tide. High storm surges that night were about 2.13 m (7.4 ft) above NGVD at Atlantic City, and up to 2.6 m (8.7 ft) at Sandy Hook, New Jersey, some 120 km (75 miles) apart. The storm lasted approximately 140 hours (or 5.8 days), affecting 12 tidal cycles. The highest surge occurred early in the storm period, associated with spring tide, but a surge of greater than 0.6 m (2.0 ft) existed for 9 successive high tides, each of which flooded the marshes and adjacent areas and reached into the upper portions of the beach.

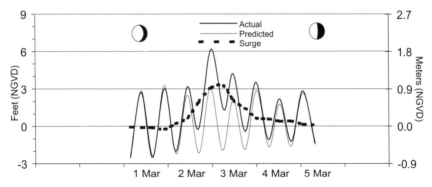

Figure 4.10. March 1994 storm. *(http://co-ops.nos.noaa.gov/)*

The storm's impact on the New Jersey coast was substantial. Beach erosion was evident all along the coast, as was structural damage to seawalls, boardwalks, and protective dunes. Flooding up to 1.52 m (5 ft) closed numerous roads, including the Garden State Parkway, and invaded many homes in low-lying areas. In the history of observations at Atlantic City, this storm surge of 2.26 m (7.4 ft) was the second highest on record, exceeded only by the storm surge of the 1944 hurricane of 2.32 m (7.6 ft) above NGVD.

MARCH 1–5, 1994

The March 1994 northeaster occurred during a relatively quiet storm season. However, this storm produced considerable storm surge despite occurring just prior to neap tide (figure 4.10). Maximum water level was 1.89 m (6.2 ft) above NGVD, with a surge of 1.0 m (3.3 ft). One surge level reached the dunes, and two others operated more than 2 feet above normal high tides. The storm caused regional damage, including minor to major beach erosion, road closures, and some flood damage to commercial and residential buildings throughout the coastline. This northeaster lasted approximately 50 hours, with the most intense surge effects occurring over 36 hours.

DECEMBER 19–23, 1995

In this storm of December 1995, a very strong low pressure system traveled along the coast and passed through New Jersey just before spring tide (figure 4.11). Strong winds propelled waves and water toward the coast during the early portion of the storm, which pushed the water level to 1.8 m (6 ft) above NGVD, a surge of 0.79 m (2.6 ft) above pre-

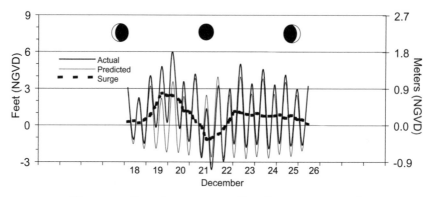

Figure 4.11. December 1995 storm. *(http://co-ops.nos.noaa.gov/)*

dicted high tide. As the center of the low moved to the northeast, the winds shifted to the offshore and the water level was quickly lowered to 1.1 m (3.6 ft) below NGVD, 0.36 m (1.1 ft) lower than the predicted low tide. Then, with the abatement of the locally strong winds, the spring-tide effects and regionally high water returned to produce a minor storm surge of 0.36 m (1.1 ft). The signature of this storm is particularly instructive; it shows the influence of wind direction in causing local water levels to change relative to onshore or offshore flows. Further, the combination of storm effects and spring tides amplified the magnitudes of both the high-water level and the low water.

JANUARY 7–10, 1996

The "blizzard of 1996" was the next big storm to be classi-fied as *the* "storm of the century." This storm system was relatively short. It lasted approximately 48 hours and produced three relatively high tides (figure 4.12); it also created hazardous conditions throughout the state of New Jersey because of significant snowfall. Wave heights reported ranged from 4.5 to 7.6 m (15–25 ft) with tides 0.9–1.2 m (3–4 ft) above normal (Lelis 1996). Water levels exceeded 1.5 m (5.5 ft) above NGVD during the height of the storm on January 7th, but its effects were reduced because the tidal range was decreasing after a spring tide earlier in the week. Heavy snow and strong winds accompanied this storm throughout the coast and added to the perception of magnitude of the event. However, it was a modest storm relative to the elevated water levels. There was severe local erosion, with scarps in the beach reaching up to 3.7 m (12 ft) (Bates and Moore 1996). Snow fences were torn away in many locations as well as sections of boardwalks. Flooding due to waves breaching sand dunes,

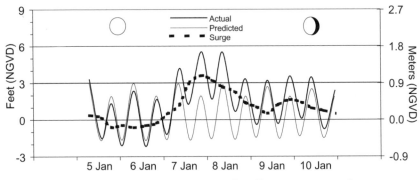

Figure 4.12. January 1996 storm. *(http://co-ops.nos.noaa.gov/)*

accompanied by significant amounts of snow and rain, made some road-ways impassable and caused water damage to houses in low-lying areas.

MARCH 19–21, 1996

In March 1996 a combination of frontal storm and spring tide once again produced an exceptionally high water level (figure 4.13). The maximum elevation of the water reached 1.92 m (6.29 ft) and was a singular peak, with a surge value of 0.94 m (3.1 ft). The maximum water level was sustained for only one high tide on March 19, but the storm surge was greater than 0.3 m (1 ft) for an additional three high tides. This storm illustrates the effect of a storm's occurrence in association with a spring tide to produce a long period of high storm surge.

DECEMBER 3–17, 1996

Some weather patterns are significant because of their duration as well as their magnitude. Such was the case during the first half

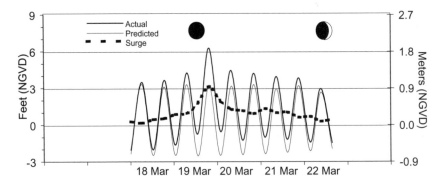

Figure 4.13. March 1996 storm. *(http://co-ops.nos.noaa.gov/)*

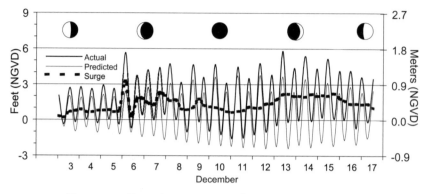

Figure 4.14. December 1996 storm. *(http://co-ops.nos.noaa.gov/)*

of December 1996 when several storms sent elevated water levels and waves to erode the beaches and dunes (figure 4.14). The first storm caused the water elevation to peak on December 6 at 1.73 m (5.67 ft) above NGVD. This was shortly after the neap tide of December 3. Another storm sent the water elevation 1.77 m (5.82 ft) above NGVD on December 13, shortly after the spring tide of December 10. Although the absolute water elevation was higher during the latter storm, the earlier storm had a storm surge of 0.99 m (3.25 ft) compared with a 0.66 m (2.16 ft) surge for the latter. Importantly, the prolonged influence of the several storms kept the water level high during this period. Sixteen of the recorded high-tide water levels were greater than the maximum predicted water level for the period. Thus there was constant flooding and beach erosion, and scarping of the dunes along the shore.

NOVEMBER 13–16, 1997

A winter storm combined with spring tide conditions pushed up the water level in mid-November (figure 4.15). The maximum water elevation of 1.81 m (5.93 ft) above NGVD occurred on November 14, with an accompanying surge of 0.63 m (2.06 ft). However, the water remained high during four successive high tides, and deep-water waves of up to 5.2 m (17 ft) eroded the shore (NOAA 1998).

JANUARY 27–FEBRUARY 8, 1998

Two storms one week apart produced among the highest water levels of the last two decades of the century (figure 4.16). The late January storm coincided with a spring tide and pushed the water to

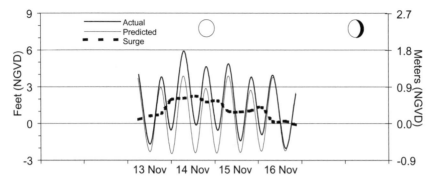

Figure 4.15. November 1997 storm. *(http://co-ops.nos.noaa.gov/)*

1.76 m (5.78 ft) above NGVD on January 28. The following week, another storm occurred very close to neap tide and raised water levels to 1.81 m (5.93 ft) above NGVD on February 5. Whereas the January storm surge measured 0.82 m (2.68 ft) on top of the spring-tide level, the February storm had a surge of 1.89 m (3.9 ft) above the neap-tide level. Each storm maintained a long duration of high water. The January storm had three very high tides and a total of eleven successive high tides that were greater than the forecast spring-tide level. The February storm had three high tides that reached to within 0.15 m (0.5 ft) of each other and a total of ten high tides that were greater than the spring-tide level. Together, these two storms pushed the water into all of the low-lying regions and kept it there. The beaches were constantly eroded and lost a great deal of sediment volume. Following these two events, the beaches were very low and the dunes were very vulnerable to succeeding storm effects.

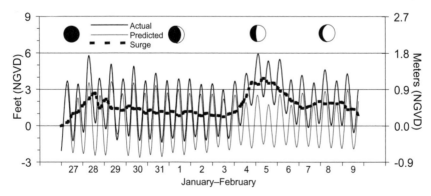

Figure 4.16. January–February 1998 storm. *(http://co-ops.nos.noaa.gov/)*

Table 4.3 ∼ HURRICANES AFFECTING COASTAL NEW JERSEY SINCE 1980

Name	Date	Wind speed kmph (mph) Atlantic City, N.J.	Category on Saffir-Simpson scale
Dean	September 30, 1983	32 (20)	Below 1
Josephine	October 13–14, 1984	40 (25)	Below 1
Gloria	September 27, 1985	128 (80)	Weak 2
Bob	August 19, 1991	40 (25)	Below 1
Felix	August 7, 1995	>32 (>20)	Below 1

SOURCE: U.S. ACOE (1996).

Other Significant Northeasters

Since 1980 there have been other storms that have affected the New Jersey coastline, although they have not had effects as significant in terms of water-level changes and damages as those just reviewed. But although their duration may have been shorter, each one of them has influenced the shaping of beaches. Less powerful storms can cause a net loss in sand and, in turn, a net loss in beach dimensions. In general, these other storms produced minimal to moderate erosion along the New Jersey coastline, some dune and structural damages, and localized flooding in low-lying areas. The December 1994 and November 1995 storms are two examples of storms with relatively short duration. Water-level elevations were significantly higher than that of predicted high-tide levels for approximately 12 hours on December 24, 1994, and 24 hours during November 15, 1995. An important element in any storm is the cumulative nature of its effects: the removal of sediment from the beach and the flooding of low-lying areas reduce the buffering capacity of the dune and exacerbate the effects of succeeding storms. The winter of 1997–98 provides a good example of this type of sequencing, when changes beyond what might be expected from an individual storm were produced because of the lack of recovery between storms.

Significant Hurricanes Affecting New Jersey, post-1980

Since 1980 there have been numerous hurricanes along the East Coast of the United States, but only a few have tracked along the coast and generated storm surges and high waves along New Jersey (table 4.3). Hurricane Felix is an example of one hurricane that was of sufficient strength to have an effect upon the coast. On August 7–9, 1995, Hurricane Felix "stalled" in the Atlantic, southeast of the New Jersey shoreline,

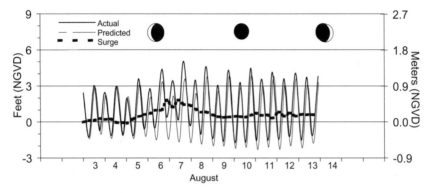

Figure 4.17. Hurricane Felix, August 1995. *(http://co-ops.nos.noaa.gov/)*

and caused some elevation of the water and accompanying high waves (Prichard 1995). Some compared the storm surges from this hurricane to those of the December 1992 northeaster (Lelis 1995). However, the storm surge levels achieved during Felix were less than 1.68 m (5.5 ft) above NGVD (figure 4.17), or lower than any of the northeaster storms already described. Water levels were consistently 0.6 to 0.9 m (2 to 3 ft) higher than the predicted high-tide levels throughout the storm's duration, but they occurred during neap tide conditions and thus the flooding effect was reduced. Damages sustained along the coast from Hurricane Felix included area flooding, strong winds that blew debris around, and severe beach erosion because of high waves and the duration of the event. Damages did not exceed those produced by the March 1984 storm or the December 1992 storm.

The biggest concern with hurricanes that reach the New Jersey coastline is their duration. In general, hurricanes move or track quickly. (The most notable exception is Hurricane Dennis, of August 26–September 6, 1999, which stalled and remained stationary off the coast of North Carolina for over a week.) Because hurricanes are intense and move rapidly, they usually cause localized flooding and erosion to the beaches. Although not an exceptionally high surge producer, Hurricane Josephine (figure 4.18) produced water-level elevations that were significantly higher than the predicted high-tide levels for over 60 hours (figure 4.19). Hurricane Bob's influence, in contrast, was barely noticed as it produced high water elevations for approximately 12 hours. In some cases, such as Hurricane Dean, water-level elevations were not affected and were fairly consistent with predicted high-tide levels. In other cases, hurricanes have produced conditions similar to those of northeasters. High water levels produced by

Figure 4.18. Hurricane Josephine storm waves overwashing barrier island. Remnants of coastal foredune. Holgate Section, Forsythe National Wildlife Refuge, October 1984.

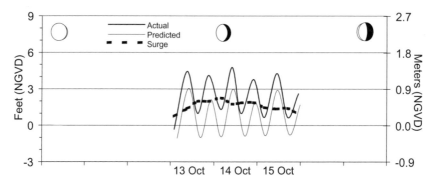

Figure 4.19. Hurricane Josephine, October 1984. *(http://co-ops.nos.noaa.gov/)*

Hurricane Gloria in late September 1985 occurred during spring-tide conditions and pushed water levels to 2.16 m (7.1 ft) above NGVD (figure 4.20). These water levels were slightly more severe than those of most of the storms of the previous two decades, and only slightly lower than the most severe storms on record (table 4.4). However, because Gloria moved through the area very rapidly, onshore winds and storm surge were restricted to one high tide. The result was minor erosion and damage from the water along the coast.

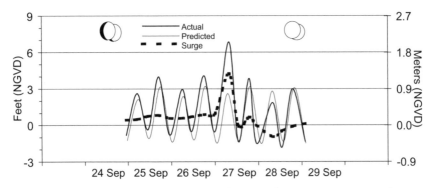

Figure 4.20. Hurricane Gloria, September 1985. *(http://co-ops.nos.noaa.gov/)*

Measures of Storm Severity and
Intervals of Recurrence

The occurrence of severe storms brings forth the notion of the "storm of the century" or a similar designation. Certainly, the erosion, damage, disruption, and casualties caused by each severe storm create an individual measure of its effect. Comparative measures are available in the Saffir-Simpson scale for categorizing the severity of hurricanes and the Dolan-Davis scale to classify northeasters.

Another method developed to compare storms involves the use of frequency curves. As we have seen, storm magnitudes and duration vary considerably. Storm water levels and storm surges can build up over several tidal cycles or pass quickly. However, one common element of storms is their increased elevation of water. This sole characteristic is directly correlated with many effects of coastal storms and thus can assist in determining the possible magnitude of erosion from waves and surge, penetration of dunes, overwash, and the flooding of barrier islands and bayside communities. In enabling an estimate or comparison of a range of storm intensities, flood-water elevations serve to reduce the characteristics of a storm to a single variable or common element, water level. Although the storm water level is only one of several storm-related variables, it is perhaps the single best descriptor of the overall effects and impact of a storm in the coastal zone.

Through time, storm events and associated water levels have been measured at many coastal locations, allowing comparisons to be made. This type of comparison organizes the observed data from specific sites so as to construct a frequency curve and to calculate probabilities of a range

Table 4.4 ⌒ STORM SURGE LEVELS AT ATLANTIC CITY TIDAL GAUGE (METERS)

National Geodetic Vertical Datum (NGVD) (1929)	Vertical Mean Lower Low Water (MLLW)	
2.89	3.37	FEMA 100-year tidal surge
.		
.		
.		
2.58	3.07	FEMA 50-year tidal surge
.		
.		
.		
2.43		
.		
.		
.		
2.34	2.83	FEMA 30-year tidal surge
2.31	2.80	Hurricane 1944
2.25	2.74	December 1992
	2.71	October 1991
2.19	2.68	March 1962, Hurricane Belle 1976
2.16	2.65	FEMA 20-year tidal surge
2.13	2.62	November 1950
	2.58	Hurricane Gloria 1985
2.07	2.55	March 1984
.		
.		
.		
	2.43	January 1987, January 1992
1.92	2.40	FEMA 10-year tidal surge (Margate Bridge begins to flood), March 1996
	2.37	March 1994
1.82	2.31	Hurricane Donna 1960, December 1995, November 1997, February 1998
1.79	2.28	December 1996
1.76	2.25	FEMA 5-year tidal surge, January 1998, January 2000
1.73	2.22	March 1993
	2.19	January 1996, February 1998, September 2000
1.64	2.13	Hurricane Felix 1995, January 1997, February 1997, August 1997
1.61	2.10	(Black and White Horse Pike begins to flood) May 1998
	2.07	November 1995
	2.04	June 1995, December 1995, July 1996, January 1999
	2.01	June 1997, July 2000, August 2000
	1.98	December 1997, June 2000

(*continued*)

Table 4.4 ~ (*continued*)

National Geodetic Vertical Datum (NGVD) (1929)	Vertical Mean Lower Low Water (MLLW)
1.52	1.95 December 1986
.	
.	
.	
1.31	1.79 Minor flooding (average 6 times per year). December 1994
.	
.	
.	
0.18	0.68 NOS Mean Sea Level

SOURCE: State Hazard Mitigation Team (1993), augmented by *http://co-ops.nos.noaa.gov.*
NOTE: 0.0 m NGVD 1929 = 0.5 m MLLW.

of storm water levels. Flood frequency curves have often been used to describe actual and predicted flood events along the coast, which, in turn, have application to public and emergency planning and even storm event categorization.

The Federal Emergency Management Agency has conducted an evaluation of the storms measured at Atlantic City and has created a classification based on the probability that certain water levels will occur (table 4.4). The data portrayed in this comparison are all related to the 1929 NGVD and the 1960–1978 tidal epoch. This means that the water elevations are standardized relative to these norms. In this comparison, the FEMA 100-year storm surge will have an elevation of 2.89 m (9.5 ft) above NGVD, the 30-year storm surge will attain an elevation of 2.34 m (7.7 ft) above NGVD, the 5-year storm surge will reach 1.76 m (5.8 ft) above NGVD, and so on. All of the major storms of the 1980s and 1990s have been placed on this scale, and many of the minor storms have also been included. Further, the major storms of the past half century have been added as a further means of comparison. The highest storm surge on record at Atlantic City is associated with the 1944 hurricane, followed by the December 1992 and the October 1991 northeasters. The March 1962 northeaster and Hurricane Belle are only a little bit lower. It is important to note that none of these storms of record exceeds the 30-year storm (probability of once in a 30-year span of time). Thus there has not been a 100-year storm striking the New Jersey coast in terms of water-level elevation within the period of record (56 years). In addition, there are an increasing number of storms toward the lower end of the scale. This is as it should be. By definition, there should be more instances of 5-year- and 10-year-storm water levels than the higher magnitude and less frequent storm events during any extended period of observation. There is a bias in the table, because it includes as much information as could be determined for the past two decades, whereas the previous information was skewed toward the higher storms. Yet the record from 1944 to the present shows that most of the highest water elevations are the products of recent storms. This may be the product of probabilities or increased storminess during the most recent decades. Regardless of the possible skewing of the data set, the lesson inherent in it is obvious: storminess is a condition of the coast, flooding and erosion and storm water penetration are conditions of the coast, and the past storms have been severe but they have not been 100-year storms. Assessing probabilities is a statistical game, but it is does provide a basis for creating alternative scenarios and it does establish a basis for planning and management.

In order to design more effective and efficient policies, policy makers and coastal decision makers at the federal, state, and local levels must determine very clearly the level of protection desired to reduce the threat and effects of the storms and storm surges likely to occur in the barrier islands and bayside communities. Options to protect reaches against 1-in-5-year storm water levels differ from strategies to protect reaches from a 1-in-50-year storm, and this difference needs to be recognized in policy design. Although protection from a 1-in-5-year storm requires less immediate investment than that of a 1-in-50-year storm, providing a low level of protection (for, say, a 1-in-5-year storm) will require periodic and at times substantial postdisaster cleanup and repairs from larger storms. Protection against a 1-in-50-year storm, in contrast, will reduce the overall probability of flooding from frequent, less severe storms, but this level of protection is not cheap. It is likely that a medium level of protection, such as providing protection from a 1-in-20-year storm, may be the most feasible option, one that will buffer communities from the general effects of these moderately severe, less frequent storms as well as the more frequent storms.

Through an understanding of recurring storm events and their effects, planning and management agencies can establish alternatives to dealing with the dynamics of coastal storms. The storms will continue to occur, and the coast will respond in both natural and cultural terms. The

Table 4.5 ~ SEVERE STORMS, 1980–1998

Storm date	Storm type	Water level (Meters [feet] above NGVD)	Duration (hours)
March 29, 1984	Northeaster	2.19 (7.2)	~36
October 13, 1984	Hurricane	1.43 (4.7)	~60
September 27, 1985	Hurricane	~2.1 (~7.1)	~24
January 2, 1987	Northeaster	1.8 (5.9)	~50
October 30, 1991	Northeaster	2.0 (6.6)	~114
January 4, 1992	Northeaster	1.93 (6.3)	~32
September 23, 1992	Northeaster	1.7 (5.6)	~36
December 11, 1992	Northeaster	2.32 (7.6)	~140
March 13, 1993	Northeaster	1.68 (5.5)	~40
March 3, 1994	Northeaster	2.3 (7.6)	~60
August 7, 1995	Hurricane	1.65 (5.4)	~96
November 14, 1995	Northeaster	1.52 (5.0)	~36
January 7, 1996	Northeaster	1.66 (5.4)	~48
March 19, 1996	Northeaster	1.91 (6.3)	~24
December 6, 1996	Northeaster	1.77 (5.8)	~65
November 14, 1997	Northeaster	1.81 (5.9)	~35
August 7, 1995	Hurricane	1.68 (5.5)	~50
January 28, 1998	Northeaster	1.76 (5.8)	~36
February 4, 1998	Northeaster	1.81 (5.9)	~72

SOURCE: *http://co-ops.nos.noaa.gov.*

question is whether anyone has learned from the events of the past to reduce the effects of the storms on the built environment in the future.

Conclusion

Coastal storms are important conditions that influence the dynamics of the New Jersey coast. Since 1980 more than 25 major and minor storms have affected the state's coastline in some manner. Some of these resulted in extreme water-level elevations above NGVD (table 4.5), such as the storms of December 1992, October 1991, and March 1984. Storms typically produce a peak storm surge at their height or center of intensity, with water levels rapidly building up and then slowly receding. The increase in water-level elevations, coupled with high winds and the erosive effects of large waves, can produce significant damage to both the natural system and the built environment and infrastructure. These elements can erode beaches and dunes; flood low-lying areas; cause damage to buildings, boardwalks, and other man-made structures; cause loss of life; and create hazardous conditions for coastal communities in general. Ominously, the total damage to communities and loss of life increased over the 1990s as more development crowded into the coastal zone (H. J. Heinz Center 2000b). An understanding of these storms and of their frequency and magnitude, therefore, aids us in determining the nature of the hazard and the efforts needed to lessen the risks to public safety at the coast.

CHAPTER 5

Sea-Level Rise: Its Dimensions and Implications

On sandy barrier coasts, rises in mean sea-level lead to overwash breaching on several time-scales and perhaps, ultimately, to overtopping and flattening of the barrier coast.

R.W.G. Carter, *Coastal Environments* (1989)

To explore the elements of sea-level rise that affect coastal management decisions, the level of inquiry must be expanded beyond short-term efforts functioning at the local level to include regional considerations. The issue of sea-level rise and its effects on the coastal system is an example of a ubiquitous problem requiring management options that involve local efforts within the scope of regional dynamics. It is especially important to be aware of the changes that sea-level rise is producing, both in the natural components of the coastal zone and in the continuing modifications to development there.

Introduction

Sea-level rise is producing a progressive submergence of the land mass within the coastal zone. This inundation is presently causing some migration and displacement of natural habitats. It is changing the position of the shoreline, it is extending the inland penetration of coastal storms, and it is having a wide variety of other direct and indirect effects upon the natural and cultural features in the zone. The process of sea-level rise and its effects produce a gradual drowning of existing fixed structures and habitats and a gradual increasing exposure to the hazards of coastal storms. Although the rates of sea-level rise are relatively slow in terms of day-to-day activity, they become important in longer-term issues of public safety and hazard management. Their effects need to be recognized in future land-use planning and in the design of future coastal engineering techniques. This chapter examines the concepts associated with sea-level rise and summarizes the evidence of, and variation in, sea-level rise along the coastline of New Jersey. These data are extended using several estimates of future sea-level rise in order to examine their significance for the state. Future scenarios form the basis for the development of longer-term management planning to accommodate the changing dynamics associated with the progressive inundation of the coastal zone.

The issue of sea-level rise has become an impetus for the consideration of a variety of coastal management strategies. The problem is not simply an increase in the water level in coastal areas, but an increase in exposure to storm effects as well. The increased inundation and penetration of coastal storms act upon higher water levels and affect the ocean and bay shorelines, the barrier islands, and the low-lying mainland.

The amount of sea-level rise along the New Jersey coast can be identified through analysis of tidal gauge records. Its effects are manifested in a variety of changes that have occurred throughout time, and as sea level continues to rise, the natural and cultural components of the coastal zone will be affected. This situation is receiving increasing attention because of the intensive urban development that has occurred along much of the coastline in the United States, including most of New Jersey. Coastal decision makers need to anticipate the effects of sea-level rise and to incorporate hazard management and emergency management strategies that enhance public safety and reduce the exposure, both direct and indirect, of the susceptible elements in the zone.

Absolute and Relative Sea-Level Rise

Sea-level rise is composed of a combination of several factors. The most basic factor is an increase in the amount of water in the ocean. The melting of mountain glaciers and high-latitude snowfields of the earth, caused by the warming of the earth's atmosphere, adds water to the oceans and causes the ocean surface to become elevated. This change is referred to as the "eustatic effect," or *absolute* sea-level rise (that is, the change in the volume of water in the ocean basin). A second factor in sea-level rise is that the land of the coastal zone is subsiding, or slowly sinking. Referred to as the "tectonic or isostatic effect," this is a characteristic of the older coastal margins of continents or locations of great quantities of sediment accumulation (for example, river deltas and barrier islands).

Thus the older and depositional margins of continents are slowly moving down relative to the ocean level. Furthermore, the relatively young sediments that make up barrier islands undergo some compaction because of their thickness and weight. As these sediments compact, the reduction in vertical dimension contributes to an overall lowering of surface elevations. The total change in sea level is caused by the combination of these isostatic factors that produce a general lowering of the land in the coastal zone plus the eustatic increase in volume in the world's oceans; this is the *relative* sea-level rise. Relative sea level is the measure of how fast the water is encroaching onto the continent and inundating the coastal margin of the land. In discussing management options concerning relative sea-level rise, whether the land is subsiding or the sea is rising is of little concern. The focus is that the water is rising and encroaching upon the elements of the coastal zone. It is a driving force that is producing changes to the natural and to the built environments at the land margin.

The Effects of Sea-Level Rise

The issue of sea-level rise is multifaceted, because so many indirect effects are associated with it. The problem is not simply an overall increase in the water level of the oceans, but is also related to a general displacement of the shoreline at all the margins of the barrier islands and bayside communities, including those on the mainland. Furthermore, as sea level rises, storm conditions are able to reach farther inland. Smaller storms, which were of little concern before, are now able to reach levels and locations that were attained only by a rare storm event in the past. As sea level rises, the effects of a diminishing sediment supply are also magnified. Although displacement of the shoreline is perceived to be erosion, it is also a product of increasing water levels. Therefore as sea-level rise contributes to the measurable displacement of the shoreline, an increasing amount of replacement sand will be necessary just to maintain a constant position in the future. Other related issues include a change in the extent of and distribution of coastal wetlands, altered estuarine habitats both within the water and in the benthic areas, intrusion of saltwater into upper portions of estuaries, intrusion of saltwater into potable groundwater supplies, increased frequency in the inundation of evacuation routes, decreased clearance under bridges, changes in the ability of stormwater drains to discharge flood waters, need for pumps to move water that is near or below the new sea-level position, and effects on hazardous waste sites located in coastal areas.

The Application of Sea-Level
Rise Information

In the *New Jersey Shore Protection Master Plan* (NJDEP 1981), the issue of sea-level rise was introduced as a variable that was changing the condition of the coast. General information was known, and there was a record of water-level changes observed in tidal gauges. However, there were few studies on sea-level change in New Jersey at the time. In addition, most of the discussion in the 1981 assessment was theoretical and focused on the problems associated with future sea-level rise. Major advances in the investigation of sea-level rise occurred in the 1980s, as additional information became available. It became possible to describe long-term records of sea-level change, immediate past conditions of sea-level positions, and to place these two pieces of information in perspective to acquire more conclusive evidence about past, present, and future rates of change.

Concern for sea-level rise and its effects has generated considerable activity on a worldwide basis. Both international and national organizations have studied the rates of sea-level rise and its associated impacts on a global basis. Some of this interest is driven by a recognition that global climatic change is causing many environmental responses. One of these is an elevation of the global sea level. Some of the early forecasts suggested extreme changes (greater than 3 m, or 10 ft) within a short time scale (less than a decade). However, newer forecasts of rates of sea-level rise are considerably lower, and the focus has partially shifted toward analyzing approaches to coastal management that incorporate the multiple aspects of sea-level rise. There remains considerable concern at the international level, however, regarding the problems that are faced by low-lying island nations that have limited areas of high land.

Pertinent Studies on Sea-Level Rise

There are two major sources of information on regional sea-level rise: (1) studies by the U.S. Environmental Protection Agency (U.S. EPA), and (2) studies by the Intergovernmental Panel on Climatic Change (IPCC) created as a joint effort of the United Nations Environment Program and the World Meteorological Organization. Both sources have addressed the issue by compiling information and analyzing the current state of knowledge. A third agency, the U.S. National Research

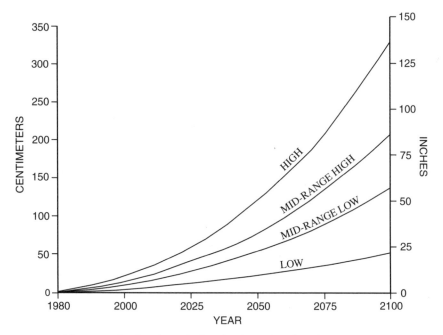

Figure 5.1. Predicted trend of sea-level rise. (Barth and Titus 1984.)

Council, has conducted a more limited inquiry and has issued a report on its findings.

The U.S. EPA Initiative

The first effort of the U.S. EPA was a comprehensive analysis of the greenhouse effect on changes in sea level (Barth and Titus 1984). The report included a methodology to identify and predict factors that influence sea level, and also incorporated scenarios of coastal and estuarine changes produced by sea-level rise. The general approach used was to consider a range of scenarios that varied from a low of 56.2 cm (22.1 in) to a high of 345 cm (135.8 in) above the sea level of 1980 and to apply them to the situation present in the year 2100 (figure 5.1). Two mid-range estimates were based on a lower value of 144.4 cm (56.9 in) to a higher value of 216.6 cm (85.3 in) above the 1980 sea level in the year 2100. These calculations only pertain to the eustatic rise of sea level, hence an adjustment to the reported figures should be made to incorporate the added effects of local subsidence and compaction. For example, at Atlantic City, New Jersey, an adjustment to include land movement in order to arrive at the relative sea-level rise would mean an increase by about another 24 cm (9.44

in) above the 1980 sea level in the year 2100. These factors would result in estimates associated with the high scenario of 369 cm (145.2 in) above the 1980 sea level in the year 2100 and estimates associated with the low scenario of 80 cm (31.9 in). In their conclusion, Barth and Titus (1984) emphasize that although sea-level rise is uneven along the seacoast (because of regional variation), evidence suggests an overall eustatic rise estimated at 15 cm for the nineteenth century with a higher rate anticipated for the twentieth and twenty-first.

Several other U.S. EPA publications expanded on the sea-level rise theme by examining the effects of environmental change for specific locations, such as Ocean City, Maryland, beaches (Titus et al. 1985), salinity levels in Delaware Bay (Hull and Titus 1986), and coastal wetlands (Titus 1988). Much of the development of the wetlands scenario was based on conditions presently occurring in coastal Louisiana where, because of the very large accumulations of the Mississippi River Delta, the relative rate of sea-level rise in some areas is on the order of 1.0 cm (0.39 in) per year (Titus 1988). Implications of these studies and their application to areas such as Delaware Bay and coastal New Jersey show that an increase in the rate of sea-level rise results in shoreline displacement, wetland loss, and inundation of land masses of barrier islands. In other words, the effects of sea-level rise will affect the entire barrier island, bay, and wetlands complex of coastal New Jersey.

National Research Council Initiative

The Engineering and Technical System Commission of the National Research Council (NRC 1987) conducted an inquiry into sea-level rise because of the range and implications of estimates produced by the U.S. EPA, other NRC reports, and several international studies. In its report the commission concluded that sea-level rise was indeed occurring and that it would continue into the future. The report cautioned that although the rates of future sea-level rise are uncertain, it is likely that the rate could increase over its current value. As a result, three scenarios of sea-level rise were developed, low (50 cm or 19.7 in), middle (100 cm or 39.4 in), and high (150 cm or 59.1 in) by 2100. It was further recognized that local subsidence and compaction would cause the land to sink and add to the rate of inundation. The overall conclusion was that sea-level rise was exacerbating shoreline erosion and that horizontal shore displacement was on the order of 100 times the vertical rise of the sea.

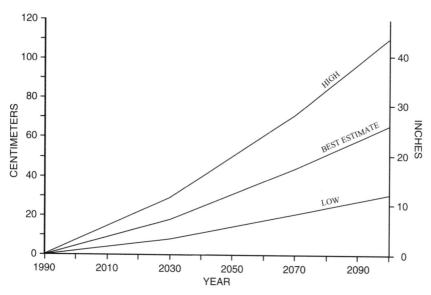

Figure 5.2. Predicted trend of sea-level rise. (Houghton, Jenkins, and Ephraums 1991.)

Intergovernmental Panel on Climatic Change

Another source of information was generated through the Intergovernmental Panel on Climatic Change as part of the World Meteorological Organization and the United Nations Environment Program. One of the objectives of this panel was to concentrate on the effects of global climate change on the coastal zone. Several workshops were convened, and information throughout the world was collected and collated to consider evidence regarding climate change, rates of change, effects of the change, and management options to deal with present, and future changes. A large variation in the assembled data was present and it was not possible to establish a consensus estimate of global sea-level rise. Rather, ranges of sea-level rise were developed (figure 5.2), based on a point estimate of a high rate of rise (110 cm, or 43.3 in), a low rate of rise (31 cm, or 12.2 in), and a reasoned "best" rate of rise (66 cm, or 26 in) over the 1990–2100 period (Houghton, Jenkins, and Ephraums 1991). Similar to the approach used by the U.S. EPA, these values refer only to absolute sea-level rise (eustatic). Also, similar to the results of the U.S. EPA, the panel found evidence for a continuing increase in sea level. Currently, the IPCC is conducting research on the present and future rate of sea-level rise and has found evidence that the rate was greater for the twentieth century than for the nineteenth and will continue to increase in the

Table 5.1 ∼ NORMALIZED SEA-LEVEL PROJECTIONS, COMPARED
WITH 1990 LEVELS (CM)

Cumulative probability (%)	Sea level projection by year			
	2025	*2050*	*2075*	*2100*
10	−1	−1	0	1
20	1	3	6	10
30	3	6	10	16
40	4	8	14	20
50	5	10	17	25
60	6	13	21	30
70	8	15	24	36
80	9	18	29	44
90	12	23	37	55
95	14	27	43	66
97.5	17	31	50	78
99	19	38	57	92
Mean	5	11	18	27

SOURCE: Titus and Narayanan (1995).

twenty-first. The panel is also examining evidence on the types of changes that may be expected at the shoreline and in the coastal habitats related to sea-level rise (Beukema, Wolff, and Browns 1990; Warrick, Barrow, and Wigeley 1993).

Recent Revisions to Estimates

Both the U.S. EPA and the IPCC have modified their estimates of sea-level rise since their initial studies. Generally the estimates have decreased slightly, as a result of adjustments in simulation models that reduce the water contributions from the melting of the icecaps at Greenland and Antarctica to the world's oceans. In 1995 the U.S. EPA released its revised estimates on a likely sea-level rise scenario to the year 2100 (Titus and Narayanan 1995). The approach was based on a survey of a panel of experts on sea-level rise, from which an estimate of the most probable rate was determined. Increased rates of sea-level rise were then paired with probabilities.

To apply the findings of the U.S. EPA effort to Atlantic City, for example, involves an estimate of the most probable outcome of sea-level rise—an estimate of a 25 cm (9.8 in) increase from 1990 to 2100—added to the existing trend for the municipality (Titus and Narayanan 1995) (table 5.1). Hence this number is simply added to the existing rate of rel-

Table 5.2 ∽ HISTORICAL RATE OF SEA-LEVEL RISE IN AND NEAR
COASTAL NEW JERSEY (MM/YR AND IN/YR)

Locations	mm/yr	in/yr
New York, N.Y.	2.74	0.11
Sandy Hook, N.J.	4.06	0.16
Atlantic City, N.J.	3.85	0.15
Lewes, Del.	3.11	0.122

SOURCE: NOAA (1987–1994).

ative sea-level rise to determine a future sea-level position. Data from the
tidal gauge at Atlantic City establish that the existing rate of relative sea-
level rise is 39 cm (15.3 in) per century (table 5.2). If the 1990 estimate is
incorporated, the most likely elevation of sea level at Atlantic City in 2100
would be 68 cm (25.6 in) higher than that of 1990. Using the data in table
5.1, we find that there is a 10% chance that the increase of 39 cm (15.3 in)
would be exceeded by 55 cm (21.6 in) in 2100, providing a total increase
of 97 cm (37 in) by the year 2100. In addition, calculations from table 5.1
indicate that there is a 1% chance that the additional rise could reach 92
cm (36.2 in) above the current trend, raising the total 2100 sea level to 134
cm (51.56 in) above the 1990 level. Note that nearly all of the estimated
probabilities point to a continuation of sea-level rise and at a higher rate
in the twenty-first century.

Scientific-technical agencies that have studied the issue of sea-level
rise thus conclude that it is occurring and will increase in the future. The
magnitude of the eustatic rise is further accompanied by the effects of
local subsidence and compaction. Calculations of future sea-level rise at
Atlantic City, for example, from various scenarios indicate a potentially
serious problem for the low-lying coastal zone (table 5.3). Even the smallest of these values implies significant changes to the system.

Rate Changes in Sea-Level Rise
in New Jersey

Several studies have looked at the rate of sea-level rise in
New Jersey from a geological perspective (Meyerson 1972; Psuty 1986;
Psuty, Guo, and Suk 1993). It was found that sea level has been rising during the past several thousand years. Although these rates have fluctuated,
they support an upward trend. An analysis of radiocarbon dates of organic
materials accumulating in estuarine sediments over the past 7500 years

Table 5.3 ～ ELEVATION OF SEA LEVEL AT ATLANTIC CITY, N.J., UNDER
VARIOUS SCENARIOS, INCORPORATING SUBSIDENCE
(ELEVATION IN CENTIMETERS, RELATIVE TO 1990)

Agency and year of study	Year 2000	Year 2025	Year 2050	Year 2100
EPA (1984)				
Conservative	4.9	19.35	36.4	81.3
Mid-range, moderate	6.9	30.55	62.9	167.3
National Research Council (1987)				
Low	7.1	25.2	43.1	78.6
Middle	11.2	39.4	67.6	123.8
IPCC (1990)				
Conservative	4.5	15.5	32.0	58.5
Moderate	5.5	20.5	44.0	93.5
EPA (1995)				
Best estimate	5.0	18.7	33.5	68.1

shows a general increase in sea level on the average of about 2.1 mm (0.083 in) per year until about 2500 years ago, when the rise slowed to an average rate of about 0.8 mm (0.032 in) per year (figure 5.3; Psuty 1986). At that time, sea level was about 2 m (6–7 ft) below today's level. It was during this slower rate of sea-level rise of the past several millennia that a greater stability of coastal forms and habitats developed, leading to the association of land and water features that presently characterizes the coast. However, recent data from New Jersey tide gauges indicate that relative sea-level is once again increasing at a fast rate.

Tidal Gauge Measurements

Mean water levels measured by tidal gauges are probably the most convincing indication of the relative rise in sea level, because of their long records of actual water-level measurements. For example, the tide gauge at the Battery in New York City has recorded water levels since 1856. The portrayal of the trend of sea-level rise is depicted on each of the tidal gauge records available in or near the New Jersey coast (figure 5.4).

In each of the stations in figure 5.4, the value that is plotted across time is the annual mean sea level (the average of the 12-monthly mean sea levels for a particular year) at a site relative to a fixed datum (an elevation). Because each station has a different datum elevation, the specific values of sea-level-rise axes are slightly different. However, the relative values of the amount or percentage increase per year is easily determined from these graphs and is not influenced by the elevation. This information from tide

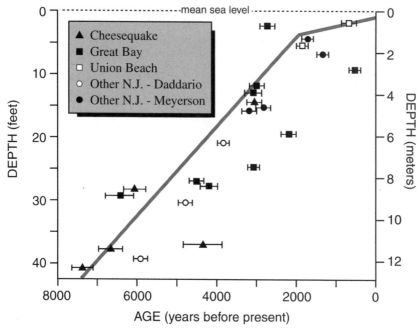

Figure 5.3. Trend of recent geologic sea level. The points on the scatter diagram are radiocarbon ages on materials taken from cores in New Jersey. The shaded line is the interpreted trend in elevation of sea level. The horizontal bars on the points represent the standard deviation in the age of determination. Note the rapid rate of rise until 2500 years BP and the ensuing slower rate of rise. (Psuty 1986.)

gauges indicates a trend of sea-level rise within which there is year-to-year variation. The fluctuations or variation in yearly mean sea level are modest compared with the longer, upward trend that exists for each station. The average rate of relative sea-level rise varies between 4.06 mm (0.16 in) per year at Sandy Hook to 3.11 mm (0.12 in) per year at Lewes, Delaware (NOAA 1987–1994). The rates of rise differ because of a combination of variables, especially tectonic subsidence and compaction of differing thicknesses of recent sediments, that affect the relative sea level.

A study of sedimentation in Great Egg Harbor (Psuty, Guo, and Suk 1993) found that the wetland surface in this estuary is being inundated at rates faster than those recorded by the tidal gauges located on the adjacent barrier island (Atlantic City). An analysis of Cesium137 isotopes in the upper foot of peat and sediment accumulations on the Rainbow Islands in Great Egg Harbor Bay determined a rate of rise of the marsh surface of about 6 mm/year (0.24 in/yr) since 1963, which is higher than the 4 mm/year (0.16 in/yr) average from nearby tidal gauges during the twentieth

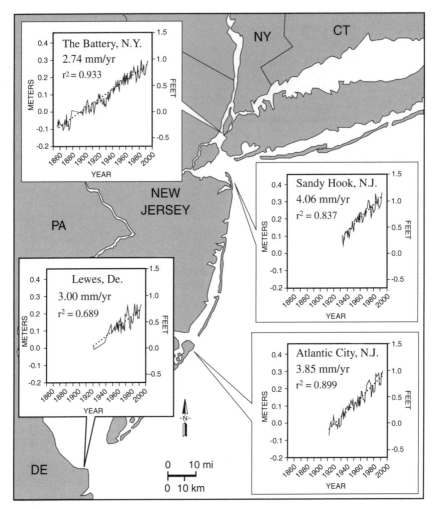

Figure 5.4. Trends of mean annual water level at tide gauges in New Jersey and nearby sites (NOAA 1987–1994; *http://co-ops.nos.noaa.gov/*)

century. It probably indicates that the wetland surface is being lowered faster than the barrier island surface, because the silts and organics of the wetlands are subject to more compaction than are the sands of the barrier islands.

The rate of sea-level rise at present is higher than at any time in the past 7500 years (figure 5.5). It is believed that the combination of sea-level rise and a limited availability of sediment in the system are causing changes in the coastal morphology and coastal habitats. With a slow rate of sea-level rise for the past several thousand years, barrier islands and

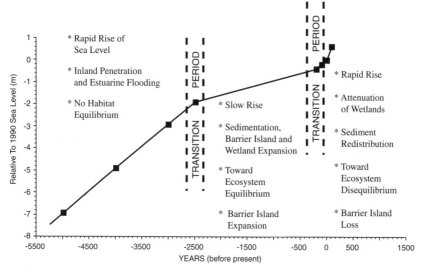

Figure 5.5. Barrier island system development in recent geologic time. (Psuty 1991.)

wetlands previously were in adjustment, but the present faster rate is causing disequilibrium conditions at the coast and in estuaries. In the past 50
years, for instance, the group of undeveloped and therefore undiked and
unditched Rainbow Islands in Great Egg Harbor, New Jersey, have lost
about 5% of their area, and one island has completely disappeared (Psuty,
Guo, and Suk 1993). Essentially, the accumulation of sediment and organic matter was insufficient to support the surface area of the island as it
was drowning. As the rate of sea-level rise increases in the future, it will
result in further adjustments to the paucity of sediment availability and
wetland maintenance, and it will force new displacements in the coastal
zone for some time. This present high rate of rise is submerging the wetlands, and this combination of submergence and a meager sediment
supply is largely responsible for the loss of wetland area on New Jersey's
Rainbow Islands. It is this rate of rise that is encroaching on the margins
of the entire barrier island and on the estuarine and bay shorelines of New
Jersey.

Sea-Level Rise and Coastal Storms

Storm surge levels generally vary as a function of a storm's
strength and duration. However, another variable also helps determine
the comparable level to which storms can raise storm water elevations and
penetrate inland. This additional factor is the change in relative sea level

over time. Because the sea level is rising, the base upon which storms occur and operate, as well as their subsequent effects or actions, is changing. This means that sea-level rise will magnify or accentuate the effects of storms. Storms in recent times have been capable of reaching higher flood levels with lower surges associated with storms of lower severity than those of the past. Furthermore, a rise in sea level can allow stronger, less frequent storm events to reach inland coastal areas that were once safe from storm activity, thus exposing more areas to the erosion and flooding effects of a storm.

When comparing storms, it is possible to describe their surge level relative to a fixed datum, such as the NGVD of 1929, the nationwide constant elevation (table 4.4), or to a changing datum, such as sea level at the time of the storm. However, when comparing storms separated by several decades, some of the differences will be due to sea-level rise. For example, although the 1991 Halloween storm and the March (Ash Wednesday) 1962 storm were nearly equal in water elevations reached, 2.225 m (7.3 ft) and 2.19 m (7.2 ft) above NGVD, respectively, the 1991 Halloween storm operated on a sea level 0.119 m (0.39 ft) higher than that of the 1962 storm. Because the March 1962 storm operated on a lower water base than the Halloween storm, the 1962 storm had a stronger storm surge than the Halloween storm in order to reach the same water levels.

Future Storm Levels

Continuation of Previous Century's Rate

The effects of sea-level rise on storm inundation levels through time are depicted in figure 5.6. Column A contains the water-level elevations of the major storms that have affected New Jersey over recent decades. Twelve storms are listed according to their peak water levels above NGVD. In addition, the elevation of the water level is referenced to the FEMA recurrence interval water levels. For example, the January 1998 storm is shown as having a storm peak water level of 1.77 m (5.8 ft) above NGVD and was classified as a 1-in-5-year recurrence interval storm. It is labeled number 2 in column A, and this storm and its number are incorporated in each of the other vertical columns in figure 5.6 (columns B, C, and D; see also table 4.4).

For each of the major storms depicted in the left column in figure 5.6, the storm water elevation is that level recorded at the Atlantic City tidal gauge at the time of the storm relative to NGVD 1929. This procedure

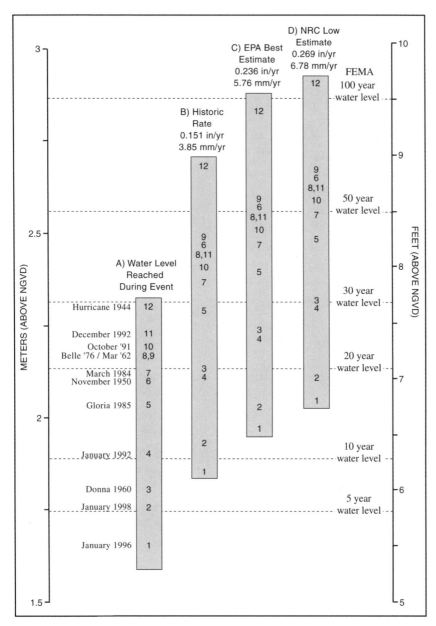

Figure 5.6. Projections of historical storm water levels at Atlantic City to the year 2050 using three sea-level-rise scenarios. (A) Actual level above NGVD. (B) Applying historical rate. (C) Applying Environmental Protection Agency best estimate rate. (D) Applying National Research Council low estimate rate.

does not explicitly identify the effects of sea-level rise during the period of record; rather, it is implicit in the storm surge measure. Consider the following scenario: both the March 1962 storm and the 1991 Halloween storm had nearly equal water levels, 2.19 m (7.2 ft) and 2.225 m (7.3 ft) above NGVD. Yet if either were to have occurred in 2000, its effective water level would have been different, because of sea-level rise. One needs to adjust a storm's surge based on the rate of past sea-level rise, 3.85 mm/year (0.151 in/yr). On this basis, had a storm equivalent to the March 1962 storm occurred in 2000 it would have been expected to produce a storm surge of 2.336 m (7.66 ft) above NGVD (compared with 2.19 m, or 7.2 ft, above NGVD in 1962) and would subsequently be classified as a 1-in-30-year storm. If an event similar to the 1991 Halloween storm were to have occurred in 2000, water elevations could reach up to 2.26 m (7.41 ft) above NGVD (figure 5.6, column B), compared with 2.225 m (7.2 ft) above NGVD in 1991. From the above application, it is clear that if a storm similar to the 1962 Ash Wednesday storm (no. 9) were to occur today, it would surpass the flood levels of Hurricane Gloria in 1985 and achieve storm water levels above those of the December 1992 storm and even of the 1944 hurricane.

Storm water levels of past storms can be projected to their equivalents in the year 2050 to further illustrate the joint-multiplier effect of sea-level rise in the future (figure 5.6). On the basis of a historical sea-level rise rate of 3.85 mm/year (0.151 in/yr) (column B), an event similar to the December 1992 storm would result in a water level 2.47 m (8.11 ft) above NGVD in 2050, or a 1-in-40-year storm. Similarly, for an event equivalent to the Halloween storm, expected water levels would be 2.45 m (8.04 ft) above NGVD in 2050. Furthermore, if an event similar to the 1962 Ash Wednesday storm were to occur in 2050, water levels could reach up to 2.528 m (8.33 ft) above NGVD, a 1-in-45-year storm.

U.S. EPA "Best Estimate" Rate

Projecting into the future involves an element of risk, but the identification of possible scenarios is appropriate for planning and public policy purposes and for raising awareness of potential situations. It is to this end that the U.S. EPA and other agencies have developed models of possible sea-level change. On the basis of probabilities, the U.S. EPA has identified a "best estimate" of the rate of sea-level rise as 6 mm/year (0.236 in/yr) to the year 2050 (Titus and Narayanan 1995). These estimates are contained in figure 5.6, column C. The predicted water level was determined from the past century's rate of sea-level rise of 3.85 mm/yr (0.151 in/yr) to the year 1990, and from the U.S. EPA rate for 1991–2050. Results from these calculations indicate that a March 1962 "equivalent" storm may be expected to reach a water level of 2.63m (8.65 ft) above NGVD or a 1-in-60-year water level, greater than any storm surge of the twentieth century. An event similar to the December 1992 storm would result in a storm surge of 2.58 m (8.49 ft) above NGVD in 2050, a 1-in-50-year occurrence interval, and if an event equivalent to Hurricane Gloria were to occur in 2050, the water level is projected to be 2.35 m (7.71 ft) above NGVD.

National Research Council
"Low Estimate" Rate

Estimated water elevations of similar events using the "low estimate" of accelerated sea-level rise of 6.84 mm/year (0.269 in/yr) to the year 2050 from the National Research Council's study *Responding to Changes in Sea Level* (NRC 1987) are depicted in figure 5.6, column D. Predicted water levels were calculated based on the past century's rate of 3.85 mm/yr (0.151 in/yr) up to the year 1990 and based on the NRC rate thereafter. From these projections, if a storm event similar to Hurricane Gloria were to occur in 2050, water levels are likely to be 2.529 m (8.3 ft) above NGVD. A storm similar to that of December 1992 would result in levels 2.647 m (8.68 ft) above NGVD, a 1-in-65-year storm. Projected water levels for a March 1962 "equivalent" storm in 2050 would be 2.70 m (8.88 ft) above NGVD, a recurrence interval that would classify it as a 1-in-75-year storm. It should be noted that the projected rates of relative sea-level rise used here are among the lowest rates reported in the literature.

One final comparison involves the 1944 hurricane, which produced a water level of 2.31 m (7.6 ft) above NGVD in 1944. This was "the" major storm, and it retains that status throughout the comparisons. Based on the NRC "low estimate" rate, calculations show a projected elevation of 2.897 m (9.5 ft) above NGVD in 2050, or a 1-in-100-year flood in present-day terms. The difference between the 1-in-30-year flood level of this hurricane and a 1-in-100-year level, over a threefold increase, is the result of sea-level rise. Although the meteorological events that produced this storm may recur at a probability of three times per century, the resulting surge and flooding will reach far greater elevations in the future.

General Options for Management

Coastal storms represent a meteorological process that will continue to affect the New Jersey shoreline. These storms often result in the loss of life, extensive damage to property, and coastal erosion (see figures 4.1, 4.2, and 4.3). As the sea level continues to rise, the effects of storms will be felt farther inland, across more of the coast, and in general will be magnified. Efforts are needed now to reduce the threat and potential losses from less frequent severe storms, as well to buffer the effects of the more frequent low-magnitude storms. Several issues should be considered in the process of developing management strategies:

- Identification of high-hazard areas
- Identification of the specific level of protection desired from coastal storms, on a reach basis
- Determination of objectives to be achieved in the future (for example, the year 2050) that incorporate rising sea levels

As sea levels rise, most effects of coastal storms will be manifested initially and to the greatest extent in the coastal areas that are currently prone to the erosion and flooding effects of present-day storms. Such hazardous areas need to be identified, and this information should be made available to the public. Further, elevated levels of storm water will begin to cause damage to barrier island and estuarine areas currently at moderate risk from flooding. To reduce further loss of life and damage to property in these areas known to be exposed and vulnerable, decisions regarding public safety must be initiated. Policy makers similarly should readdress the issues of shoreline stabilization and the increasing trend to develop and live in exposed areas. A variety of management tools exist from different disciplines that are being applied to coastal management and hazard mitigation issues in the United States, and there is a need to introduce more flexibility into the design of land-use plans, building codes, and population concentrations. Zoning and boundary lines should be flexible and capable of being adjusted periodically to reflect changes in exposure and risk from the rising sea level and the increased severity of coastal storms.

Coastal officials must determine, preferably on a regional basis, the level of protection desired to reduce the effects of the probabilistic occurrence of storm surge and storm-induced waters to the shorefront and bayside communities. Options to establish buffers in reaches or regions against a 1-in-5-year storm are different from strategies to buffer or protect reaches from a 1-in-50-year storm. Although buffering from a 1-in-5-year storm requires less immediate and total investment than from a 1-in-50-year storm, its limitations will result in more frequent and larger costs associated with postdisaster cleanup, repairs, and subsequent mitigation. To provide buffering against a 1-in-50-year storm will reduce and possibly prevent flooding from less frequent severe storms. However, it may require expensive structural solutions, such as dikes, that are not economically feasible or realistic. In any case, some level of buffering is needed against the effects of storms and storm surge. One possible option is to adopt an intermediate approach that could provide protection from a 1-in-20-year storm. By providing protection from such a storm, communities could be protected from the effects of both the moderately severe, less frequent storms and the more frequent storms of lower intensity. Regardless of the degree of buffering selected, the creation of a buffer will require a commitment of space for its construction. Its location will need to be adjusted, moreover, as the sea level rises.

In coastal areas with low-density development, it may be possible to consider options that allow for the relocation of structures both inland and out of low-lying or highly eroded portions of the coast that are severely affected by sea-level rise. However, in densely developed coastal locations, little space is available to relocate buildings and/or infrastructure. Three alternative strategies show the diversity of possible policies to reduce the public's exposure to the encroaching sea:

- Utilize seawalls, beach nourishment, dikes, and other constructed barriers to prevent the rising water from penetrating inland and to defend the existing shoreline.
- Allow the shoreline to migrate inland and accept the losses of property and infrastructure at the water's edge.
- Use a combination of several approaches, involving short-term hard structures or nourishment along with dune creation, in conjunction with planned changes in land uses for the most exposed areas.

The Intergovernmental Panel on Climatic Change of the UN looked at several options that could be applied to coastal management in the face of sea-level rise. No new revelations were put forward. Options considered were to build dikes (similar to the Dutch approach), allow the natural system to function while the populace moves back from the rising water (most of the developing world), or perform short-term holding actions while a longer-term solution is sought (delaying the hard choices).

The Engineering and Technical System Commission of the NRC effort on sea-level rise (NRC 1987) recommended only two management options: (1) to stabilize the coast, or (2) to retreat. No one option was preferred because of the wide range of sea-level-rise scenarios and the need to gather site-specific data to assist in determining possible structural solutions. A structural solution was thought possible, although the report indicated that it might be too expensive to apply. The commission concluded that the evidence for an accelerated sea-level rise into the twenty-first century was well established, and that it and its effects should be incorporated into sound planning and design.

The Atlantic City Tide Gauge Revisited: Reason to Pause

As new data from the Atlantic City tidal gauge are verified at the NOAA data center, they are combined with the annual sequence of water-level measurements to extend the record. Results from statistical analysis of measurements to 2000 have differed in comparison with the outcomes of previous years. One difference has been a net increase in the average annual rate of sea-level rise for the entire period of record. Whereas the previous averages for the entire data record into the early 1990s varied from about 3.8 to 4.0 mm/yr (0.150–0.158 in/yr), the new average rate for the entire data record was 4.2 mm/yr (0.165 in/yr), an increase of 5.3% to 10.4% (figure 5.7). Another statistical analysis revealed that distribution of the mean annual water level values was being displaced upward at an increasing rate. Evidence from the statistical test indicates that the distribution of the points along a slightly upward curving line was a better fit than the distribution of points along a straight line. The data have a slightly better statistical fit (that is, a higher coefficient of determination, based on $r2$ on the figure) with the nonlinear upward-sloping relation (exponential trend) versus a straight line for the period of record (figure 5.7). Extending both estimated relations into the future emphasizes an increasing separation in trend rates. If the linear rate were continued to the year 2100, relative sea level would be 42 cm (1.38 ft) higher at that time than in 2000. If the exponential rate were to apply, the relative sea level in 2100 would be 50 cm (1.65 ft) higher, a difference of almost 20% over the linear rate.

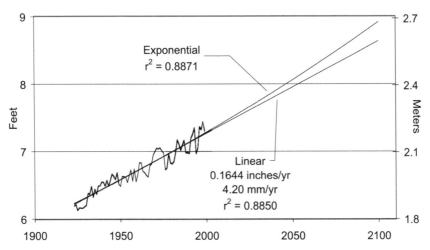

Figure 5.7. Linear and exponential trends of sea level rise, Atlantic City, N.J. (www. nos.noaa.gov)

Statistical analyses of the data from the Atlantic City tide gauge thus suggest for the first time the increased possibility of an exponential rate of sea-level rise over a linear rate. Although more data are needed to provide conclusive evidence, it is valid at this time to consider both these possibilities in planning scenarios. Such a planning exercise will likely show that it is more important to anticipate the impacts of the increasing rate and develop appropriate strategies.

Challenges for New Jersey

A central issue raised in the discussion of sea-level rise along with the increase in exposure of people and infrastructure in the coastal zone is that of public safety and protection. Retention of necessary waterfront development is also important. Public policy can help strike a balance between public protection and planned development of the coastal zone, but the necessary decisions are not easy, popular, or inexpensive. For example, dikes that keep the sea out are extremely costly. Other solutions, such as beach nourishment, are short term and do not address the problem of inundation by a rising sea level; they merely postpone policy decisions addressed at managing the most hazardous coastal areas and making large public investments. Regardless of the choices made, each is expensive and will demand periodic investment as the coastal system changes. Approaches that address the effects of sea-level rise will probably require greater expenditures of public funds. To ensure that these funds are spent wisely and support regional and state objectives, policies should be developed in advance that recognize the need to respond to a changing situation and to determine longer-term objectives associated with a range of scenarios about what the shore may look like in coming decades. If changes are required, it is necessary to identify them early and create procedures by which they are incorporated into the planning and redevelopment process.

Both the structural and the retreat options represent extreme positions and must be selectively applied. An intermediate approach may be the most feasible and the most acceptable. Regardless of the approach selected, each will require a process involving the development of state policy that must treat the broad ramifications of sea-level rise. The following steps and considerations can assist in such a process:

- Initially, policy should require that the investment of public funds in the coastal zone in post-storm situations should include efforts to mitigate the effects of a rising sea level.
- Incentives and disincentives via public funds should relate to state policy to defend the shoreline or to retreat from designated high-hazard areas.
- Public policy should be developed to assist states in determining goals regarding the impact of a higher sea level and a modified coastal zone. For example, what objectives should be achieved in 50 years as sea level rises and inundation and displacement occur?
- Risk reduction and management of the effects of sea-level rise relative to public safety and exposure to hazards should become part of public policy approaches that manage the coastal zone.
- There should also be a recognition that the coastal system is highly dynamic and that any means to interact with a changing system must also be dynamic, especially in the case of construction lines and land-use boundaries on maps and planning documents. These lines and zones should be periodically reassessed and revised, particularly after major storm events.
- There should be flexibility in the design process to permit changes in land use and distribution of facilities. Policies and public officials need to address the issue of what options are appropriate regarding redevelopment in areas damaged or previously destroyed rather than automatically assigning the same land uses and structures, or allowing the same densities, as in pre-storm conditions.

Figure 5.8. Water level very near to infrastructure on bayside, Seaside Park, N.J.

- Public policies should take a balanced, flexible approach to the inter-related issues of coastal land use, hazard reduction, and public safety.

Nearly all of the management options discussed have incorporated the effects of sea-level rise at the ocean shoreline. However, the effects of sea-level rise will also be manifested on the shorelines of bays and estuaries in the coastal zone, and they will likely be without a sizable protective buffer of beach or dunes. These locations are not only typically low-lying, initially, and very exposed to the effects of flooding, but may also contain a large proportion of the local infrastructure (figure 5.8). Bayside development is expected to experience increasing episodes and severity of flooding as the sea level continues to rise. As a result, these bay margins should be the first locations for the application of state policy that recognizes the effects of sea-level rise.

Conclusion

Relative sea level in New Jersey rose about 39 cm (15.3 in) in the twentieth century and the consensus is that the rate of rise will increase in the twenty-first. Sea-level rise is a multifaceted issue involving global and regional efforts. There is measurable evidence that water levels are rising and that the coastal zone is becoming inundated. One result is that beaches have migrated inland and barrier islands have become narrower due because of the displacement of land at the water's edge. Low-lying bay shorelines are especially vulnerable. Both residents and development at the shore are at risk, and these risks will continue to increase as the sea level rises, probably in a nonlinear manner. Given this background, policy makers must develop likely scenarios of the future direct and indirect effects of the rising sea level and formulate policies that will enhance public safety and reduce the threat of loss.

Although beach replenishment or the construction of seawalls addresses some of the issues associated with sea-level rise, neither can eliminate flooding or the high risk of damage from storms operating on elevated sea levels. A range of policies is needed. Some of the actions may direct the investment of public funds into projects to enhance areas of adequate elevation to accommodate the effects of sea-level rise in the short term; others may reduce public expenditure in locations of high hazard that require continuous repairs to both development and infrastructure, thereby removing the subsidies awarded for the occupation of exposed locations. Furthermore, flexibility needs to be introduced into design of existing land-use practices, building lines, and densities. The idea of flexible zones and boundary lines in flood-prone areas will allow for adjustments of their boundaries on a periodic basis to reflect changes in exposure and risk from a rising sea level.

CHAPTER 6

Hard and Soft Approaches to
Coastal Stabilization

Structural protection against natural disasters comes at a high cost in dollars and does not provide complete protection. . . . People and businesses tend to view the structures as providing complete protection from loss, when in fact they provide only partial protection up to the limits of some storm event, such as a storm that happens, on average, once in 50 or 100 years. . . . Because people do not understand that structural protection has limits, however, structures have been found to actually induce development in hazardous areas and to increase, not decrease, the likelihood that when a large flood or hurricane does occur, losses will be truly catastrophic . . .

R. J. Burby, "Introduction," in *Cooperating with Nature* (1998)

Too many of the accepted methods of coping with hazards have been shortsighted, postponing losses into the future rather than eliminating them. People have sought to control nature and to realize the fantasy of using technology to make themselves totally safe.

D. S. Mileti, *Disasters by Design* (1999)

The primary goal of coastal stabilization throughout most of the twentieth century and into the twenty-first has been to reduce or prevent the effects of chronic shoreline erosion. The combination of the limited sediment supply available to beaches and the gradual drowning by the rising sea level means that the shoreline in New Jersey and elsewhere is being displaced inland. As this occurs, the shoreline begins to encroach upon the human features of roads, houses, utilities, hotels, and so on. Eventually the width of the beach is narrowed, and the development and accompanying infrastructure become exposed to damage during storm events. In an effort to provide an adequate buffer, a number of approaches to shoreline stabilization have evolved. They range from attempts to slow erosion by hardening the shoreline, to balancing the effects of erosion by replacing the sand that has been lost, to erecting dikes that define the water-land contact and prevent the sea from encroaching farther inland. The approaches have been applied at a very local level (for example, a single oceanfront property lot) as well as to an area encompassing an entire sand-sharing reach, in concert with state and federal programs. Some approaches can be very short term in nature (as in a response to a single event); others strive for a longer-term balance over several decades.

This chapter considers the primary engineering approaches of modern-day coastal stabilization, summarizing the principles of coastal management and the shoreline-stabilization tools in use. It discusses the engineering practices prevalent in New Jersey to explore the dimensions of the problem and to demonstrate how these techniques have been implemented and modified over time.

Because of differences in shoreline type, coastal configuration, past practices, level of development, and long-term management objectives, a variety of approaches to managing the stability of the shoreline exist. There is no universal approach, and hence no "best" management practice. Rather, the physical characteristics and socioeconomic characteristics of a specific location will help determine an appropriate, practical approach for that area. Even then, conditions change, objectives change, and new techniques evolve to meet a changing coastal system.

Earlier emphasis on "hard" barriers to shoreline stabilization, such as seawalls, groins, bulkheads, and so on, has given way to "softer" approaches, such as beach nourishment and coastal dune development. This change in management practice has occurred not only at the state level, as in the case of New Jersey, but at the federal level (NRC 1995). New management attitudes have resulted in reduced destruction of wetlands and a prohibition of construction in designated dune zones. Changes in coastal land-use zoning have also been a means in more recent times to reestablish boundaries of development, to identify development zones, and to reduce the threat of and loss from natural hazards. In some areas, buildings have been elevated above predicted storm surge levels. The Rutgers reassessment (Psuty et al. 1996) of the *New Jersey Shore Protection Master Plan* (NJDEP 1981) recommends that approaches to coastal and hazard management be directed toward specific objectives and applied in a regional context. A similar conclusion was reached by a large panel of experts involved in the nation's Second Assessment of Natural Hazards (Burby 1998a; Mileti 1999).

Since the turn of the twentieth century, methods to control beach erosion in New Jersey have concentrated on the construction of permanent structures such as jetties, groins, seawalls, and breakwaters to protect shorefront property and the physical structures of residential housing, buildings, and other coastal development. Stabilization structures have been built along 165.2 km (102.66 miles), or 80.8%, of the New Jersey coastline. These structures are located in 41 of the 45 (91%) shorefront communities. Table 6.1 summarizes these methods and their relative costs and life spans.

The history of shoreline stabilization in the state is a long narrative of attempts to maintain a shoreline position. Early reports from the New Jersey Board of Commerce and Navigation and the New Jersey Beach Erosion Commission describe ongoing construction of barriers to reduce wave energy or to retain sand at the beach (NJBCN 1930; NJBEC 1950). Numerous publications from the U.S. Army Corps of Engineers chronicle the construction of groins, jetties, and bulkheads, providing photographic records of the placement of these structures and summary characterizations (such as U.S. ACOE 1964, 1990). This collection of continuing efforts at shoreline stabilization via structural means is impressive. Both the state and the U.S. ACOE reports illustrate the central problem: the lack of sufficient sediment input to the coastal zone with which to build beaches and to buffer the effects of waves and currents. The result is continued erosion of the beach and the stranding of structures in the sea in various states of disrepair.

Instead of "hard" engineering, however, both the *NJSPMP* and previous versions of the New Jersey Coastal Management Program indicate that the preferred approach to managing shoreline erosion is nonstructural, such as beach nourishment and land-use options, although some structural approaches may be conditionally acceptable (NJDEP 1980, 1981). The rationale for favoring beach nourishment since the late 1970s was that it addressed the central problem of inadequate sediment supply and was also the preferred approach at the federal level, although it was recognized that the technique was expensive. Beach nourishment was further viewed as interacting with the natural dynamics of coastal processes, a perspective that represented a shift in philosophy away from static approaches of the past. An important consideration was the ability of beach nourishment to provide a recreational beach, which could better accommodate anticipated increases in recreational use (NJDEP 1977). Thus a change in policy in the late 1970s occurred that was to deter any

Table 6.1 ～ COMPARISON OF STABILIZATION TECHNIQUES

Method	Physical description	Approx. costs (linear foot)	Life expectancy (years)
Shore-parallel approaches			
Revetment	Interlocking concrete blocks or rip-rap built on a slope	$1000–$1500	30
Bulkheads	Vertical walls constructed of steel, concrete sheet piling, creosote-treated lumber, aluminum, plastic, or timber	$600–$1000	30
Seawalls	Vertical wall constructed (sometimes with a curved face) of stone or concrete	$1000–$2000	50–100
Shore-perpendicular approaches			
Groins	Concrete blocks, steel, or timber built to varying lengths with a range of profile shapes (constant top or, for low-profile groin, sloping top)	$1000–$2000	30–40
Sand bypassing	Hydraulically passing sand across an inlet with a pump or trucked fill	$2/cu yd and cost of pump station	Sand is transferred periodically
Detached/attached breakwaters	Offshore structures constructed of stone and/or concrete armor units (CAU)	$1500–$2500	30
Beach nourishment	Process of bringing quantities of sand from an outside source onto an eroding beach area	$4–$8/cu yd	4–7
Dune stabilization	Creation and stabilization of artificial dunes with natural dune vegetation, I5 fill, or sand fences	Per square yard: $2 (vegetation). Per linear foot: $6–$120 (I5 fill); $1 (sand fencing)	5–10
Geo-tubes	Large, filled bags covered with sand to create a dune/dike ridge	$60–70	5–10

SOURCES: John Garofolo of the Coastal Engineering Division of NJDEP and Anthony Ciorra of the U.S. ACOE.

NOTE: Linear and volume measurements were originally taken in English units and are maintained in those units. I5 is a construction term describing coarse fill, used as the core of constructed dunes to create artificial stability.

further construction of hard structures, because of their long history of localized interference with the natural movement of sediment in the beach zone and the additional erosion problems associated with them. State policy now emphasized support of existing beaches on a regional basis and discouraged any actions that would disrupt the processes of sediment transport and accumulation in the existing beach system.

Beach nourishment addresses the basic problem of insufficient sediment supply in the coastal zone, a major advantage of the technique. It replaces lost sediment and rebuilds the beach to some previous position. In theory, beach nourishment should contribute to the maintenance of the beach zone, the offshore zone, and the foredune zone. Furthermore, it is understood that the new material will be lost over time. In essence, the addition of sediment buys time, that is, it provides some time before the beach is again eroded to its prenourishment position. According to the *NJSPMP*, beach nourishment should be planned to respond to the concerns of an entire reach and not an isolated erosion problem (NJDEP 1981). There should be an opportunity for the emplaced sediment to move downdrift and buffer other beaches in the reach. Moreover, according to the policy articulated in the *NJSPMP*, beach nourishment should no longer be used to respond to emergency cases of erosion, because it is designed to address problems on a much different time scale.

Application to New Jersey

The New Jersey shoreline is an area of continuous and at times dramatic change. Atmospheric disturbances generate winds, which in turn create waves, which eventually break at the shoreline, resulting in a release of energy. This energy mobilizes and transports sand. Most of the sand is transported alongshore and some is transported offshore. Storm waves result in net offshore transport, whereas calm-weather waves cause the sediment to return to the beach. The onshore transport rates are much slower, however, and full recovery may not occur between frequent storms. Furthermore, storm conditions may transport sediment sufficiently far offshore to depths where wave action during periods of calm weather is unable to reactivate and return this sand to the beach. If the sediment is replaced with equal quantities of beach sand from other areas or returned from the same area with no change in sand volume, then the beach is said to be in "dynamic equilibrium." If less sand replaces that which is lost in natural coastal processes, erosion occurs, leading to an overall loss of sand volume and displacement of the beach. Erosion is a natural process that shoreline communities in New Jersey and around the world have attempted to slow or stop for the last century. Desiring to protect beachfront homes and businesses, municipalities have responded to erosion problems with structural engineering approaches to retain the sand or with nonstructural solutions such as beach nourishment. These structural and nonstructural approaches have usually been applied independently of each other, but sometimes are used in combination.

The multitude of factors that affect the short-term and long-term benefits of any structural approach include regional weather conditions, storm events, and the maintenance schedule adopted by the community for upkeep of the structural approach(es) applied. Most experts believe that structural methods are only short-term adjustments to the natural, continual loss of sediment. These techniques do not add new sand to the beach, but attempt to reduce the rate of loss or to cause sand to concentrate in one location at the expense of another. The resulting changes to a shoreline's sediment budget can have both positive and negative consequences for a host community and its neighbors. The negative side effects that occur downdrift can be recast as a negative externality and treated formally in economics within externality theory, as is commonly done in the case of pollution. The basic solution to this traditional problem is that communities must account for the social costs they impose on other communities by the use of structural tools. For example, the community that causes the externality could forfeit a portion of its beach revenues (from beach-badge sales) or tourism revenues (from taxes on tourists' spending) to the affected communities equal to the amount of negative effects caused. Another possibility is that the state could tax the community causing the problem equal to the social costs or damages downdrift and provide income transfers to businesses in the affected communities. However equitable this approach might be, it is not in pratice.

Structural "Hard" Approaches

Shoreline-Parallel Structures:
Sand Retention and/or Wave Interference

Traditionally, four major types of shoreline-parallel approaches have been used to stabilize the shoreline against the effects of wave action. Three types are placed in the beach (revetments, bulkheads, and seawalls), and one is offshore (breakwaters). There are always a vari-

Figure 6.1. Revetment composed of rip-rap stone blocks against headland. Revetments attempt to stabilize the slope. Deal, N.J.

ety of modifications, such as sediment-filled fabric tubes referred to as "geo-tubes," and different materials, such as wood, concrete, or metal, may be used. Nevertheless, most structures can be classified in one of the above four categories because of their interaction with the incident waves, their location in the beach profile, and their interaction with the mobilized sediment. In each of these approaches, the consistent variable is that the structure extends along the shoreline (parallel with the coast and water's edge).

REVETMENTS—TRADITIONAL

Revetments are by far the simplest of shoreline-parallel structures. They are placed on the seaward face of a sloping beach and are designed to stabilize an eroding shoreline in areas of light wave action. Revetments can be made of a layer of stones or interlocking concrete-stone blocks called "rip-rap," and they line or cover the seaward face of the exposed slope (figure 6.1). Over the short term, revetments stabilize the slope and reduce wave run-up, thereby reducing the risk of direct wave attack on inland structures and development. However, an increase in turbulent energy caused by waves reflected off the hard surface and associated with an increase in local erosion may threaten the long-term survival of this approach. Local beach dynamics and storm events also affect the life expectancy of revetments. With careful thought given to the

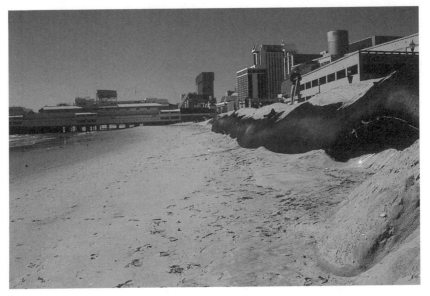

Figure 6.2. Exposed geo-tube core of artificial dune. Tube also forms a seawall, protecting the boardwalk. Atlantic City, N.J.

structural design, a revetment may last an average of 30 years or more. The cost of a revetment is approximately $1000–$1500 per linear foot (in 2000 dollars) and can vary depending on materials used and the elevation or height desired. Along the coastline of New Jersey, there are 5 hard revetments that were built during the 1952–1962 period.

REVETMENTS—SEDIMENT-FILLED
FABRIC TUBES (GEO-TUBES)

Elongate fabric tubes, known as geo-tubes, are an approach resurrected in the 1990s as a barrier to coastal processes, and they function in much the same way as traditional revetments. These fabric tubes have a diameter of 0.5 m to 2.5 m (2–8 ft) and are usually filled with local sediments. The tube envelope is made of a soil-tight geotextile fabric and is hydraulically filled with a sand-water mixture. In the past this technique was used at locations of severe erosion in New Jersey as a form of sandbag revetment extension. More recently it has been used in two other situations. In Atlantic City and Sea Isle City, geo-tubes have been placed on the upper beach in front of the boardwalk, filled with sand pumped from the beach, and covered with sand to create a dune (figure 6.2). In Avalon, a geo-tube has been placed at the low-tide line at the inlet to form a type

Figure 6.3. Wavebreaker pilings in beach. Strathmere, N.J.

of low revetment. In both cases the filled fabric tubes encapsulate some of the local sediments and form a hardened barrier to the sediment-transporting processes, a soft type of wall. In areas of modest erosion, geo-tubes can protect the areas directly inland. In locations of high erosion, however, geo-tubes can be undermined and may collapse seaward. Although these tubes are resistant to ripping and tearing, once they are torn the sediment will eventually wash out.

WAVEBREAKERS

Wavebreakers are structures that are in the beach, interfering with the passage of the wave across the beach. They are lines of closely spaced pilings, sometimes connected by metal rods to a timber bulkhead on the shore face (figure 6.3). There are 5 wavebreakers (2 in Long Beach, 1 in Ventnor, and 2 in Strathmere) recorded in the "Inventory of NJ Coastal Structures" (U.S. ACOE, 1990, appendix B). Their construction dates are largely unknown, with the exception of the wavebreaker at Ventnor, built in 1920.

BULKHEADS

Bulkheads are vertical walls constructed of steel or concrete sheet piling, creosote-treated lumber, aluminum, plastic, or timber (figure 6.4).

Figure 6.4. Vertical bulkhead, with rip-rap at toe. Bulkheads defend a line. Strathmere, N.J.

These structures extend from below low water to above high tide and are usually in the upper portion of the beach, at or within the dune zone. They are not usually exposed to direct wave action except during storms or other high-water events. In the short term, bulkheads preserve adjacent landward property. However, direct wave action can undercut and undermine the bulkhead, flank its end, and lead to the construction of seawalls in its place. The cost of constructing bulkheads is dependent on the chosen material. Materials such as aluminum and creosote cost approximately $600–$1000 per linear foot ($50–$100 per ton) (2000 dollars). The life expectancy of bulkheads depends on the material and the conditions associated with the location. Material such as creosote-treated lumber has lasted approximately 30 years when properly maintained. Aluminum has also been used experimentally in bulkhead construction and is thought to have an extended life expectancy. There are 82 bulkheads along the New Jersey shore, built from 1905 to 1989.

SEAWALLS

Seawalls may be constructed of stone or concrete, and are sometimes built with a curved face to dissipate wave energy and prevent undermining (figure 6.5). They are designed to sustain the full force of

Figure 6.5. Seawall with waves breaking directly on the wall. Small pocket beach has formed on updrift side of groin in background. Sea Bright, N.J.

wave action, and are often used in conjunction with the structures already described. Seawalls as well as bulkheads are designed to protect only the land immediately behind them and function as a dike to keep water from penetrating inland. Beach erosion continues both in front of and downdrift of the structure, and the presence of seawalls may locally intensify erosion. This can result in scouring, or erosion, at the base of the seawall and may lead to the eventual collapse of the structure. Costs of constructing seawalls depend on the material but generally range from $500 to $1000 per linear foot ($50–$100 per ton) (1996 dollars). Seawalls may be expected to last 50–100 years, depending on their construction, maintenance schedule, and local conditions. In New Jersey there are 14 seawalls of varying length, constructed from 1898 (at Sandy Hook) to 1980 (in Avalon).

BREAKWATERS: AN OFFSHORE BARRIER

Breakwaters are offshore structures constructed of stone and/or concrete units. They are designed to protect shore areas from direct wave action and to create littoral sand traps. In a detached breakwater, a structure not connected to the mainland, sand accumulates inland of the breakwater, forming a seaward projection of the shoreline (figure 6.6). The structure shields a portion of the beach, but can lead to

Figure 6.6. Detached breakwater offshore from beach. Sand accumulation extends seaward toward breakwater. Beach is scarped to either end. Cervia, Italy.

downdrift erosion. These structures are used on the western coast of the United States and, more commonly, in the Mediterranean and Australia. The attached breakwater, which is connected to the mainland, functions slightly differently. Its objective is to shield an area from direct exposure to waves (figure 6.7). However, it will also intercept a longshore sediment transport, causing accumulation on the updrift side and accelerating sediment loss downdrift. In New Jersey an attached breakwater, constructed in 1965, exists at Ocean City, shielding the music hall, and there is another in Asbury Park, with an unknown construction date, which shields the theater (figure 6.7). Over the long term, breakwaters increase beach accretion and shoreline stabilization immediately landward of the structure. However, breakwaters accumulate sand at the expense of the downdrift area. Both attached and detached breakwaters can be installed at a cost of approximately $1000 per linear foot (1996 dollars) and may be expected to last an average of 30 years with proper maintenance.

SUBMERGED BREAKWATERS— ARTIFICIAL REEFS

The submerged breakwater, or artificial reef, is constructed parallel to the shoreline to shield the beach and adjacent infrastructure from the effects of large waves. The structure may occupy about half of the water depth at a location and permits the passage of small waves. In 1993–94, three submerged breakwaters were installed along the shoreline

Figure 6.7. An attached offshore breakwater shields an area from direct wave attack. It also intercepts littoral transport of sediment. Asbury Park, N.J.

at Avalon, Cape May Point, and Belmar–Spring Lake. These reefs are composed of interlocking concrete units that measure approximately 4.5 m (15 ft) in length. The structural design of the unit (figure 6.8) consists of a ribbed seaward face and a more steeply sloping landward face, leading to a slotted opening that runs along the slightly curved crest of the reef. These units were engineered to theoretically reduce the offshore loss of sand during storm events by reducing the incident wave height and creating a vertical current from deflection of the bottom flow upward through the slotted openings at the crest.

Each installation occupies a different location relative to the ambient processes and geomorphology. Both the Avalon and the Belmar–Spring Lake projects included nourishment on the landward beach. The Avalon reef was constructed immediately adjacent to Townsend's Inlet and was the only structure of the three to have an open end. Both the Belmar–Spring Lake and the Cape May Point installations involved a complete spanning of groin compartments. In all cases the reefs were connected to the groins or jetty, with additional stone at the junction with the reef structure. All reefs were placed with different top elevations (that is, the depth of water at mean low water). The Cape May reef was placed in the most shallow water, followed by the Belmar–Spring Lake and then the

Figure 6.8. Breakwater reef cross-section configuration. (Photo courtesy of Breakwaters International.)

Avalon reef. Geotextile fabric or mattresses were placed beneath each unit to provide scour protection.

The submerged breakwaters were evaluated through a state-sponsored monitoring project conducted by engineers from the Davidson Laboratory at Stevens Institute of Technology in Hoboken, New Jersey (Bruno, Harrington, and Rankin 1996). The monitoring effort included beach surveys, offshore bathymetric surveys, wave and current measurements, dye-release studies, and visual (scuba) inspections of the structures. The three reefs varied in their success. The Avalon reef was considered a success in limiting the loss of the renourished beach. The northern portion of the project site appears to protect the area from the strong tidal currents in the Townsend's Inlet channel. The reef has not, however, provided the same degree of protection at the southern portion of the installation. This area has experienced the same rate of erosion as that of the unprotected coastline to the south, along with a local scour zone just landward of the southern end of the structure. This loss of sediment is thought to be caused by the open end at the southern limit of the installation.

Because the area is not sheltered by a groin installation, it is believed that the reef causes enhanced wave activity in the area immediately shoreward of the structure, resulting in sediment loss equivalent to an unprotected or natural shoreline.

Both the Belmar–Spring Lake and the Cape May Point installations have experienced general accretion landward of the reef (Bruno, Harrington, and Rankin 1996). Of some concern, however, is the presence of local scour zones immediately landward of the reef in enclosed situations. These scour zones do not appear to extend any great distance landward. This feature is thought to be caused by the deflection of the bottom flow by the reef, thereby causing locally intense bottom shear stress and sediment removal. The Belmar–Spring Lake installation experienced a redistribution of sand to the offshore bar. Beach erosion and bar formation were due to several major coastal storms associated with a very active hurricane season in 1995. The artificial offshore breakwater reefs lose their effectiveness in the vicinity of an open end, and should therefore either be employed only in the protection of natural or man-made cells (for example, groins) or be redesigned to minimize the end effects. Observations have also found no evidence of adverse impact on adjacent beaches.

Shoreline-Perpendicular Structures

In general, two shoreline-perpendicular approaches have been used to maintain shoreline positions, groins and jetties.

GROINS

A groin is a stone, concrete, steel, or timber structure placed perpendicular to the shoreline out into the ocean (figure 6.9). It is designed to slow the rate of littoral drift of sand and to capture sand along the updrift side, causing accretion, or an accumulation of sand. Groins have been built at varying lengths and with a range of profile shapes, for example, constant-top elevation or sloping-top elevation (low-profile groin), T-shapes, L-shapes. Groins initially trap sand moving along the shore, causing the beach to build seaward on the updrift side. However, a groin may also accelerate erosion in the downdrift direction to such an extent that it becomes necessary to construct a second groin, then a third, and so on. Downdrift erosion, occurring in a domino pattern, can become an urgent problem requiring more elaborate stabilization measures. Groin installations do not create new sand but are intended to slow the

Figure 6.9. Numerous groins projecting seaward from upland. Pockets of sand indicate accumulation associated with northerly drift. Long Branch, N.J.

rate of loss created by alongshore transport of sand. Groins function best when there is an adequate supply of sand and are not effective when the littoral or nearshore materials are finer than sand. The problems associated with downdrift erosion may be partially corrected by cutting notches into the groin or using gaps when constructing it, in order to allow some sand to pass across or through the structure. The top elevation of the groin may be designed so that the seaward end slopes downward or is lower than the waterline, allowing some sand to pass across. When constructed and maintained properly, these structures may last 30–40 years at a cost of $1000–$2000 per linear foot ($75–$100 per ton) (2000 dollars). There are 368 groins, built between 1942 and 1967, within the four coastal counties of New Jersey (U.S. ACOE 1990). A few new groins were constructed in Sea Isle City after this time.

JETTIES: INLETS AND SAND BYPASSING

Jetties are constructed of rock and concrete at inlets to confine tidal flow and to prevent littoral drift from shoaling navigation channels within an inlet (figure 6.10). They rise above water level and are often constructed in pairs. They are designed to help stabilize the depth and location of channels. By their very nature and function, jetties are directly at odds with beach stabilization. A stable, jetty-protected navigation

Figure 6.10. A pair of jetties at Barnegat Inlet are controlling inlet width for navigation. Shoals on the bayside are part of the flood-tide delta. Fill dredged from the inlet has widened beach to left (downdrift) of the jetty.

channel prevents or minimizes shoaling of the channel, in turn creating a barrier to littoral transport. This causes accretion on the updrift and erosion on the downdrift sides of the channel. Natural inlets (that is, without structures) exhibit natural sand exchange across an ebb-tide shoal. Most often, controlled inlets (that is, inlets with jetties) limit the sediment transfers across the navigable channel and reduce the dimensions of the associated submarine shoals. An immediate consequence of this restriction is a loss of sediment supplied to the downdrift side of the channel and an increase in shoreline erosion. This problem can be addressed by transporting the accumulated sediment mechanically, by sand bypassing, via dredge, pipeline, or periodic truckload, thus minimizing erosion on the downdrift side (figure 6.11). Sand bypassing may be conducted hydraulically at an estimated cost of $2 per cubic yard (2000 dollars), excluding capital costs. Sand bypassing is conducted on a regular basis at Indian River Inlet, Delaware, where the capital costs (above annual maintenance costs) of constructing the system were approximately $1.7 million dollars (1996 dollars). Within New Jersey, there are 24 jetties that were initially constructed between 1908 and 1967, with significant maintenance and reconstruction as recently as 1990 (U.S. ACOE 1990).

Figure 6.11. Sand bypass system (intake hose at crane in background) transfers sand hydraulically across inlet through pipeline. Indian River Inlet, Del.

Nonstructural "Soft" Approaches

Beach Nourishment

Beach nourishment involves bringing quantities of sand from an outside source onto an eroded beach area to counter the losses from that beach. The scale of this activity can range from short-term replacement of sand lost during a specific storm event (figure 6.12) to extensive placement of sand designed to create a beach with a life expectancy of about five years (figure 6.13). New sand may be pumped from an offshore location (known as an offshore borrow site) or may be trucked in and deposited on the beach. Beach nourishment projects are usually intended to broaden and heighten the beach surface, producing a berm 3 m (10 ft) above mean low water and 30 m (100 ft) wide. The offshore slope of the berm is usually a ratio of about 1:30, that is, for every one-meter rise in height, there is a change in length or width of 30 meters down to the intercept with the preexisting surface. Sometimes an artificial dune ridge is incorporated at the inner margin of the constructed beach.

Past beach nourishment has seen sand placed on the beach by various means and from various sources. Small amounts of sand (thousands of cubic yards)[1] have frequently been transported by truck and deposited on the beach in areas of severe local erosion that endangers or threatens to endanger property or infrastructure. Larger amounts (hundreds of thousands to millions of cubic yards) are usually deposited on a beach site by

Figure 6.12. Beach nourishment in small section of shoreline, emplaced by truck and earthmoving equipment. Brant Beach, N.J.

Figure 6.13. Pumping sand onto beach via pipeline. Ocean City, N.J.

hydraulic pipeline. The sand is typically dredged from an offshore site or inlet and is transferred directly or via some intermediate handling process to the beach. Because of the large volumes involved with pumping, nourishment projects try to combine building the beach at a selected site along with the emplacement of sand updrift (that is, on a feeder beach) so that the natural processes will eventually transport the sand to downdrift locations, thus benefiting many beaches along a transport continuum. In this way the use of fill can maintain the sediment budget throughout a reach rather than at a single location.

In the short term, beach nourishment creates a wide beach surface and provides for better shore maintenance and recreational use. However, beach nourishment is understood by all to have comparatively short-term effects on erosion; sandfill must be replaced periodically to maintain a shoreline position. There are many factors to consider in a nourishment project, including the rate of loss of beach material in the region, the predominant direction of littoral drift and its interaction with all existing engineered structures, the availability and suitability of beachfill material, the method of beachfill placement, the physical effects of fill removal at offshore sites, and the impact on the ecological systems at both the borrow and the fill sites. It is important to select beachfill material that closely matches the average natural grain size of the beach. Fill that is too fine, as from the back bays, will be rapidly eroded. Upland sources tend to have a silt component that will wash out quickly and may create a yellowish stain for a short time.

The use of nourishment as a shoreline-stabilization measure generally requires a continuous commitment to financial and material maintenance. The costs associated with beach nourishment must be balanced against the value in protection of infrastructure, property, and development. Sandfill may cost $600 per linear foot ($2–$8/cubic yard) (2000 dollars) and is usually replaced on a two- to six-year cycle. The fill cycle is highly dependent on local beach dynamics. Beaches on the downdrift side of inlets are especially dynamic, and these beaches, in Ocean City and Avalon, for example, have been the sites of multiple nourishment projects.

Coastal Dune Development and Enhancement

Dunes are natural features of the coastal landscape, the product of many years of interaction between waves, wind, and sand. Dunes exist in conjunction with the beach and are part of the sand-sharing system that actively exchanges sand between the dune, the beach, and the offshore bars. Dunes can provide a protective buffer, and their presence and function in the beach zone can reduce the rate of shoreline erosion (displacement) and can serve as a barrier to storm surge, in turn offering protection to properties directly inland of the dune zone (figure 6.14). In areas of adequate sand supply, coastal dunes achieve their full form. Where the sand supply is limited, however, dunes are small and narrow and can be frequently overwashed. In areas of very meager sand supply, dunes will not exist.

Figure 6.14. Artificial sand dune, created by bulldozing into ridgeform, augmented by sand fences, stabilized by planted vegetation. Lavallette, N.J.

In culturally developed coastal zones, dunes perform their natural function as sites of sand storage along with the added roles of providing a natural barrier to storm surge and flooding and contributing an important niche to the coastal ecology. As a result dunes are often valued by coastal communities as a natural, esthetic, ecologic, and protective feature of the landscape. Although in New Jersey dunes have been recognized as a form of coastal protection since the early 1930s, in it was not until 1984 that shoreline communities and the state took an active role in maintaining them. Through these efforts, the value of the coastal foredune as a barrier against coastal storm surges and waves was generally established. Providing technical and financial support, New Jersey's Department of Environmental Protection and Office of Emergency Management (NJOEM) continue to encourage communities to restore and improve their dunes.

A common dune-building method consists of erecting wind obstructions, such as picket-style sand-snow fences, that break the wind and capture sand particles. Another approach is to create a dune ridge by direct placement of sand on the upper beach, as a short-term stopgap. In the United States generally and in New Jersey, many additional materials have been used as wind-flow barriers to trap sand in transport (across the beach), such as fences of cloth or plastic. Old Christmas trees have often been recycled as brush in the dune to help trap sand, but can be unsightly if they are not buried or covered. The use of sand fencing in dune construction is relatively inexpensive at $1 per linear foot (1996 dollars).

Natural coastal foredunes principally function as sites of sediment accumulation and storage. As long as sufficient sand is present, dunes will exist at the inland margin of the beach. However, if erosion is too severe or if there is no space for the dune to shift further inland as the shoreline erodes, dunes will not persist. Dunes are treated in more depth in chapter 7.

An Overview of Beach Nourishment in New Jersey

Large-scale beach nourishment was applied to New Jersey beaches following the March 1962 Ash Wednesday storm that caused extensive damage and erosion, with overwashes occurring at many sites. In 1962 and 1963, sand was pumped onto the beaches in quantities reaching approximately 7.5 million cubic yards and 4.65 million cubic yards, respectively (U.S. ACOE 1990, augmented by data from NJDEP). Each of the coastal counties was part of this statewide nourishment project and received sand to rebuild its beaches and dunes. Most of the subsequent projects in the communities over the next two decades can be characterized as small, local episodes, except for several projects in Atlantic City and Ocean City. In the mid-1990s several very large beach nourishment projects were initiated and continue to the present (table 6.2). The cost of the projects became a significant constraint. For most small shore communities, the cost of beach nourishment was beyond their financial capacity and thus required substantial state aid. During the 1980s increasing amounts of beachfill and sand were placed on New Jersey's beaches (figure 6.15), sometimes in response to storm erosion (for example, in 1984 and 1993) or as part of a longer-term program. There were several substantial beach nourishment projects during this time, including an extensive federally sponsored project at Sandy Hook in 1983 and 1984, and repeated in 1989. In addition, beach nourishment projects were carried out at Ocean City and Atlantic City (largely state supported). In the 1990s the amount of beachfill continued to increase, with efforts concentrated at the municipalities of Ocean City, Cape May City, and throughout Monmouth County. A strong catalyst for this more recent beach nourishment was the creation of a stable funding source for beach protection approved by the state legislature in 1992. A pool of $15 million

Table 6.2 ~ BEACHFILL PROJECTS IN NEW JERSEY, 1980–1999

Project number or sponsor	Location	Dates	Amount of fill (cubic yards)	Site	Cost	Other information
NJDEP	Allenhurst	1985	35,000–40,000			
Federal	Asbury Park to Shark River Inlet	1994–1999	3,100,000		$17,000,000	
174a	Atlantic City	1/83–6/83	75,000	Mass. Ave.	$358,250	Trucked
575	Atlantic City	9/86–2/87	1,000,000		$7,000,000	
576	Avalon	1987	1,300,000	8th to 30th St.	$2,873,940	
1219	Avalon	12/87–1988	158,945			
NJDEP	Avalon	1989	60,000			
Local	Avalon	1990	404,000		$600,000	
NJDEP	Avalon	1992	410,000		$1,188,000	
Federal	Avalon	1993	239,000		$1,777,193	
567	Avon	1982	136,000		$352,240	
NJDEP	Barnegat Light	1991	75,000		$326,087	Dredged from Barnegat Inlet
State/local	Brigantine	1996	1,200,000		$6,000,000	
171	Cape May	12/1981	36,000		$93,000	
583	Cape May	1991	770,000		$1,017,501	
578	Cape May, U.S. Coast Guard base	1989	465,000		$3,158,000	100% USCG
583/584	Cape May City	1991	900,000		$4,380,000	
587	Cape May City	1992	500,000		$2,232,143	
587/Federal	Cape May City	4/1993	415,000		$4,561,000	Storm rehab.
Federal	Cape May City	9/1993	300,000		$2,135,000	1st nourishment cycle
Federal	Cape May City	9/94–2/95	330,000		$2,605,000	2nd nourishment cycle
Federal	Cape May City	1997	366,000			

(*continued*)

Table 6.2 ~ (*continued*)

Project number or sponsor	Location	Dates	Amount of fill (cubic yards)	Site	Cost	Other information
NJDEP	Cape May Point	1992	42,000		$187,500	Dredged from Cape May Canal
1217	Cape May State Park	1/85–86	15,000	Sand dune construction	$272,618	
176	Cape May State Park	6/1992	200,000		$261,905	
581	Harvey Cedars	1990	27,300		$34,957	
2092	Harvey Cedars	1/1992	110,000	Between 85th and 82nd	$491,071	
4005	Harvey Cedars	2/1994	485,000		$3,700,000	Trucked
NJDEP	Long Beach Twp.	1990	175,000		$36,000,000	Dredged from Barnegat Inlet
Federal	Long Branch	1999	4,300,000			
582	Longport	1990	250,000		$949,000	Dredged from Great Egg Inlet
NJDEP	Lower Township	1986	87,000		$337,209	Dredged from Cape May Canal
Federal	Monmouth Beach	1994	800,000	3.1 miles		
Federal	Monmouth Beach	1995	4,600,000		$19,600,000	Dredged Herford Inlet
2086	North Wildwood	1989	190,000		$875,000	
566	Ocean City	1980	150,025		$647,147	
1062	Ocean City	1982	1,149,683	Morningside Rd. to 13th St.	$4,885,000	
1235	Ocean City	5/87–1988	Dunes, 190,000 Fill, 40,000		$2,847,086	
579	Ocean City	1989	250,000		$717,236	
1250	Ocean City	1990	256,000		$1,207,250	Emergency fill
585	Ocean City	1991	100,000		$130,840	
586	Ocean City	1992	2,617,000		$10,915,970	
Federal	Ocean City	1993	2,700,000		$14,571,908	
Federal	Ocean City	1993	845,000		$2,915,132	

Federal	Ocean City		1994	607,000	$3,217,825
Federal	Ocean City		1995	1,411,000	$5,746,992
State/local	Ocean City		1995	360,000	$1,232,572
Unknown[a]	Ocean City		1997	800,000	
571	Sandy Hook		1983	2,370,000	$10,236,161
572	Sandy Hook		1984	800,000	$3,968,965
1228	Sandy Hook		1989	3,302,273	$1,350,000
Federal	Sea Bright		1996	3,800,000	$16,300,000
1055	Sea Isle City		1981	20,880	$54,080
569	Sea Isle City		1983	45,000	$194,294
1061	Sea Isle City		1984	800,000	$3,652,500
577	Sea Isle City	South of 78th St.	1987	150,000	$528,244
NJDEP	Sea Isle City	Between 77th and 82nd	1992	375,000	
1614	Sea Isle City	Between 2nd and 10th	3/1992	20,000	
Federal	Shark River Inlet to Manasquan Inlet		1997–1999	4,100,000	$27,000,000
4009	Spring Lake/ Belmar	Between 19th Ave. and Pitney Rd.	1/1994	70,000	$347,199
568	Strathmere		1982	45,000	$90,000
1080	Strathmere		1984	450,000	$2,986,679
574	Strathmere (upper)		1984	592,000	$3,929,142
1056	Upper Township		12/1981	36,000	$93,240
1201	Upper Township		1984	120,000	$2,453,600
573	Upper Township		1984	1,600,000	$6,451,613
NJDEP	Upper Township	Whale Beach	1992	23,000	$102,679
Federal	Wildwood		1991	100,000	$434,783

SOURCES: Uptegrove et al. (1995); U.S. ACOE (1995); NJDEP files; *www.geo.duke.edu/research/psds/njersey.htm; www.nan.usace.army.mil.*

a. Incomplete information.

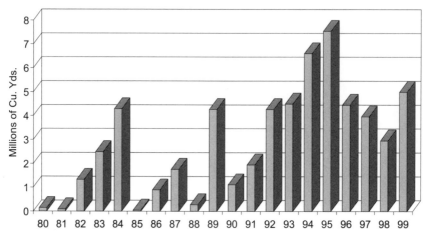

Figure 6.15. Beach nourishment volumes by year, 1980–1999 (millions of cubic yards). (Uptegrove et al. 1995; U.S. ACOE 1995; NJDEP files; *www.geo.duke.edu/research/psds/ njersey.htm; www.nan.usace.army.mil.*)

annually for a Shore Protection Fund, later raised to $25 million, was to provide most of the nonfederal share (35% in 1992–1999) and was an important factor in the increase in nourishment projects in the post-1991 period.

Allocating the Cost of Beach Nourishment

Support for beach nourishment was recommended in the *NJSPMP*, but with several important qualifications. Initially the plan suggested that beach nourishment be used to maintain existing recreational beaches and indicated that this approach should not be applied primarily as an emergency measure. Furthermore, the plan indicated that the use of beach nourishment should be a reachwide activity, suggesting that the benefits must apply to the reach and not to a smaller locality. This important consideration appears to have been neglected in the applications of beach nourishment in the 1990s, except for part of the project design in the Sea Bright to Barnegat Inlet project. Significantly, all of the federal Army Corps of Engineers projects in the state that were listed as in the reconnaissance, feasibility, or operational phase in the 1990s were applied to entire reaches.

Public concern about the appropriateness of stabilizing the shoreline and about specific techniques used in the process has generated consider-

Table 6.3 ~ INITIAL COSTS AND MAINTENANCE FOR AUTHORIZED AND
PROPOSED PROJECTS IN NEW JERSEY, 1994–2000

	Initial cost, total project	*Initial cost, nonfederal share*	*Maintenance, nonfederal share, 50 years*	*Maintenance per year, 50 years, nonfederal*
Sandy Hook to				
Manasquan Inlet				
Section I: Sea Bright to				
Ocean Township	$140,000,000	$25,000,000	$45,000,000	$5,957,700
Section II: Asbury Park				
to Manasquan Inlet	$70,000,000	$45,000,000	$25,000,000	$4,211,200
Cape May City	$10,526,000	$2,149,000	$10,600,000	$212,000
Ocean City	$33,195,000	$10,482,000	$201,518,000	$4,030,360
Brigantine	$8,558,000		$12,728,800	$254,576
Long Beach Island	$35,794,000		$60,055,450	$1,201,109

SOURCES: Uptegrove et al. (1995); U.S. ACOE (1995); NJDEP files; *www.nan.usace.army.mil.*

able interest and exchange of views. Some see the emplacement of sand on beaches as an unrealistic and expensive attempt to stabilize an extremely dynamic system, one that has changed greatly in the past and that will continue to evolve. Others see beach nourishment as a necessary process to provide stability and protection for economic development at the shore. Because beach nourishment must be repeated at some time interval to replace the lost sediment and to rebuild the shoreline, it is usual to design projects that incorporate renourishment or maintenance over a multiyear period (for example, 50 years). These projects involve an initial investment and periodic costs over time that are budgeted for. What length of maintenance to commit to is an issue for state policy, to be decided by New Jersey citizens and elected officials. In the past, the cost of maintenance has been shared between the federal government and the state on a 65%–35% basis, with the cost of maintenance efforts typically several times greater than the amount of the initial project cost. A partial listing of recent projects and their initial as well as maintenance costs demonstrates a considerable debt for future support (table 6.3). Concern over this issue has increased as a result of decisions at the federal level to either reverse the federal share of maintenance costs or eliminate its share entirely.

Generally, renourishment maintenance occurs on a three- to five-year cycle, with storm rehabilitation an additional expense. Although this relatively brief cycle is a general rule of thumb used by coastal engineers, economists and public policy analysts may view the lifespan or effective life of a nourishment project differently. A basic problem with the planned maintenance period of 50 years used by the Army Corps of Engineers is that the variability of the effective life of a project and of the associated erosion rates is not explicitly treated. In effect, all projects in all locations are treated equally. One way to incorporate temporal variability in a project's effective life is to designate the final year of a nourishment project as the year preceding the one in which a new supply of beachfill is needed. The new supply should involve an amount of sand roughly equivalent to that initially deposited there or at a similar location. This is a fairly standard application of the cost-benefit analysis (CBA) used by economists, in the sense that a new project is initiated when an equivalent amount of beachfill is placed in the same location. Hence over a period of time (decades) we may observe that beach nourishment projects last different lengths of time, and thus evaluative procedures should provide a measure of the variability in their effective life.

As a result the effective life for a given nourishment project can be determined, along with a measure of the variability in the effective life of projects. This is a necessary variable in applying dynamic decision-theoretic techniques that we can learn from and incorporate into CBA procedures. Other factors that have been handled poorly or not at all in CBA analyses of shoreline-stabilization projects are the influence of exogenous risk, such as erosion rates on the effective life of projects that can vary in both a temporal and a spatial sense, and the tradeoff between projects where the flow of net returns introduces variability across projects, and hence risk. For example, it should be possible to compare two projects, Project A, which is less variable and less risky because the flow of net returns ($5000) over five years is evenly distributed ($1000/year), and Project B, a more variable and riskier project with an uneven distribution of net returns ($2000 in the first year, $1000 in the second year, $500 in the third year, $1500 in the fourth year, and $0 in the fifth year). The family of decision-theoretic models from the field of finance and portfolio theory can handle risk and uncertainty and offers new insights to improve procedures for evaluation. From these models one can assess the tradeoffs between risk and uncertainty and define a portfolio of projects that meet appropriate criteria. It is possible that these insights can be introduced into the planning and provision of future shoreline-stabilization projects. They are explored further in chapter 8.

A Brief Evaluation of Beach Nourishment:
NRC Assessment

Promoted by public debate over the effectiveness of beach nourishment in treating the problem of an eroding shoreline, the Marine Board of the National Research Council established a Committee on Beach Nourishment and Protection. Its mission was to "conduct a multidisciplinary assessment of the engineering, environmental, economic, and public policy aspects of beach nourishment to provide an improved technical basis of judging the use of beach nourishment and protection technology in shoreline stabilization, erosion control, recreational beach creation, dredged material placement, construction of coastal storm barriers, and protection of natural resources" (NRC 1995). Among the conclusions of the committee was that beach nourishment is an appropriate technique to buffer the erosional effects of storms. It is an approach that attempts to interact with the natural processes and rebuild a beach in areas of erosion, and it addresses the problem of loss of sediment directly. However, committee members conclude that beach nourishment is not a panacea. Projects should be designed for specific areas, with a solid foundation in science and engineering. Areas exhibiting high erosion rates may not be good locations for the application of beach nourishment, because of the rapidity of sediment loss. Further, beach nourishment is effective in human time scales (decades, not centuries). It should be applied in concert with established goals and a method of quantifying the measure of success. It is important that maintenance be part of the original plan and that sources of sediment for renourishment be identified at that time.

A brief follow-up to the 1995 NRC report offered a small caveat to the endorsement of beach nourishment as an appropriate means for shoreline stabilization (Seymour 1996). The chair of the committee noted that federal public policy toward beach nourishment changed as the report was in its final stages. At the time, there was a decrease in the availability of federal funding and a modification of the federal-state partnership upon which all economic assessments had been based. Although this shift in federal policy brought various aspects of the economic viability of beach nourishment into question, it was felt that the report's review of techniques and applications of beach nourishment remained valid regardless of any new policies or partnerships reducing the federal share of the costs. Furthermore, because states and local governments would have to be increasingly responsible for the planning, financing, and execution of beach nourishment projects in the future, the assessment information would become more useful in evaluating the full direct and indirect costs of beach nourishment.

Policy Evolution

Today's policy climate contains a heightened concern for financial constraints on public expenditures, more economic accountability, and fewer federal funds available to support beach nourishment projects. Prior to 1995, beach nourishment and shoreline stabilization was under the purview of the U.S. Army Corps of Engineers, with cost sharing set at 65% federal and 35% nonfederal. The U.S. ACOE provided the engineering expertise that designed, implemented, and supervised most projects. In fiscal years 1995 and 1996, funding for beach nourishment projects by the U.S. ACOE was disapproved by President Clinton, and the budget item for shoreline stabilization was subsequently removed. As a result, support for shoreline stabilization was only possible by direct congressional appropriation. The overall view of the Executive Office at that time was that shoreline stabilization was something that states, counties, and communities had primary responsibility for and that federal support would no longer be available. In the state of New Jersey, this message came at a time when the local availability of shoreline-stabilization funds had increased following the approval by the state legislature of an annual fund of $15 million for such projects, with further efforts to raise this amount to $25 million annually. Hence the state was in a position to bear more of the responsibility for shoreline stabilization. However, $15 million or $25 million could support only a limited number of projects without matching federal funds (figure 6.15). With the reduction of federal support, the state's citizens and legislature have had to reevaluate the financial commitments required into the future.

Continuing Issues

The NRC report (1995) noted that although the application of beach nourishment points to the symptoms of the problem, it does not address the basic conditions responsible for the erosion of the coast, which are sea-level rise and the increasingly limited supply of sediment at the shore. In the case of barrier islands, as the sea level rises each entire island becomes slowly submerged, and the placement of sand on the beachfront does not reduce the rate of inundation or the narrowing of the island. Nor can beach nourishment change the decreasing elevations of the islands or the increasing exposure to storm damage as a result of sea-level rise. Sea-level rise is a continuing process, and it will change the conditions of exposure and modification due to storms. The cause of sediment deficits is another issue not directly addressed by beach nourishment, and it too will persist as a result of the absence of sediment inputs by rivers leading to the beaches of New Jersey. The effective life of a nourishment project is largely related to the existing sediment deficit as well as to location, beachfront exposure, presence of other protection structures, and storm events.

Conclusion

The long-term trend of the shoreline position along the Atlantic coast and New Jersey's coastline is one of inland displacement caused by a loss of sediment and rising sea level. Major factors leading to this situation, as we have seen, are the general absence of new sediment supplies to replace the material that is being eroded and the slow inundation of the ocean, bay, and estuarine shoreline associated with sea-level rise. Most of the engineering approaches discussed in this chapter have been applied as an attempt to stabilize portions of the New Jersey shore and defend "the line." Hard or armored shore structures (groins, jetties, bulkheads, offshore breakwaters, and so on) are abundant along the state's coast and are usually directed toward local erosion or stabilization problems. Generally, they attempt to restrict the movement of sediment either by slowing its rate of transport through an area or by shielding the sediment from direct wave attack. The structures do not create any new sources of sediment, so they do not change the balance of sediment available in a region. They do, however, redistribute locally existing sediment supplies. Most structures have downdrift effects because of local interference with sediment transport. If they are used, it is advised that they should be part of a regional program concerning sediment and shoreline management. Walls and barriers should be regarded as the last resort in attempts to stabilize the shoreline.

The "soft," nonstructural approach of beach nourishment places sand in problem areas and creates a temporary beach with a limited life. As coastal management strategies strive to enhance public safety, community-based programs for coastal dune development and maintenance can easily be incorporated. Because coastal dunes buffer the effects of storms and storm surges and add some sand to the local sediment budget, they assist in the reduction of damage. Therefore policies that promote coastal dune protection and enhancement are consistent with both federal and state hazard mitigation objectives. If the federal government decreases financial support for beach nourishment projects as a type of coastal stabilization, the buffering abilities of coastal dunes may become the primary means of stabilization. All of the approaches outlined in this chapter are costly and may not be appropriate in all situations.

Choosing an appropriate solution to shoreline problems depends on the location, severity of the erosion, local coastal processes and storm history, availability of materials, and associated costs as well as environmental, esthetic, and social concerns. In the application of management and policy decisions, local resource managers need access to accurate and current information on appropriate strategies and their relative effectiveness in order to manage the effects of severe storms and shoreline erosion and to approach the management of natural hazards rationally.

CHAPTER 7 ∽

Coastal Dunes: A Natural Buffer

Coastal barrier dunes are valuable to coastal protection . . . as flexible barriers to storm surges and waves, and . . . by providing stockpiles of sand to nourish the beach during storm attack. Coastal dunes are also used for water storage and recreation, and are highly valued as wildlife habitats.

W. W. Woodhouse, Jr., *Dune Building and Stabilization with Vegetation* (1978)

The dunes should provide an uninterrupted barrier and a source of sand to mitigate the effect of storm waves for the benefit of the entire Borough. . . . Dune Areas are vulnerable to erosion by wind, water, and the absence of good husbandry by those responsible for their maintenance and preservation.

Borough of Mantoloking, N.J., Ordinance no. 348 (1995)

Dunes are natural features of the coastal landscape. They are products of the ambient coastal processes and they exist as part of the dune/beach-dune landform system (Psuty 1988). In the 1990s several publications highlighted the emerging knowledge regarding the processes of coastal dune development and erosion (Nordstrom, Psuty, and Carter 1990; Carter, Curtis, and Sheehy-Skeffington 1992; Pye 1993). Many types of dune forms exist within the coastal zone. From a planning and management perspective, however, the foredune is the most important geomorphological feature. The foredune or the primary dune is the first sand ridge inland from the beach (figure 7.1). It exists in conjunction with the beach and is part of a sand-sharing system that actively exchanges sand between the dune, the beach, and the offshore bars. In areas of adequate sand supply and where beaches are sufficiently wide, coastal foredunes can

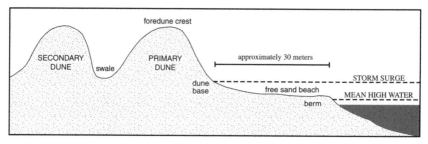

Figure 7.1. Coastal foredune on beach-dune profile.

achieve their full form. If sand supply is limited, however, the dune form becomes more restricted and may result in small, narrow features subject to frequent scarping (erosion that steepens the face of the foredune) (figure 7.2), breakthroughs, and overwashing. In areas of very meager sand supply or where beaches are not of sufficient width, dunes will occur as a series of small isolated hummocks or not at all.

Coastal foredunes, hereafter usually referred to as coastal dunes, occupy a transitional zone between marine and continental processes and mark the seaward extent of continental vegetation. Coastal dunes can be thought of as occupying an ecological niche that involves a transition between the harsh, salty, open environment of the beach and the less harsh, protected environment directly inland or to the lee of the dune crest. Natural dunes are replete with hollows and knolls, ridges and swales (lows), wind-created blowouts, and vegetated slopes. The location and dimensions of dunes on the beach profile are related to the historical development of the shoreline. Under conditions where sand is continuously accumulated (accretional conditions), older lines of foredunes are stranded inland and are succeeded by newer foredunes created in the active upper beach zone. Under conditions of active erosion, the dune form shifts inland as the beach retreats or is displaced inland. If erosion is too severe or there is no space to shift inland, however, dunes will not survive. Thus dunes as well as the beach erode and migrate when there is a shortage of sand in the system. Dunes do not stop erosion; they share in the loss of mass and volume as the beach erodes.

Coastal dunes in developed areas continue to store sand, but they also serve as a natural barrier to storm surge and flooding. Hence dunes are often highly sought by shoreline communities, because they offer a natural esthetic and protective presence. Valued for their coastal buffering since the early 1930s in New Jersey, dunes began to be actively restored,

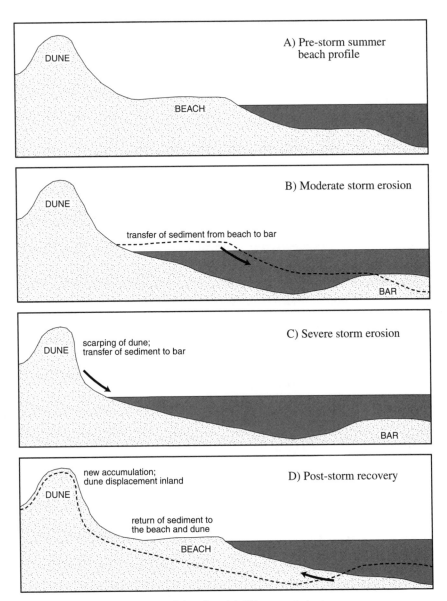

A) Pre-storm summer beach profile

DUNE

BEACH

B) Moderate storm erosion

DUNE

transfer of sediment from beach to bar

BAR

C) Severe storm erosion

scarping of dune;
transfer of sediment to bar

DUNE

BAR

D) Post-storm recovery

new accumulation;
dune displacement inland

DUNE

return of sediment to
the beach and dune

BEACH

Figure 7.2. Stages of dune/beach/offshore sand-sharing system.

repaired, and maintained by local communities and the state only in 1984. The actions grew out of a period of considerable erosion and a state report, *Assessment of Dune and Shore Protection Ordinances in New Jersey* (NJDEP 1984a). Concern for the creation and enhancement of coastal dunes subsequently became part of a management program to provide a protective buffer for the communities inland from the beach. Through a combination of local and state efforts, the public's perception of the value of coastal dunes was strengthened and this attitude continues to the present day.

A particularly attractive aspect of the preservation and restoration of coastal dunes is that the effort can be carried out at the community level and become a community responsibility. Furthermore, good management practices regarding dunes are also good management practices for the community. It is necessary to understand the processes that create, augment, and later destroy dunes in order to develop an effective dune management program. This chapter seeks to provide guidance to coastal communities both inside and outside of New Jersey as they develop strong dune enhancement and maintenance programs. It describes the processes that govern the development of New Jersey's coastal dunes and offers management strategies directed at the community level to maximize the function and effectiveness of dunes. Elements of a model ordinance developed from ordinances from many of New Jersey's coastal communities are also discussed (see also appendix A).

Characteristics of Coastal Dunes

An understanding of the processes that influence the presence and attributes of coastal dunes will greatly assist efforts related to their management. Coastal dunes are part of the natural beach system. Dunes that are located in dry sandy areas or areas not near seacoasts (for example, in deserts) are mostly shaped by wind, whereas dunes in the coastal zone are shaped and formed by the actions of wind as well as waves. In the context of a beach-dune profile, the coastal dune consists of a ridge of sand that accumulates above the high-tide line and is inland of the extremely mobile, bare (sand) beach surface (figure 7.2). The coastal dune is a site of sediment accumulation and storage. It exchanges sand with the beach and offshore during storms, and is built up during times when sand is blown inland to accumulate in the vegetation. It exists because more sand is deposited in its location than is removed. As a result, it develops a physical form with height, width, and mass characteristics. The foredune is in dynamic interaction with the beach and the processes that move sand in the beach, including waves, currents, and winds.

The coastal dune is the site of certain types of vegetation that can tolerate the harsh conditions of heat, aridity, high salinity, and low nutrient availability found at the shore. These plants are referred to as "pioneer" plants and comprise the dune grasses and other plants that spread from the interior and colonize the seaward face or crest of the foredunes. Pioneer plants are natural stabilizers for the accumulated dune sand. As the wind blows across the bare beach surface, sand can become airborne and transported inland to the location of the pioneer vegetation. As the wind makes contact with the rough surface of the vegetation, it meets resistance and its velocity is decreased, causing sand to be deposited around the vegetation and accumulate in this zone. The accumulation will eventually build a mound or ridge, thus starting the natural cycle of dune formation. The sand from the beach contains nutrients, usually bound by water around the individual sand grains, and sand accumulations become a source of nutrients for the plants as well as mass to build the dune form. Over time the vegetation serves to anchor the dune below the surface through its root mass and organic matter. The dune form eventually becomes a well-defined ridge and a natural barrier to storm waves and surge. Sediment may stabilize on the dune surface, in turn providing a natural barrier to incoming waves. Without the stabilizing effect of their vegetative cover, dunes become extremely vulnerable to the forces that create them, and can be easily mobilized and destroyed.

The location of the foredune on the beach profile is inland and above the sandy beach slope that is continually changing under storm conditions. Foredunes develop in this upper-beach zone because it is the area where vegetation can persist. Generally, the coastal foredune is not located seaward of the pioneer vegetation line, because then it would lack the stabilizing grasses and other plants it needs in order to accumulate sand transported inland from the beach. On accreting shorelines, pioneer vegetation can extend seaward from the foredune face to eventually establish a new line or ridge of sand accumulation. This type of broadening shoreline situation gives rise to a series of low, small ridges (figure 7.3). Such features are present at the northern tip of Sandy Hook and at the southern, accretionary tips of New Jersey's barrier islands. With the exception of these ends of the barrier islands, however, most of New Jersey's beaches are eroding, and the seaward edge of the foredune is located at or close to the storm waterline and, hence, is frequently attacked by waves.

On many beaches the foredune can continue to exist in the beach-dune profile even though the shoreline has a long history of erosion. This is possible where foredunes are able to migrate or move inland as the beach erodes. Hence dunes can exist as the beach profile loses sand to off-

Figure 7.3. Accretionary foredune ridges, with positive sediment budget, downdrift end of barrier island. Corson's Inlet State Park, N.J.

shore and alongshore locations, because space allows them to migrate inland. Foredunes are capable of migrating as a result of natural processes that cause some of the eroded sediments to be transported into the dunes during high water levels, storm surges, and high wind, and where development (for example, residential housing) is not present or is situated in a way that allows inland migration of the dune form. It should be noted that the existence of dunes on an eroding shoreline indicates that sand is being transported into the dunes intermittently, thus adding mass to the crest of the dune and to its landward side even as it loses sand on its seaward side. After a storm it is common to identify sand that has been transported into the dune, onto the crest of the dune, and into the topographic lows, or swales, inland of foredunes, as well as to observe fresh layers of sand blanketing the vegetation (figures 7.4 and 7.5).

The coastal foredune accumulates sand as pioneer vegetation traps sand blown across the beach into the ridge. It loses sand when wave action erodes the beach, reaches the foredune, and erodes or scarps the foredune. Scarping by waves is the process by which the sand held in storage in the foredune is returned to the beach for subsequent transfers offshore or alongshore (figure 7.2). If the amount of sand lost by waves scarping the seaward margin of the foredune is replaced later by sand moved by wind and water from the beach to the dune, the net effect of the exchange is considered to be in balance. The dune will be reduced in size if the

Figure 7.4. Sand transported inland covering vegetation, associated with storm. Island Beach State Park, N.J.

Figure 7.5. Sand transferred inland during and after snow storm. Snow at base of accumulation. Lavallette, N.J.

amount of sand removed is greater than that replaced. Conversely, the dune will grow larger if the amount of sand replaced is greater than that removed by the scarping process. Holding the dune in place and retaining its size is accomplished by a healthy, dense, dune vegetation cover, which serves to anchor the sand during erosion and fosters accumulation during recovery.

The foredune forms a natural barrier to inland penetration of high water from storm surges. Thus dunes can restrict the effects of storm waves and currents to the beach and the foredune face. As the dune buffers the effects of these forces, its sand may be released by the mobilizing processes of waves and moving water. Because the amount of protection is directly related to the mass of the dunes, higher and wider dunes will provide more buffering than lower and narrower dunes. However, the buffering effect of the coastal dune will be diminished when the dune crest is overtopped and becomes eroded. Overwash may sometimes be so severe that it completely removes or levels the dune form and transports much of the sand inland, as occurred during the March 1962 and March 1984 northeasters in New Jersey and in the 1987 blizzard at Cape Cod National Seashore, where dunes up to 4.5 m (15 ft) high were completely leveled.

Thus far, this discussion of dune development has focused on the processes by which natural dunes exist in the beach-dune profile and their interaction with waves, winds, and currents. Most of the dunes in New Jersey, however, have been modified by cultural activities that both diminish and enhance the remaining forms. In a few places, some natural or near-natural dunes still exist because of the absence of human interference. These dunes are located on the northern portion of Sandy Hook, much of Island Beach State Park, northern Barnegat Light, southern Sea Isle City, southern Stone Harbor, the shoreline portion of the Forsythe Wildlife Refuge, and the central portion of Avalon. Dunes located at Island Beach State Park and Avalon are important features of the New Jersey coastline, because they show the overall conditions of the barrier islands and the coastal dunes prior to extensive development of the shore (figures 7.6 and 7.7). These locations demonstrate that even though erosion has occurred, primary and secondary dunes still survive. Furthermore, they reveal the variety of dune forms, their relationship to the position of the storm waterline, dimensions that can occur within short distances of the storm waterline, and the interdependence of dunes with the ambient processes and sediment availability in the beach.

Figure 7.6. Primary and secondary coastal dune formation. Island Beach State Park, N.J.

Figure 7.7. Natural dune field retaining predevelopment characteristics. Avalon, N.J.

Interest in Dunes and Dune Management

An early state of New Jersey study, the *Report on the Erosion and Protection of the New Jersey Beaches,* recognized the importance of dunes and emphasized that coastal dunes should be given more consideration for their protective qualities (NJBCN 1930). It was not until 1972, however, that the attributes of coastal dunes were recognized at the federal level with the passage of the Coastal Zone Management Act (P.L. 91-583). Among its basic tenets was the desire to encourage the maintenance and enhancement of coastal dunes as a natural protective feature. Under the act, states could receive federal funding to develop and implement the act's objectives, and for the first time there was a national incentive for coastal states to promote dune restoration and maintenance. Yet it was not until after the March 1984 storm, which destroyed most of New Jersey's dunes, that the state implemented a Federal Emergency Dune Restoration Program. At that time state coastal management strategy began to emphasize dunes as a preferred form of coastal erosion buffering and storm surge protection through technical and financial support to communities for dune restoration. Further, the State Department of Environmental Protection and Office of Emergency Management have continued to encourage communities to restore, improve, and maintain dunes with guidance and support.

In 1984 NJDEP completed a report, *Assessment of Dune and Shore Protection Ordinances in New Jersey,* which involved an evaluation of municipal management of coastal dunes. The report concluded that state expenditures for shore stabilization would be most cost effective if they were associated with programs that protected and created dunes, and that future shore-stabilization expenditures by the state should be conditioned upon adoption and enforcement of an effective dune management program by municipalities (NJDEP 1984a). As a by-product of the Section 306 dune restoration program, NJDEP produced a manual of guidelines for the restoration and creation of coastal dunes in 1985. It was designed to "list recommended dune restoration and creation techniques, help municipalities plan effective and environmentally sound dune projects, and explain what information municipalities should submit to complete their applications for dune funds" (NJDEP 1985b). Further, NJOEM also continued to support a coordinated government program of dune creation and expansion via the state Hazard Mitigation Plan recommendations (NJOEM 1994). In addition, through NJOEM coastal communities could qualify for funding from a state Hazard Mitigation Grant Program to restore, develop, and repair their dunes after storm damage. More recently some of the beach nourishment projects have also incorporated a modest dune ridge in the project design.

As New Jersey's coastal management strategies evolve to support the concepts of mitigation and enhancement of public safety, coastal dunes

Figure 7.8. Community dune enhancement program.

can greatly assist this objective, and dune maintenance can easily be incorporated into such efforts. Because coastal dunes buffer the effects of storms and storm surges, they can prevent subsequent damage. Therefore policies that promote coastal dune maintenance and enhancement are consistent with both federal and state hazard mitigation objectives. If the federal government decreases financial support for beach nourishment projects as a type of coastal stabilization, the buffering abilities of coastal dunes may become the primary means of damage reduction. Furthermore, through dune maintenance programs such as annual beachgrass plantings and the placement of sand fences, municipalities can bring various resources together for a common cause and in turn create a feeling of community (figure 7.8).

Dune Management Objectives

Communities should determine the level of buffering protection they wish to achieve from their dunes. That is, in order to achieve their goals, communities should establish preferred dune dimensions (buffering capacity) that is consistent with beach width and height, space for dunes, maintenance program, and other variables that could affect foredune creation and stability. The spatial dimensions of the foredune that can protect against a 1-in-5-year storm differ from those required to buffer a 1-in-50-year storm. Although a municipality may wish to provide protection from the effects of a 1-in-50-year storm, the beach-dune area may only provide adequate space for a foredune that can buffer a 1-in-20-year storm. For communities that desire to develop dunes to protect against higher-magnitude storms but do not have the desired space, a possible strategy is to consider rezoning beachfront areas. Rezoning could allow communities to develop or expand existing dunes as space becomes available. Unless a community is willing to designate adjacent landward property for use as a dune area, a limited space will result in smaller dunes that offer a low level of protection at best.

The concerns expressed regarding dune preservation are largely directed toward maintaining storm-buffering capabilities. But there are other attributes of coastal dunes, such as their esthetic value, that communities may wish to foster as well. For example, the city of Brigantine incorporates ordinance regulations to preserve habitats for bird-nesting sites (Brigantine 1986). When developing strategies, it is often possible to incorporate the various attributes of coastal dunes and to define steps to enhance them.

Another factor that communities should consider when choosing objectives is the temporal component of dune preservation. Through time, the ability of coastal dunes to buffer storms becomes compromised as the shoreline erodes and sea levels rise. As shorelines erode, unless coastal dunes are able to shift inland to adjust to these changes, they will become closer to the waterline and more subject to scarping and overwash; eventually they may become completely eroded. In the process of determining the level of buffering protection to be provided by dunes, municipalities may therefore want to incorporate a setback area that will allow a coastal dune's position to be adjusted landward in response to the dynamic nature of the coastline (figure 7.9).

Placement and Dimensions of a Dune

Coastal dunes are in continuous interaction with their adjoining beaches, where the beach is a source of both sand and protection from small storm events (figure 7.1). Generally, a dune located 30 m (100 ft) inland from the high waterline on a beach with a berm 1.5 m (5 ft) high will survive a 1-in-5-year storm. However, more severe storms will erode the protective berm and eventually erode the dune.

Figure 7.9. Dune buffer/setback area. South Seaside Park, N.J.

The amount of dune erosion from a storm is dependent both on the storm conditions and on the sand reservoir that existed in the beach-dune area before the storm. The Federal Emergency Management Agency applies the concept of the sand reservoir to the portion of the dune that is in front of the dune crest and above a 100-year flood level, that is, having a probability of surviving floods that occur once in a hundred years (figure 7.10). Obviously, larger dunes have a greater cross-section and can survive larger storms. FEMA suggests that in the event of a 100-year storm (that is, a storm surge that reaches a 100-year-storm water level), the dune would require a sand reservoir area of 50 m² (540 ft²) above the 100-year flood level to prevent overwash and to survive (FEMA 1995b).

Few dunes along the developed coastline have such sizable dunes, and it may be unreasonable and is probably economically unjustifiable to try to build dunes 40 m (130 ft) wide and 4.5 m (15 ft) high above a beach 3 m (10 ft) high to protect against the once-in-a-century storm. Psuty and Tsai (1997) have applied the dune-erosion simulation model developed by Kriebel and Dean (1985) and Kriebel (1995) to higher-frequency storm events, which are much less severe. They establish nine different beach-dune profiles and simulate the effects on them of a 10-year, 20-year, and 30-year storm. Using scenarios of different wave sizes in the storms,

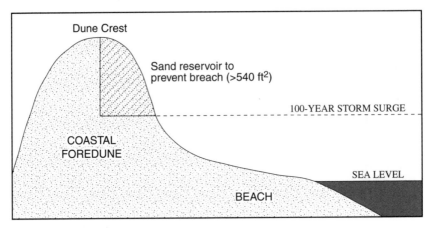

Figure 7.10. Schema of FEMA's sand reservoir in the coastal foredune.

responses can be modeled to determine their protective capacity (figure 7.11). These simulation models are used to show the response of existing profiles or to establish design profiles for management.

Dune Placement

Dunes should be placed sufficiently inland from the storm-tide position to provide a buffer and to protect the dunes against erosion during frequent storms. Otherwise, the dune can become substantially eroded during low-magnitude storms and will not be capable of providing any protection against larger events. In order for a dune to provide buffering from larger storms, the specific dimensions of a dune need to be maintained. As storms of greater magnitude occur, they will reach higher into the dunes and have a greater impact on dune forms. Hence the size and position of the dune is directly related to its ability to survive major storms.

Dune Height

A dune's height varies with weather conditions, availability of sand, shoreline erosion rate, vegetative cover, and in cases of human intervention, the methods used to build the dune. Throughout much of New Jersey, dune heights reach 2.5–4.5 m (8–15 ft) above the beach. For artificially created dunes, the initial accumulation and the growth may be rapid at first (Hamer, Cluster, and Miller 1992). As the dune height increases, however, the rate of increase in elevation slows, because of the

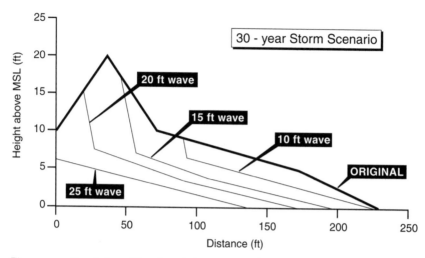

Figure 7.11. Simulation of beach and dune erosion during a 30-year storm surge with waves of differing sizes.

greater mass needed to raise the entire surface of the foredune, the variation in vegetation cover, and other perturbations on the dune face. Under ideal conditions and a combination of sand fencing, vegetation, and ample quantities of dry sand, the rate of increase in dune elevation can reach up to 1.3 m (4 ft) in one season (Hamer, Cluster, and Miller 1992), although rates of less than 0.6 m (2 ft) in one season are more common.

Dune Slope

Although there is a limiting slope for the natural accumulation of sand, a number of variables combine to affect this slope on a dune face, such as rate of sediment delivery, vegetation, presence of living organisms, fences, salt content, and other disturbances. One measure of the relationship of dune slope was reported by Gares, Nordstrom, and Psuty (1983) from a survey of New Jersey dunes, which included the effects of erosion and accretion on natural dunes and altered dunes. In this survey, they determined that New Jersey dunes were up to 20 times as wide as they were high (table 7.1). This estimate includes the effects of combining multiple dune lines and is thus overstated. Other measurements from coastal New Jersey and from Long Island, New York, show that human-constructed dunes are 7 to 10 times as wide as they are high (Psuty and Piccola 1991). This means that the slope could be expressed as a ratio of 1 to 3.5–5.0.

Table 7.1 ～ RELATIONSHIP BETWEEN DUNE
HEIGHT AND DUNE WIDTH

Dune height	*Dune width*
5'	35'
8'	56'
10'	70'
12'	84'
14'	98'
16'	112'

SOURCE: Adapted from Psuty and Piccola (1991).

Building and Enhancing Coastal Dunes

After a municipality has decided that coastal dunes can be a viable option to protect a developed area, it should identify objectives regarding the level of storm buffering provided by the created coastal dune (that is, its level of protection), and it must also determine whether there is adequate space to achieve these goals. The next step is to implement the plan and begin dune enhancement or construction. Several different techniques are available. Dune enhancement may entail mechanical manipulation, planting appropriate dune vegetation, erecting sand fences, or a combination of these methods. If no dune exists, dune creation can employ the same techniques.

Mechanical Manipulation

A simple method of creating a coastal dune is by bulldozing sand into a shoreline-parallel ridge. An advantage to this method is that dunes and their buffering-protective features are instantly achieved. However, the sediments that make up mechanically made dunes are more easily mobilized or are unstable during coastal storms, because they lack the internal organic binding provided by vegetation at the core. Therefore, once the appropriate dimensions of a dune are established, vegetation and fencing should be applied and maintained to improve its stability.

Dune Vegetation

Another approach is to mimic the natural processes and plant dune grass, "Cape" American beachgrass, *Ammophila breviligulata,* or other pioneer dune vegetation at adequate distances inland from mean

Figure 7.12. Dune created using American beachgrass, *Ammophila breviligulata*. Ocean City, N.J.

high water (figure 7.12). Unlike bulldozed dunes, this method takes some time before sizable dimensions are achieved. Once established, beachgrass will continue to accumulate sand to add to the dune form, and its extensive root system helps to bind the sand in place. American beachgrass will continue to grow relative to the supply of nutrients provided. With adequate sand transport from the beach into the vegetation zone, the nutrients will be delivered on the sand grains and in the propelling wind. If the sand supply is low, however, nutrient delivery will likely be reduced and the vegetative cover may become thin. Dune growth and stabilization are therefore dependent on sediment delivery and the trapping effect of the vegetation. Well-fed vegetation will increase plant density, in turn becoming a more effective sand trap and adding to the foredune system.

Although most attention has been directed toward dune grass, *Ammophila breviligulata,* as the principal pioneer species, the coastal sand ridge has many exposures and niches, and it is currently thought that a mixture of species is desirable to support biodiversity in the habitat. Other primary vegetation such as sea rocket, dune cordgrass, seaside goldenrod, and dusty miller are suitable primary vegetation that communities can plant. After primary dune vegetation has been established, woody plants adapted to coastal climates can be planted in secondary coastal dunes for added stabilization and biodiversity. Hamer, Cluster, and Miller

(1992) recommend bayberry, wax myrtle, beach plum, Japanese black pine, salt spray rose, and "Emerald Sea" shore juniper as secondary dune vegetation.

Sand Fences

Sand fencing is also an effective method of trapping sand, but it is a fairly slow process and more expensive than vegetation alone. Sand fencing accumulates sand in the same manner as dune vegetation. As windborne sand travels from the beach to the backbeach area, the fencing creates resistance to the airflow and wind speed decreases locally, with the result that sand is deposited and accumulates to the lee of the fence line. As sand accumulates, additional fencing can be placed over the filled areas until the dune reaches a desired level of buffering or height. Hamer, Cluster, and Miller (1992) found that sand will generally reach a level about three-fourths of the exposed height of the sand fence. Fences should be at least 30 m (100 ft) from the mean high-tide line, and the general fence line should be parallel to the shoreline. When a dune is built with fencing, vegetation should be planted in the sand accumulations to hold the sand in place.

There are several different configurations of sand-fence placement, each creating a somewhat differently shaped sand mass or sand ridge. The most cost-effective type of sand fence used to create a foredune is a single line of fence placed parallel to the shoreline. Hamer, Cluster, and Miller (1992) recommend that two parallel fences be erected 10 13 m (30 40 ft) apart instead of a single fence (figure 7.13). Parallel sand fences can also be used to restore or expand existing dunes, and should be placed no more than 15 feet seaward from the base of the old dune (Hamer Cluster, and Miller 1992) (figure 7.14). An advantage of parallel fence lines surrounding the dune form is their control of pedestrian traffic through the dune zone. This is especially important in reducing the number of access paths through the area.

Coastal Dune Restoration and Maintenance

Once the dimensions of the enhanced or constructed dune are established, the next challenge facing a municipality is to ensure that the dune will be maintained. Maintenance of the physical dimensions of the foredune will require a rigorous program. Even the best-vegetated

Figure 7.13. Combination of parallel fencing and dune vegetation. Ocean Grove, N.J.

Figure 7.14. Sand fencing used to expand dune farther seaward. Beach Haven, N.J.

dune may need attention to support a continuous vegetative cover in order to enable it to retain the integrity of the sand ridge. Beachgrass must be planted and possibly fertilized, and broken fencing must be replaced on a continuing program. Additionally, any blowouts or scarping of the dunes may need attention.

Limiting Human Impact

Dune vegetation cannot tolerate much trampling. Even light foot traffic can cause breakage, churn up the roots, and eventually

Figure 7.15. Sand fencing installed to prevent indiscriminate passage across dunes. Laval-lette, N.J.

destabilize the dune if it is allowed to continue (NJDEP 1984a). Dune vegetation, therefore, must be protected from foot and vehicular traffic. By installing either elevated walkways or fenced paths, traffic can be controlled across the dune and access points can be limited. Fences can also be installed in the front and back faces of a dune as borders to further prevent indiscriminate passage across the sand ridge (figure 7.15). Signs should also be placed along dune areas and access points to educate and remind the public not to walk across the sensitive, vegetated dune areas.

Pathways

Street ends are often weak links in the coastal dune ridge, becoming sites of overwash and breaches. A major concern of beach access through the dune area is the elimination of low straight-through gaps at street ends. Pathways should be oriented on an angle to the shoreline to reduce the ease of storm surge penetration. Offsets in pathways or a curving pathway will serve the same function in limiting overwash. Another concern is the low elevation of pathways. Elevated crossovers can be installed to reduce vertical erosion of the dunes (figure 7.16). If used, elevated walkways should be high enough so that they exceed the design height of the intended dune. Otherwise, roll-up sidewalks or some other material that can be laid on top of the sand surface to prevent vertical destruction by foot traffic should be used.

Figure 7.16. Elevated cross-over installed to reduce vertical erosion of dune. Surf City, N.J.

Repairing Dunes

Municipalities should establish a periodic dune maintenance and monitoring program. Any changes in a dune system, such as blowouts, extensive scarping, or human-induced damage, require immediate attention if dunes are to provide their maximum level of protection. Blowouts can easily be repaired by installing a single shore-parallel sand fence at the gap between the existing dunes. Following major coastal storms, dunes often become scarped from storm waves (figure 7.17), a natural process of cutting back the face of the dune as the beach is eroded. Dunes that are located too far seaward will experience scarping from storms of low magnitude, with a subsequent reduction in the size of the dune (figure 7.18). Although the general post-storm reaction of communities has been to restore the scarped dunes to their pre-storm dimensions, there may be a need to reevaluate the location of the dune area and hence dune maintenance efforts. All repairs to scarped dunes and to eroded dunes in general should continue to observe the dune's placement at 30 m (100 ft) from the mean high waterline. If the dune zone is maintained sufficiently inland, the beach can function as a buffer and protective area and reduce future scarping. However, continued beach erosion and sea-level rise will gradually narrow the protective beaches.

Following high-magnitude storms, even well-maintained dunes may experience some degree of scarping. Where dunes are situated at the

Figure 7.17. Scarped dune. Ocean City, N.J.

recommended minimum distance from the mean high waterline, the scarped dunes can be repaired by placing short perpendicular spurs into the scarped dune followed on the seaward side by zig-zag fencing (NJDEP 1985b). After a wedge of sand begins to accumulate, American beachgrass can be planted. Within a matter of months, the dune may regain most of its natural profile. If dunes are located closer to the waterline (MHW) than the minimum recommended distance, it will be necessary to relocate the position of the created foredune or to build another dune line at an inland site to attain the desired objective of storm protection.

Dune Buffer Areas

In an effort to maintain a stable shoreline position, coastal dunes have often been used as a type of seawall in an attempt to intercept waves and to prevent storm surge penetration. These management practices have often resulted in dunes built within the reach of minor storm waves, subjecting them to frequent scarping and eventual destruction. An alternative approach is to place the dunes farther from the water and to manage the coastline in a manner that is more compatible with the natural forces that shape it. As New Jersey's coastline continues to erode as a result of a rising sea level coupled with sediment loss, coastal dunes will be subject to increased scarping and erosion if they are not

Figure 7.18. Dunes scarped by wave action, insufficient beach width to protect dune. Cape May Point, N.J.

permitted to shift inland naturally and maintain an appropriate distance from the mean high waterline. Because the basic dune processes often cause transport of sediment inland of the dune crest, the creation of a buffer zone at the inland margin of the foredune supports an inland continuity of dune processes, dune forms, and habitats as the dune shifts inland (figure 7.9).

Developing a Dune Protection and Maintenance Ordinance

Most coastal communities in New Jersey have adopted a dune protection ordinance (table 7.2). Many of these ordinances were passed after the devastating 1962 nor'easter that destroyed almost every dune along the New Jersey coastline (NJDEP 1984a). Following a 1984 storm, local governments, with federal and state support, rebuilt the dunes, and additional municipalities subsequently passed in the 1990s, ordinances to protect and preserve them. Prior to the spate of beach nourishment projects most of Monmouth County's beaches were not wide enough to support coastal dunes, and therefore these municipalities had not adopted dune protection ordinances.

Table 7.2 ～ COASTAL MUNICIPALITIES WITH DUNE ORDINANCES
THROUGH 1995

Municipalities	*Date of last revision*	*Municipalities*	*Date of last revision*
Monmouth County		Harvey Cedars	1989
Sea Bright	No ordinance	Surf City	1972
Monmouth Beach	No ordinance	Ship Bottom	1994
Long Branch	No ordinance	Long Beach	1994
Deal	No ordinance	Beach Haven	1994
Allenhurst	No ordinance	Dover Township	1981
Loch Arbour	No ordinance	*Atlantic County*	
Asbury Park City	No ordinance	Brigantine City	1986
Neptune (Ocean Grove)	No ordinance	Atlantic City	1989
Bradley Beach	No ordinance	Ventnor City	1989
Avon-by-the-Sea	No ordinance	Margate City	1991
Belmar	1992	Longport	1996
Spring Lake	1993	*Cape May County*	
Sea Girt	1994	Ocean City	1994
Manasquan	1989	Upper Township	1975
Ocean County		Sea Isle City	1987
Point Pleasant Beach	1994	Avalon	1970
Bay Head	1993	Stone Harbour	1985
Brick	1988	North Wildwood City	No ordinance
Mantoloking	1995	Wildwood Crest	No ordinance
Lavallette	1985	Wildwood City	No ordinance
Seaside Heights	No ordinance	Lower Township	1988
Seaside Park	1988	Cape May City	1995
Berkeley Township	1994	Cape May Point	1974
Barnegat Light	1994		

Coastal dunes are managed and protected on a local level, and consequently dune ordinances vary considerably from community to community. Although municipalities have created them with good intentions, many of these ordinances have failed to prevent dune damage or to provide clear guidance on the proper maintenance and construction of dunes. Lack of scientific knowledge on dune dynamics has at times been a contributing factor (NJDEP 1984a). Several communities have begun to remedy this problem by strengthening and amending their ordinances as further knowledge has become available. The result is that their dunes have provided enhanced buffering from the effects of subsequent storms. A model dune ordinance is included as appendix A in this book. It incorporates the best elements of existing ordinances found along the New Jersey. Examples of its composition follow below.

Definitions

Defining the parameters of what is being regulated is a significant aspect of any effective dune ordinance. Without good definitions, what is being regulated can become ambiguous. For example, most ordinances have legally defined dune areas (or building lines), set by a fixed line. However, dunes naturally migrate in response to wind, water, and other elements. When dunes migrate out of the legally defined dune area, communities may become unable to prevent future construction in the new location because the dunes now exist landward of a fixed building line. The result is a narrow dune ridge that is unable to provide much protection from storm surge and overwash. Because communities are unable to redefine and map dune areas without the threat of a "takings" issue (similar to eminent domain, when private property is taken by public governments; see Merriam, and Frank Meltz, 1998), many pass the burden of defining the dune zone to higher administrative levels, such as the county or the state.

Following dissemination of the New Jersey Shore Protection Master Plan (NJDEP 1981), a few communities have revised their ordinances to provide a more scientifically defined setback line that acknowledges dune migration. The communities of Mantoloking, Bay Head, and Point Pleasant use a case-by-case review for the construction or renovation of residences in order to keep development away from the backslope of foredunes. The 1984 assessment of dune ordinances by NJDEP also acknowledged Long Beach Township's attempts to define a beach-dune district 45 m (150 ft) wide, but a clause in the ordinance has nullified this effort by allowing houses to be built upon the dunes as long as they are 20 feet behind a bulkhead line and the dune is 4.8 m (16 ft) high at the oceanfront building line (NJDEP 1984a).

In 1993, the New Jersey State Legislature redefined a dune in amendments to the Coastal Resources and Development Policy, a law regulating certain development in a defined area (Coastal Area Facilities Review Act II) (N.J.A.C. 7:7E-3.16). Presently only a few communities have revised their definition to incorporate CAFRA II's new definition. One of these, the township of Berkeley, has revised its definition as follows:

> Dunes. A coastal dune is a wind or wave deposited or man-made formation of sand (mound or ridge), that lies generally parallel to, and landward of, the beach, and between the upland limit of the beach and the foot of the most inland dune slope. "Dune" includes the foredune,

secondary, and tertiary dune ridges, as well as man-made dunes, where they exist.

 1. Formation of sand immediately adjacent to beaches that are stabilized by retaining structures, snow fences, planted vegetation, or other measures are considered to be dunes regardless of the degree of modification of the dune by wind or wave action or disturbance by development.
 2. A small mound of loose, windblown sand found in a street or on a part of a structure as a result of storm activity is not considered to be a "dune."

Regulation of Activities

Because dune vegetation is easily disturbed by foot traffic or other activities, dune ordinances regulate all activities in the dune areas that may disturb the condition of dunes. Typically ordinances regulate construction, public access, and maintenance in the dune area. Construction on the dune itself is prohibited by every community. The only exceptions are shore-stabilization projects, such as the construction of bulkheads, jetties, or groins approved by the state (NJDEP), the federal government (U.S. ACOE), and the municipality. Mantoloking and Brick have included clauses to the effect that if any of these projects impedes the natural flow of sand supplementing dunes, it will not be permitted (Brick 1988; Mantoloking 1995).

In its 1984 assessment of dune ordinances, NJDEP concluded that municipalities that permit the use of brush-type fencing (trees or shrubs) as a means of trapping sand had unintentionally transformed dune fields into dumping grounds. As a result of this finding, many municipalities eliminated this form of sand fence from their ordinances. However, several communities still allow dead trees and other shrubs to be used as drift fences.

Almost every community restricts public access over dunes to prevent the destruction of dune vegetation or the lowering of dune height. Only a few, however, have addressed the problem of overwash through walkways at street ends during periods of high water. The communities of Mantoloking and Bay Head recommend the use of elevated walkways for both public and private access to the beach. Ocean City also tries to prevent overwash by recommending that pathways should be constructed as zig-zag pathways angled to the southeast or in the direction that is at a right angle to approaching waves in the area (Ocean City 1994). In addition, many communities are now placing an artificial surface on pathways, such as roll-up sidewalks, plastic mats, or cordgrass, to prevent the vertical destruction of the dune.

Dune Maintenance

A principal purpose of dune maintenance is to protect and enhance the elevation and breadth of the foredune to provide buffering from the effects of storms. Most dune ordinances do not permit the lowering of dune height below an elevation set by the community through either the direct or indirect action of any person. Only a few municipalities state directly in their ordinance the ideal height of dunes above mean sea level. For example, the municipality of Bay Head recommends 16 feet as its ideal dune height (Bay Head 1993). Mantoloking sets a minimum height of 18 feet at the prevailing dune crestline (Mantoloking 1995), and Long Beach recommends a minimum height of 16 feet (Long Beach Township 1994). Mantoloking, Ocean City, Point Pleasant, Dover, Berkeley, and Brigantine are some of the communities that mandate specific requirements for the annual maintenance of dune vegetation, to be conducted by either the individual homeowner or the town. The appendix to Mantoloking's 1995 ordinance is adopted from the U.S. Natural Resource Conservation Service's (1992) *Restoration of Sand Dunes along the Mid-Atlantic Coast* and is a good reference for dune maintenance.

Conclusion

Communities benefit from the preservation and enhancement of coastal dunes, which are an important component of the natural coastal system. Although dunes are valued for their function as a natural barrier, they have many other roles and functions. By acting as a natural storage area for sand, coastal dunes actively exchange sand within this system. Additionally, coastal dunes provide a habitat for diverse plant life and offer various esthetic qualities. Standardized procedures to assist in the general maintenance and enhancement of coastal dunes can be developed at a community level. Dune protection ordinances provide a legal mechanism for communities to develop programs to maximize the function and effectiveness of dunes. It is important that these ordinances, in turn, reflect the objectives of the community's philosophy and efforts toward dune preservation.

Although there is considerable interest in coastal dunes, it must be stressed that dunes do not prevent erosion or reverse an erosion trend.

Dunes offer buffering from storm surges and contribute sand to buffer the rates of shoreline displacement. Dunes act as a protective barrier on the seaward side of coastal communities and can reduce exposure to storm effects. However, dunes have a finite capacity. They can be eroded and overwhelmed by waves and winds, and they may be overtopped by very high storm surges. Although communities can accentuate the protective capabilities and other qualities of dunes through maintenance programs and ordinances, dunes should be considered a short-term protective strategy that is within the capabilities of a community to pursue. It is likely that dunes will have to be rebuilt and relocated as well as reassessed in the future.

CHAPTER 8

Coastal Economics: Application to Coastal Management, Shoreline Stabilization, and Tourism

> Because people do not understand that structural protection has limits, . . . structures have been found to actually induce development in hazardous areas and to increase, not decrease, the likelihood that when a large flood or hurricane does occur, losses will be truly catastrophic.
>
> R. J. Burby, "Introduction," in *Cooperating with Nature* (1998)

One goal of this chapter is to introduce readers to the concepts and decision models commonly used in economics, so they can begin to assess coastal management and beach-stabilization strategies in the context of economic criteria. This understanding will help to distinguish between what is referred to as the "economic value" of a beach or beach nourishment project and its "economic importance," or expenditure impact. These two basic measures are conceptually different. The chapter contains background about the nature of the issues involved; an overview of pertinent economic measures, methods, and models; and a summation of the literature as it pertains to (1) the economic value of beach use and shoreline stabilization, (2) the role of beaches in tourism and economic development, (3) policy on shoreline stabilization, and (4) New Jersey's Division of Travel and Tourism findings concerning the impact of expenditures from tourism in the New Jersey coastal zone. It closes with issues for future thought.

Background

It is important to quantify the beneficial effects of shoreline management approaches such as beach nourishment, which provide

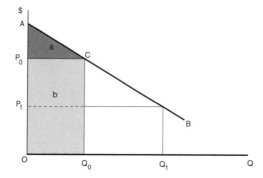

Figure 8.1. Demand curve

wider beaches over short periods of time. Wider beaches can encourage greater use and, for return visitors and adjacent property owners, produce additional enjoyment and greater property protection, all contributing to increases in economic benefits. Increases in economic benefits or value are composed of added satisfaction from new and previous users, and from property owners who know that an increased buffer between their property and the water's edge is now in place. Such beneficial effects are treated as increases in economic surpluses (that is, increases in the areas of surplus under the respective demand curves; see figure 8.1) and have been estimated in numerous studies discussed below. In figure 8.1, a downward-sloping demand curve, AB, represents demand for various beach services (beach trips, amenities provided by beach nourishment, and amenities of property in close proximity to the beachfront). Given a price paid for these services, there exists a surplus above and beyond the price paid (that is, expenditures represented by the shaded area *b* in figure 8.1) that accrues to the consumer, the shaded area *a*, and represents economic welfare or enjoyment from consumption of this good. A money measure of this economic welfare or surplus is represented by the consumer surplus, the shaded area *a*. Similarly, figure 8.2 shows a surplus that accrues to businesses and firms that provide various beach services. In this figure, an upward-sloping supply curve, CD, represents the supply for various beach services and, for a given price-quantity, the area under the supply curve represents the total costs of supplying these goods, shown by the shaded area *b*. Given a price paid for beach services, a surplus exists above the supply curve and below the price paid, the shaded area *a*, and is referred to as producer surplus, a monetary measure of economic welfare that accrues to producers.

In addition, claims have been made by various groups and by the

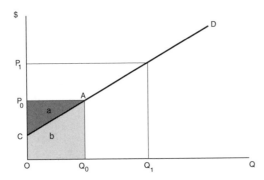

Figure 8.2. Supply curve

business community regarding the beneficial effects of beach-stabilization practices on business activity expressed as economic impacts, for example, increased sales, customers, and business; this is termed the "economic importance" of a beach. These expenditure effects can be envisioned as the shaded area *b* in figure 8.1. And one can then think of expenditure impacts as expansions of this shaded area *b* via impact multipliers to represent the effects of respending throughout an economy. These expenditure effects have received little attention and have not been examined by statistically rigorous studies that incorporate appropriate experimental designs to isolate such potential effects; they remain an area for future research. Several studies discussed below attempt to develop estimates of economic activity linked to shoreline stabilization that are not based on appropriate statistical designs, a limitation that undermines the estimates and the conclusions based upon them. The need for measures of coastal economic benefits and activity is very high, and it is necessary that researchers focus on adequately designed studies to explore and isolate specific effects rather than focusing on the end result.

Furthermore, a state's coastal tourism industry depends closely on the actual and perceived condition of the coastal zone, in terms of both erosion and marine pollution. For example, Ofiara and Brown (1999), using the number of beach tags sold as a proxy for beach attendance, found a decrease in beach tag sales at public beaches in New Jersey over the 1986–1988 period that ranged from 9% to 44% (average of 26%) in connection with 1988 marine pollution events. The processes of both coastal erosion and marine pollution, in turn, can adversely effect the state's coastal economic activity. Ofiara and Brown (1999) estimated that direct economic losses from the 1988 marine pollution events ranged from $379.1 million to $1,597.8 million for the state of New Jersey. Thus any positive actions regarding the preservation of water quality should also serve to maintain the economic activity of a state's coastal tourism industry. By extension, positive action toward beach quality in the form of stabilization should have a similar effect.

Another set of issues arises for policies concerning the management of coastal erosion. Evaluations of these policies can become increasingly complex given real-world scenarios. Shoreline stabilization is complicated because of the risks associated with the expected life of a stabilization project, which can vary, and with the expected returns associated with each project, which can also vary. Furthermore coastal shorelines, and hence shoreline-stabilization projects, are also subject to exogenous risk stemming from factors such as the geographic location, the physical characteristics of the shoreline, weather patterns, and storm activity and intensity—which over a sufficiently long period can be identified and categorized as being of high risk (that is, high storm activity) to low risk (low storm activity)—as well as the effects of sea-level rise. As one might expect, a stabilization project undertaken during periods of low risk should have the longest expected life (defined as the difference between the projected life, measured in years, and the actual time when a new project must be undertaken for the same coastal location) and the greatest economic return compared with initiating the same project during a period of high risk. However, even to simply maintain the coastal area, shoreline stabilization must be in place before periods of high risk are expected to occur; these projects could be treated as one-time emergency projects that only last the life of the current storm. Sea-level rise magnifies the above risk factors and the effects of erosion and storm damage.

Ideally, to evaluate these policies and projects, an investigator would prefer a welfare analysis (that is, a first-best analysis) of alternative policy options of shoreline stabilization based on benefit data (that is, measures of the values of society's welfare associated with different shoreline-stabilization policies), but such an approach becomes intractable. The data are not readily available; and in regard to recreational use and nonuse values, costly field surveys are necessary and sometimes involve highly controversial economic techniques. Furthermore, aggregation problems exist that involve many individuals with different tastes, and determining some type of benefit function that would depend on utilization by these individuals given alternative policy options becomes highly complex.

In the absence of a "first-best" approach, the most common technique used in economic analyses of shoreline-stabilization policies is cost-benefit analysis (CBA), a "second-best" approach, for a variety of reasons. Using such an approach, the investigator can determine the *ranking* of each project in a particular year with the decision rule to fund those projects that yield the greatest net economic benefits-returns until the program funds are exhausted; a similar criteria was used in the *New Jersey Shore Protection Master Plan* (NJDEP 1981). This process could be repeated each year. However, CBA has several limitations that may prove unrealistic. It assumes that benefits are measurable and can be accurately measured; for nonmarket goods such as beach use, benefits usually are not explicitly measurable and are subject to measurement error. Comparisons can only be made across projects that yield equal net benefits, and CBA is further limited by its association with welfare criteria; that is, projects that yield a "pareto improvement" (that is, a change in economic welfare where at least one individual is made better off and no one is made worse off) will unanimously be deemed superior to all other projects, but one can say little about any two projects that yield the same social welfare without further assumptions about distributional aspects of the project's benefits to members of society (that is, assumptions regarding equity across society). Furthermore, the treatment of risk and uncertainty creates additional complications for CBA (Ofiara and Psuty 2001). However, in spite of its shortcomings, and in lieu of better approaches, CBA is widely used and provides practical decision rules for public officials who face public policy matters.

And last, recent developments in the property casualty insurance industry stemming from recent catastrophes are a concern for public policy (Kunreuther 1998). The property devastation caused by Hurricane Andrew in Florida resulted in the failure of many private insurers to meet the claims submitted and forced them into bankruptcy. Those that survived found themselves in a very weakened financial position, subsequently forcing them to withdraw from the property protection area. The state of Florida had to intervene and began to provide flood protection insurance to affected homeowners (Lecomte and Gahagan 1998). Given the prospect of a rising sea level, any significant coastal storm could have the potential to wreak the havoc of a Hurricane Andrew, a Hurricane Hugo, or a similar event such as the Ash Wednesday storm of 1962. Implications are that states may have to assume financial responsibility in insurance coverage, on the one hand, and devise programs that both reduce the threat of natural hazards and address shoreline stabilization in a dynamic framework, on the other. Public policy makers will require improved information and decision models that are capable of handling tradeoffs involving risk and returns and are ultimately able to take account of the complex, dynamic problem of coastal hazard management and shoreline stabilization. Further limitations of the CBA are discussed in this light below and potential alternative decision models are explored.

Economic Principles of Shoreline-Stabilization Management: An Introduction

A number of different economic measures exist that are conceptually separate. From a national income accounting perspective, measures of economic activity, such as gross domestic produce (GDP) measures, represent total expenditures on final goods and services produced by a country and are usually separated into distinct sectors (that is, consumption, production, government, and foreign trade). The literature of input-output analysis treats economic measures as the change in the dollar amount of output (sales), income (wage income), the number of individuals (employment levels), and the dollar amount of tax revenue (taxes) that results from a proposed outside change. In economic welfare theory, economic benefits and losses are other types of economic measures that represent gains or losses, respectively, in economic welfare or surplus (that is, net economic value). From the CBA literature, economic benefits and losses are measures that represent gains or losses in economic surplus and are equivalent to the measures from economic welfare theory.

Several terms and concepts have been widely used by economists and in the popular press to represent the above measures. *Net economic value* refers to the monetary value of changes in economic surplus to consumption and production activities (that is, the sum of changes in consumer surplus and producer surplus—these are economic *welfare* measures) following a change in the quality of the environment (for example, storm damage and coastal erosion), and when losses of economic surplus occur it is referred to as *lost net economic value*. But losses in economic surplus are not the only adverse change in economic activity that can occur from coastal erosion and storm damage. From the public's perspective there are other impacts that should also be counted. Economists refer to such effects as *adverse economic effects* or *impacts* following degradations of the environment, and they can be thought of as decreases in measurable economic activity (for example, sales, output, employment, income, and so on—

these are economic *nonwelfare* measures) and conceivably could include all possible economic effects that can be quantified. These measures of effects are sometimes referred to as *economic activity measures*; in this context *economic losses* are thought of as decreases in measurable economic activity resulting from coastal erosion and storm damage. *Economic losses* have also been used to represent *lost net economic value* (that is, the value of lost economic surplus or welfare—welfare measures) from erosion and storm damage. *Economic losses* and *lost net economic value* represent different concepts; the former represents a nonwelfare measure and the latter, a welfare measure. *Economic impacts* also used in another context, is used, in conjunction with input-output (I/O) models to examine proposed policy and other changes in economic activity (that is, structural change, a casino closing). In this context *economic impacts* refer only to those changes in economic activity (for example, the sum of value-added throughout the economy where sales-output is the relevant variable) that are estimated from an I/O model where impacts to output or sales, income, and employment can be measured. Economic surplus or welfare is ignored.

Benefits and *economic benefits* suffer from a similar problem; they have different meanings and usage in various contexts. A strict meaning of *economic benefits*, from economic welfare theory, is gains in *net economic value* (that is, gains in money measures of economic surplus or welfare—welfare measure). From the perspective of economic impacts (for example, back-of-the-envelope impact analyses of proposed policies, and so on), *benefits* are gains in economic activity that can be quantified *and can include* gains in economic welfare or surplus. In a CBA context, *economic benefits* represent all gains in *net economic value* (that is, welfare measure) that directly and indirectly result *with* a proposed project or policy as compared to a situation *without* the project or policy, and are sometimes referred to as *net economic benefits*. In some cases economic benefits have been loosely used to include changes in economic activity (for example, changes in sales, income, employment—that is, nonwelfare components); this is not appropriate in a CBA context. Within CBA, then, *economic benefits* represent the same measure as in an economic welfare context. Each of these concepts is summarized below.

Economic Measures

National income accounting measures of aggregate economic activity are usually referred to as gross domestic product (GDP) or gross national product (GNP). These measures represent the total value of all final goods and services produced in an economy for a given time period, usually a year. Because prices can change over time, the above measures must be adjusted to yield real GDP (gdp), real GNP (gnp), which then represent the value of all final goods and services produced in an economy in terms of constant dollars. As such, changes in gdp/gnp will now reflect changes in real output. Because only final goods and services are considered rather than including sales of used goods and sales of intermediate goods, measures of GDP avoid the problem of double counting economic activity, a main advantage. This book treats national accounting measures as representing measures of "true" economic activity.

Economic impacts are not to be confused with economic benefits; they represent different measures of economic activity. Impacts measure the amount of aggregate economic activity, usually in terms of sales, income, employment, and sometimes tax revenues, that are associated with some type of change to the economy (structural change—a casino opening or closing, reduced demand) or policy change (proposed regulation). Many times they are based on multipliers from input/output models that account for not only direct effects (effects on primary and secondary sectors only) but also indirect and induced effects. Most individuals are familiar with "expenditure impacts," such as the impact of consumer-tourism spending. To avoid the problem of double counting one must be careful with the application of I/O models. The multipliers represent all activity generated from an initial change and capture the sum of the value-added at each stage of the analysis and for each sector of the economy without double counting. The overall impact represents the total change in value-added for the entire economy, a measure that is equivalent to the one based on the value-added approach from national income accounting. In principle, the impact from a change in a given economy due to a policy change based on a I/O model should yield identical measures of the change to value-added if based on a national accounting approach.

Economic benefits in the context of economic welfare theory are associated with the monetary value of specific measures of changes in economic surplus or welfare. These measures are associated with losses or gains in economic welfare and represent lost *net economic value* and gains in *net economic value*, respectively. To producers (that is, private firms, manufacturers, commercial fishermen), the measure of lost (gains in) *net economic value* is producer surplus, a surplus that accrues to the producer over and above the variable costs of producing the good. In practice this is measured as the area above a supply curve (that is, that portion of the marginal cost curve above the minimum point on the average variable cost curve) or above the average variable cost curve depending on the context, up to the price received. To consumers these losses and gains are defined in terms of changes in consumer surplus measures (*net economic value*), and represent changes in the monetary surplus that accrues to the consumer over and above the cost of buying the good. Any reduction (gain) in it can be considered a loss (gain) in *net economic value*. Technically, consumer surplus or *net economic value* is derived as the area under a downward-sloping demand curve above the cost to the consumer. In practice, *net economic value* to consumers is measured from *changes* in demand curves in which the utility to consumers from a good is held constant; these are known as *compensated* demand curves. Aggregate economic benefits are the sum of changes in the value of economic welfare to all agents involved (for example, consumer and producer). Net benefits are the difference between economic benefits (both direct and indirect) that would accrue to a shoreline stabilization project (that is, *with* the project as opposed to *without* the project) and the costs of implementing the project.

Economic benefits in the context of cost-benefit analysis represent the same measure compared with those in the context of welfare economics. Here benefits refer to *all* gains in economic value (welfare part) that directly and indirectly result *with* a proposed project as compared to a situation *without* the project (a *with* and *without* rule is used to isolate such effects). In some cases, studies purporting to use CBA measure economic benefits to include changes in economic activity (for example, changes in sales, income, employment—nonwelfare component) in addition to changes in *net economic value* (welfare component); this is not appropriate in a CBA context.

Economic Methods

Cost-effectiveness analysis concerns the minimum cost option to achieve a given objective. It ignores benefits and does not address economic rationales to achieve a given objective. It is appropriate when considering how a shoreline-stabilization project can be implemented in the least expensive way. The procedure is to estimate all costs for a particular option over time, discount these costs, and then sum the discounted costs (discounted costs represent the total cost in today's dollars); the sum of discounted costs is referred to as the present value of costs. The decision criterion is to select that project with the smallest present value of costs over time.

Cost-benefit analysis (CBA) is the primary method with which both the benefits and the costs associated with a shoreline-stabilization project are considered. It is based on economic justifications in determining the implementation of a project—that is, whether the outcome of a project is worth the costs of achieving it. The analyst must identify and then measure all possible benefits and costs associated with the presence of the project as opposed to a situation without the project. Two variations of this technique are commonly used. One is to examine the difference between benefits and costs (benefits less costs) for each time period, discount it, and then sum it, giving the present value of net benefits over time. The present value of net benefits is the appropriate measure for comparing projects over time given equal scale (size) and time period. The decision criterion is to select that project that yields the maximum present value of net benefits over time. The second version is the benefit/cost (B/C) ratio, where the discounted sum of benefits is divided by the sum of discounted costs. When benefits equal costs, this ratio will equal 1. The decision criterion is to select the project that yields the maximum B/C ratio. The use of this ratio is quite controversial, even among economists. Most would agree that selecting a project solely on the basis of the B/C ratio is inappropriate.

Economic impact analysis (EIA) also needs clarification. Many analysts of applied policy problems and proposed federal regulations use variations of EIA, where it is referred to as *public policy analysis.* Here the analyst seeks to determine the effects (impacts) of proposed policy changes via an analysis in which the economic effects associated with the policy are identified and quantified. These effects are not the same as measures of *net economic value.*

Economic input/output analysis (I/O) is a specific technique based on an economic input-output model. It uses aggregate measures of economic activity such as sales revenues, income, and employment related to an economy defined by geo-political boundaries (a state, region, nation). A main feature of this technique is to determine "multipliers," which can be thought of as indicators of how changes in primary economic activity translate into final economic activity. With this information one can examine how changes in specific sectors (manufacturing, services, and so on) affect the entire economy in question. I/O analysis was primarily developed to address policy questions such as determing the effects on sales, income, and employment of some type of change to the economy (structural change—a casino opening or closing, reduced demand) or policy change (proposed regulation).

Simulation models are hypothetical computer models written in either primary computer code or a simulation language to represent or mimic an actual situation and then to simulate the specific application and changes to it. They have been used in coastal sciences to simulate sediment transfers, dune development, and storm effects. They have also been used in epidemiology to simulate the spread of an actual disease epidemic, and in population ecology to simulate population dynamics, the actual spread of an insect population outbreak, and the effects of different control strategies. Some applications have been based on bio-economic models of fisheries.

Risk-return models come from the field of finance and consist of the applications of portfolio theory, risk-mean variance models, and variations of the capital asset pricing model. They are used to decide among tradeoffs between risk and return in order to determine an efficient portfolio of holdings (least risky collection of assets that yield the greatest return) for various risk levels. These models are highly complex and indispensable to analysts and researchers in financial markets.

Economic Aspects of Beach Use, Shoreline Management, and Coastal Tourism

The literature pertaining to coastal economic issues can be separated into several groups, focusing on the value of beach use, the value of shoreline stabilization, the impact of coastal tourism and the potential effect of beach nourishment, cost-benefit analyses of specific beach nourishment projects, and policy-oriented studies of shoreline management such as the *NJSPMP*.

Economic Value of Beach Use and Shoreline Stabilization

The literature on the economic value of beach use and the economic value of shoreline stabilization are combined here, because several studies have obtained both estimates and determined the economic value of shoreline stabilization as

$$NEV_{stabilization} - NEV_{recreational\ use}$$

where NEV is net economic value.

The research pertaining to the economic value of beach use is meant to provide estimates of the added *net economic value* from an additional beach trip to beach users. Concerning the economic value of beach-stabilization, researchers measure the additional *net economic value* (money value of added satisfaction) from shoreline-stabilization or beach nourishment projects that provide extra short-term beach width. Both estimates are usually based on field surveys of a random sample of beach users or on mail surveys of a random sample of individuals (users and nonusers) using a contingent valuation (CV) approach.

One of the first studies of the value of recreational beach use, focusing on Delray Beach, Florida, was based on a CV method, found that residents were willing to pay (WTP) $1.88/person per day for beach use, whereas tourists were willing to pay $2.15/person per day in 1981 (table 8.1) (Curtis and Shows 1982). In 1983 a similar study at Jacksonville Beach, Florida, found that residents were willing to pay $4.44/person per day and tourists $4.88/person per day for beach use (table 8.1) (Curtis and Shows 1984).

Another study of the value of beach use in Florida in 1984 found, using the CV method, that residents were willing to pay an average of $1.31/day per person and tourists, $1.45/day per person for beach use and, using the travel cost approach, that residents were willing to pay $10.23/day per person and tourists, $29.32 per person (1984 dollars, table 8.1) (Bell and Leeworthy 1986). Other results from this study indicate that residents spent an average of 14.68 days/year per person on beach visits compared with nonresidents, who averaged 8.64 days/year per person for beach visits. One would expect benefits of beach use for residents to be smaller than for nonresidents because residents are located closer to the beaches, costs of travel to the beach are lower, they have more access to beaches, and they have more alternative beaches to choose from. That is, residents are presented with a rather unlimited supply of local beaches compared with tourists and accordingly will take advantage of this and use the resource more.

In a study of the value of beach use, nourishment, and litter cleanup to beach users in New Hampshire and Maine in 1988 based on a CV method, Lindsay and Tupper (1989) found that the estimated value of recreational use averaged $47.40/year per person, the value for beach erosion control averaged $30.80/year per person, and an estimate of the value for beach cleanup averaged $26.40/year per person (1988 dollars, table 8.1). If one were to assume a 92-day beach season (June–August), an estimate of the *daily* economic value from erosion control (beach nourishment) would be $0.33/day per person. Interestingly, this study found that the mean estimates of economic value varied across all four beaches, which could reflect the difference in beach characteristics or difference in tastes, and so on. A limitation of this study is that the estimated mean benefits exhibit high variability: in the case of WTP for erosion control, over twice the estimated mean value (that is, the coefficient of variation was 2.233, which measures the relative dispersion of the mean WTP bids). Possible reasons for this could be the small sample size and/or that the survey used an open-ended WTP question. Furthermore, because respondents were asked to assess three types of values in successive order, an upward bias in the estimates of the second and third successive values could result.

A study of the economic value of beach nourishment in New Jersey (that is, the economic value attributable to beach nourishment projects within the Sea Bright to Deal area, a 12-mile stretch of the northern New Jersey coastline) in 1985 found that beach users were willing to pay an average of $3.90/day per person for beaches that would receive beach nourishment (B_w) compared with an average of $3.60/day per person for beaches that did not receive beach nourishment ($B_{w/o}$), (1985 dollars, table 8.1) (Silberman and Klock 1988). The difference, an average of $0.30/day per person ($B = B_w - B_{w/o}$), was interpreted as an estimate of economic value attributable to beach nourishment. Before continuing, one should note the similarity in the estimated economic value from beach nourishment in this study ($0.30/day per person, 1985 dollars) with that of the Lindsay and Tupper study of erosion protection using a less sophisticated research design ($0.33/day per person, 1988 dollars). This similarity suggests it is possible that bias from successive order is not measurable in small samples.

In a second study, Silberman and Klock (1988) estimated existence value (that is, the value placed on the knowledge that beaches exist that are newly nourished compared with beaches that are eroded *for the segment of the population that would never visit or use these newly nourished beaches*) using a CV approach at an average one-time contribution of $16.31 (1985 dollars table 8.1). If this represents an annual contribution, an estimate of nonuse value per day could be $0.177/day per person (based

Table 8.1a ⁓ ANALYSES OF THE ECONOMIC VALUE OF RECREATIONAL
BEACH USE

Study	Year/area	Sample design	Economic method
Curtis and Shows (1982)	1981 Fla.	*N.A.*	Face-to-face: CV-OE
Curtis and Shows (1984)	1983 Fla.	*N.A.*	Face-to-face: CV-OE
Bell and Leeworthy (1986)	1984 Fla.	RS of 911 residents; 4333 tourists contacted, 826 surveyed	Telephone: CV-OE Face-to-face: CV-OE
Lindsay and Tupper (1988)	1988 N.H., M.E.	RS of 1100 users	CV-OE, face-to-face
Silberman and Klock (1988)	1985 No. N.J.	Split RS	CV-IB; face-to-face

	Residents	Tourists
WTP	$1.88/trip	$2.15/trip

	Residents	Tourists
WTP	$4.44/trip	$4.88/trip

	Residents	Tourists
WTP	$1.31/trip	$1.45/trip
SD	(3.23)	(2.57)
n	804	968
%0 bids[a]	29%	38%
TCM:[b]		
CS	$10.23/trip	$29.32/trip
CVar	$10.31/trip	$29.45/trip
EV	$10.18/trip	$29.29/trip
n	870	1051

	All	*Old Orchard Beach*	*Pine Pt. ($/year)*	*Ocean Park*	*Seabrook, N.H.*
WTP	47.40	51.15	46.15	52.40	41.03
SD	(53.00)	(49.13)	(52.88)	(60.34)	(52.35)
n	934	316	173	164	281
WTPEC[c]	30.80	20.08[a,b]	27.37[a]	34.30[a]	40.45[b]
SD	(68.80)	(29.72)	(39.69)	(54.28)	(83.76)
n	834	248	153	156	277
WTPLP[c]	26.40	26.31[a]	18.50[a]	20.16[a,b]	34.47[b]
SD	(65.30)	(70.55)	(29.82)	(31.69)	(85.13)
n	834	248	153	156	277

	WTP w/o project	*WTP w/ project*	*Existence value*
Mean	$3.60/trip	$3.90/trip	$16.31
SD	(1.14)	(1.18)	(19.62)
n	462	445	822
%0 bids	17%	–	13%
% protest bids[d]	–	–	38.4%

(*continued*)

Table 8.1a ⁓ (*continued*)

Study	Year/area	Sample design	Economic method
Silberman, Gerlowski, and Williams (1992)	1985 No. N.J.	Split RS	Face-to-face: CV-IB
	Staten Is.	RS = 500	Telephone: CV-OE
U.S. ACOE (1986)	1985 No. N.J.	RS of 2917 users	Face-to-face: CV-IB
Koppel (1994); Kucharski (1995)	1994 So. N.J.	RS of 1063 users	Face-to-face: CV-CE CV-OE

Results

		Existence value: total sample	Existence value: nonzero bidders
Face-to-face			
Will use in future	Mean	15.21	23.59
	SD	(20.91)	(21.92)
	n	1177	
%0 bids		35.5%	
Will not use	Mean	9.34	21.02
	SD	(16.04)	(18.28)
	n	754	
%0 bids		55.6%	
Telephone			
Will use in future	Mean	19.65	31.98
	SD	(38.37)	(44.00)
	n	83	
%0 bids		38.6%	
Will not use	Mean	9.51	23.87
	SD	(17.49)	(20.67)
	n	138	
%0 bids		60.1%	
Tobit model			
Will use in future		15.10	
Will not use		9.26	
WTP w/o protection		$3.67/trip	
WTP w/50' berm		$3.89/trip	
WTP w/100' berm		$3.93/trip	
Midpoint WTP w/protection		$3.91/trip	

	w/o $0 bids	*w/$0 bids*
WTP–beach use	$5.04/trip	$4.22/trip
WTP–wider beach	$5.41/trip	$4.59/trip
Wider beach		
% WTP more	16%	
% WTP less	3%	
% WTP same	81%	

(continued)

Table 8.1a ⁓ *(continued)*

Study	Year/area	Sample design	Economic method
Falk, Graefe, and Suddleson (1994)	1993 DE	S of users	Face-to-face: CV-OE and CV-CE
		RS of 200 property owners RS of nonresidents	Mail survey: CV-OE and CV-CE

SOURCE: See references in chapter 8.
NOTE: CS = consumer surplus; CV = contingent valuation; Cvar = compensating variation; CV-CE = contingent valuation–closed ended; CV-IB = contingent valuation–interative bidding; CV-OE = contingent valuation–open ended; Evar = equivalent variation; n = number of observations; N.A. = not available; RS = random sample; S = sample; SD = standard deviation; TCM = travel cost model; WTP = willingness to pay; WTPEC = willingness to pay for erosion control; WTPLP = willingness to pay for litter program.
All estimates of economic value are in terms of $/person per trip or $/person per day (trip), except for

Results

	All	No outliers	No protest	n	% WTP
Beach use		Face-to-face survey			
All	$3.01	$2.80	$3.85	562	77%
Rehoboth Beach	$3.22			129	77%
Dewey Beach	$3.21			118	79%
Bethany Beach	$2.73			115	75%
So. Bethany	$2.59			96	74%
Fenwick Is.	$3.20			104	78%
Wider Beach					
All	$3.70	$3.46	$4.63		30%
Rehoboth Beach	$3.90				29%
Dewey Beach	$3.99				35%
Bethany Beach	$3.36				25%
So. Bethany	$3.12				23%
Fenwick Is.	$4.02				38%
Beach protection					
All	$63.69/yr	$55.74/yr	$78.65/yr		79%
Rehoboth Beach	$55.34/yr				77%
Dewey Beach	$42.40/yr				78%
Bethany Beach	$95.44/yr				73%
So. Bethany	$65.59/yr				81%
Fenwick Is.	$61.42/yr				81%
		Mail survey			
Beach use					
All	$2.85			348	76%
Non–property owners	$2.79			150	81%
Property owners	$2.97			141	74%
Non–beach users	$3.10			46	74%
Wider beach					
All	$3.50				21%
Non–property owners	$3.23				13%
Property owners	$3.98				30%
Non–beach users	$3.77				23%
Beach protection					
All	$26.60				34%
Non–property owners	$10.99				31%
Property owners	$51.64				42%
Non–beach users	$ 4.93				23%

estimates of existence value, which were converted to $/person/day(trip) on the basis of a 92-day season. Because of the high economic value estimates associated with the N.H.-Me. study, the net value of beach protection was estimated from the bid value for erosion control divided by 92 days.

a. %o bids represent percentage responding with zero willingness to pay.

b. Estimated from the area under an estimated linear demand curve evaluated at mean values of the independent variables.

c. Differences significant at the .05 level.

d. % protest indicates bids that signal a respondent's protest to the item being assessed.

Table 8.1b ~ ANALYSIS OF THE ECONOMIC VALUE OF RECREATIONAL BEACH USE: SUMMARY

		1985$	1992$
Northern New Jersey			
Silberman and Klock (1988)	*Use values*		
Silberman, Gerlowski, and Williams (1992)	WTP w/protection	$3.90/trip	$5.08/trip
	WTP w/o protection	$3.60/trip	$4.69/trip
	Net value of protection	$0.30/trip	$0.39/trip
	Non-use values		
	Existence value	$9.26/92 days = $0.1006/trip	$12.07/92 days = $0.13/trip
	Sum: Use and nonuse	$0.4006/trip	$0.52/trip

		1985$		1992$	
		w/o zero bids	w/zero bids	w/o zero bids	w/zero bids
Southern New Jersey					
Koppel (1994)	WTP–wider beach	$5.41/trip	$4.59/trip	$5.12/trip	$4.35/trip
	WTP–beach use	$5.04/trip	$4.22/trip	$4.77/trip	$4.00/trip
	Net value of protection	$0.37/trip	$0.37/trip	$0.35/trip	$0.35/trip

		1988$	1992$
New Hampshire–Maine			
Lindsay and Tupper (1988)	WTP–beach use	$47.40	$56.21
	WTP–protection	$30.80/92 days = $0.33/day	$36.53/92 days = $0.39/day

Northern New Jersey

	1985$	1992$
U.S. ACOE (1986)		
WTP–w/o protection	$3.67/trip	$4.78/trip
WTP–w/50' berm	$3.89/trip	$5.07/trip
WTP–w/100' berm	$3.93/trip	$5.12/trip
WTP–w/protection (midpoint)	$3.91/trip	$5.10/trip
Net value: w/protection	$0.24/trip	$0.315/trip
w/50' berm	$0.22/trip	$0.29/trip
w/100' berm	$0.26/trip	$0.34/trip

Delaware

	1993$		1992$	
	All	w/o outliers	All	w/o outliers
Falk, Graefe, and Suddleson (1994)				
WTP–wider beach	$3.70/trip	$3.46/trip	$3.59/trip	$3.36/trip
WTP–beach use	$3.01/trip	$2.80/trip	$2.92/trip	$2.72/trip
Net value–protection	$0.69/trip	$0.66/trip	$0.67/trip	$0.64/trip

Note: WTP = willingness to pay.

on a 92-day season, June–August). A shortcoming of this component of the study was that surveyed individuals were not given any alternative beach nourishment projects in other areas along the New Jersey coast to choose from; this fault could have caused an upward bias in the reported estimates. Furthermore, the sampled beach users expressed difficulty in understanding the concept of existence value, which could contribute to measurement error. A more detailed analysis of existence values was based on a statistical model (Tobit model) and found that estimates of existence value were an average of $15.10/year from the on-site survey and $9.26/year from the survey of residents who were nonusers (Silberman, Gerlowski, and Williams 1992). It was concluded that the estimate from surveyed residents ($9.26) was a more appropriate estimate of the average economic benefit based on the existence of a beach environment preserved with nourishment (1985 dollars, table 8.1).

On the basis of the New Jersey studies, if economic benefits are to be based on the sum of use and nonuse values, then benefits would be estimated as the sum of $0.30/day and $0.1006/day ($9.26/92 days in the season) per person or $0.4006/day per user; but if benefits only represent use values, then benefits would be $0.30/day per person. One must keep in mind the limitations of this study in any interpretation. It was based on a relatively small sample size, causing a great deal of variation in the benefit estimates; survey respondents had difficulty in trying to assess an amount for the existence value of beach stabilization; and a choice of other beach nourishment projects located elsewhere in New Jersey was not offered as an alternative, all of which can introduce an upward bias in the reported estimates.

Further studies of the economic value of beach use come from U.S. Army Corps of Engineers unpublished surveys based on the New Jersey data from Silberman and Klock (1988) and Silberman et al. (1992). Results of use values indicated a mean WTP bid of $3.67/person associated without beach nourishment, $3.89/person with shoreline stabilization incorporating a 50-foot berm, and $3.93/person with shoreline stabilization incorporating a 100-foot berm (1985 dollars, table 8.1). Results pertaining to existence value, the value of preserving the beach, was estimated at an average of $16.41/year in 1985. From these results, an estimate of the value from beach nourishment is $0.22/person per day associated with a 50-foot berm, and $0.26/person per day associated with a 100-foot berm; an average of both sizes based on the midpoint of the two WTP bids pertaining to beach nourishment is $3.91/person [(3.89 + 3.93)/2], hence a net effect attributable to beach nourishment can be estimated at $0.24/person per day ($3.91 less $3.67) in 1985 dollars.

A U.S. ACOE reconnaissance report (1993c) for the Raritan Bay–Sandy Hook (Reach 1) area in New Jersey used a unit-day-value method to obtain estimates of recreational benefits of $2.88/day trip with beach nourishment and $2.40/day trip without the project, or a net difference attributable to beach nourishment of $0.48/day trip (in 1982 dollars).

A study conducted at southern New Jersey beaches examined the economic value of beach use, beach stabilization, and erosion control in 1994 from a survey of beach users, business owners/managers, and home-owners (Koppel 1994; Kucharski 1995). The study found that the recreational value of beach use averaged $5.04/person per day when bids of zero dollars ($0 bids) were excluded and $4.22/person per day when $0 bids were included in 1994 (table 8.1). Concerning the value of a wider beach (that is, beach nourishment), 81% of the respondents indicated that they would pay the same, 16% were willing to pay more, and 3% indicated they would pay less (Koppel 1994: 26). Of those willing to pay more, the average use value (that is, consumer surplus) was estimated at $2.72/person per day above the initial value estimate, and for those willing to pay less, the average value was estimated at $1.68/person per day below the initial value estimate in 1994. On the basis of this information a weighted average of the economic value of a wider beach was estimated to be $4.59/person per day with $0 bids (that is, $6.94*165 + $2.54*33 + $4.22*865 = $4879.22/1063 = $4.59) and $5.41/person per day without $0 bids (that is, $7.76*165 + $3.36*33 + $5.04*865 = $5750.88/1063 = $5.41) (our calculations). On average, these beach users were willing to pay an additional $0.37/person per day for a wider beach for the cases with $0 bids and without $0 bids in 1994 dollars (that is, $5.41 − $5.04 = $0.37, and $4.59 − $4.22 = $0.37). Concerning existence value, results indicated that the estimated median value was $50 in 1994 dollars, and assuming a 92-day season, nonuse value is estimated at $0.5435/day. This would yield an *overall* economic value (sum of use and nonuse value) of $0.9135/day.

Koppel (1994) also obtained economic value estimates from business owners (that is, producer surplus) based on a CV approach, the only study to do so. Results found that business owners were willing to pay almost 20% more in taxes (19.95% more) for a wider beach, and an average $181/year per business with $0 bids or $256/year per business without $0 bids in 1994 (our calculations). But because of the small sample size

($n = 156$), little confidence can be placed on these responses and they will not be considered for further analysis.

Koppel (1994) also obtained economic value estimates from homeowners. Results indicated that 80% of the homeowners were not willing to pay more in taxes/payments for a wider beach (that is, beach nourishment), 17% were willing to pay more, and 2% indicated they would pay less. This finding is surprising, because homeowners would benefit from beach nourishment efforts. If the sample were stratified on the basis of proximity to the beach, or rental property versus personal use, responses might have been different. The midpoint of the estimated median value of households willing to pay for a wider beach was $35.50/household with $0 bids ($25–owners surveyed at home; $46–owners surveyed at the beach) or $229.50/household without $0 bids ($380–owners at home; $79–owners at the beach) in 1994. This also represents a relatively low sample size; 1.5% of all available homes in these communities. Again, little confidence can be placed on these estimates and they will not be considered for further analysis.

Subsequently, Kucharski (1995) used the estimates cited above and projected them to obtain estimates in 1994 for: (1) all beaches in the five communities, (2) all homes in the five communities, and (3) all businesses in the five communities. Projected estimates in 1994 were almost $101 million for the economic value of beach use, $4.8 billion in lost property value from beach erosion to homeowners, $2.5 billion for existence value of beach nourishment to homeowners, and $1.4 million for existence value of beach nourishment to businesses (all in 1994 dollars). But mainly due to the overall small sample size and the lack of estimates of variability, little confidence can be placed on such projections and on sample results.

Last, Falk, Graefe, and Suddleson (1994) conducted an overall study, based on a CV approach, of the attitudes and perceptions of Delaware's beach users and the economic value of recreational beach use, sand replenishment (that is, beach nourishment), and maintaining beach width by the use of an "annual beach protection fund." Results for "on-site" users found that the estimated WTP bid for recreational beach use averaged $3.01/day per person ($2.80/day without statistical outliers, $3.85/day without protest bids), estimated WTP for a wider beach averaged $3.70/day per person ($3.46/day without statistical outliers, $4.63/day without protest bids), and estimated economic value for maintaining beach width averaged $63.69/year per person ($55.74/year without statistical outliers, $78.65/year without protest bids) (1993 dollars, table 8.1). From a mail survey, results indicated that the estimated WTP bid for beach use averaged $2.85/day per person ($2.64/day without statistical outliers, $3.66/day without protest bids), estimated WTP for a wider beach averaged $3.50/day per person ($3.24/day without statistical out-liers, $4.29/day without protest bids), and estimated economic value for maintaining beach width averaged $26.60/year per person ($23.75/year without statistical outliers, $44.55/year without protest bids) (1993 dollars, table 8.1).

As in the Lindsay and Tupper (1989) study, Falk, Graefe, and Suddle-son (1994) also found that mean estimates of economic value varied across all five ocean beaches, which could reflect differences in physical features, socioeconomic use, tastes, and so on. An issue not examined concerns the assessment of values due to successive order (where benefits for successive levels of a good are asked in succession in surveys), as in the Lindsay and Tupper study, which could have introduced an upward bias in the esti-mates of the second and third successive economic values.

Summary: Beach Use and
Stabilization Values

For the purposes of this book, the following studies are used to derive a range of estimated average values associated with beach preser-vation/nourishment: (1) northern New Jersey (Silberman and Klock 1988; Silberman, Gerlowski, and Williams 1992); (2) southern New Jersey (Kop-pel 1994); (3) New Hampshire–Maine (Lindsay and Tupper 1988); (4) northern New Jersey (U.S. ACOE 1986); and (5) Delaware (Falk, Graefe, and Suddleson 1994). The *net economic value* of recreational beach use pertaining to beach nourishment is estimated as follows: from (1) $0.30/ person per day in 1985 ($0.39/person per day in 1992 dollars); from (2) $0.37/person per day in 1994 ($0.35/person per day in 1992 dollars, com-pared with [1]); from (3) $0.33/person per day in 1988 ($0.39/person per day in 1992 dollars); from (4) $0.24–$.026/person per day in 1985 (or $0.32–$0.34/person per day in 1992 dollars); and from (5) $0.65–$0.69/ person per day in 1993 ($0.61–$0.65 in 1992 dollars). The Falk, Graefe, and Suddleson (1994) study found the highest value of recreational beach use, twice that of all other estimates, and further research is necessary to identify reasons for such a difference. Based on these estimates a low esti-mate of *net economic value* is about $0.35/person per day in 1992 dollars,

242 ~ COASTAL HAZARD MANAGEMENT

from (2) and (4), and a high estimate of $0.39/person per day in 1992 dollars, from (1) and (3). Hence the *net economic value* associated with beach nourishment for recreational use is estimated to range from $0.35/person per day trip to $0.39/person per day trip in 1992 dollars; and based on (5) the high estimate could range up to $0.61–$0.65/person per day.

Beaches, Tourism, and
Economic Development

Research in this area attempts to provide evidence to support claims of the additional tourism and economic activity generated from or related to beach nourishment projects. The basic theme of the literature is that tourism expenditures in beach communities are attributable to the presence of the beach and that spending there can significantly contribute to local, regional, and possibly national economies. Although one can find little to debate about the general nature of tourism in beach communities, two related research claims have not been settled: (1) whether spending in beach or coastal communities contributes significantly to local economies and/or to state, regional, and national economies; and (2) whether *all* tourism expenditures are directly related to the presence or proximity of a beach (or beach project).

There are many external influences that can also affect tourism patterns, such as weather, relative cost of beach rentals, beach access and fees, availability of other amenities, marine pollution, and debris washups (plaster, litter, sewerage, tarballs). These external influences must be controlled for in order to pinpoint the specific contribution the presence of a beach nourishment project adds to tourism spending and related economic activity, if any. Research design is of critical importance, or measurements of added tourism spending and so on will contain many contributing influences or added effects that together can overshadow the contribution of beach nourishment projects. Hence, when reading this literature one must pay special attention to research design and, in particular, to how external effects were controlled for. Without recognition of all potential effects in its design, a study has not isolated the true effects of beach nourishment on tourism. In fact we don't really know what such studies are measuring or what effects they are capturing.

Another issue of concern pertains to the level of aggregation of tourism expenditure data. The issue is whether county-level estimates accurately represent spending in the coastal zone. County-level data have often been used to generate coastal economic data and evaluations of coastal regions (University of North Carolina 1991). In New Jersey, counties such as Monmouth and Ocean extend well inland from the shoreline, in excess of 48 km (30 miles) in some areas (this is also true of some coastal counties in Maryland, Delaware, Florida, Connecticut, and Massachusetts, for example). Within coastal counties, one can argue that tourism spending in areas that are not located in close proximity to the shoreline and beach are probably not influenced by the beach, that is, there is no beach effect in these cases, and expenditures are independent of location. The use of county-level data can present a misleading picture, in that they contain an unknown portion of economic activity that is located well inland and has no "beach effect." Including these data introduces an unknown, upward bias in county-level statistics. For example, should sales made at shopping malls located 30 miles inland while on a visit to the beach be included as beach-trip expenditures? Or should economic activity from all business units located inland and included in county-level data represent coastal economic activity? As another example, Long Island is composed of two boroughs (Queens and Brooklyn) and two counties (Nassau and Suffolk). In this case the entire area classifies as a coastal area if based on county-level-equivalent analysis, and the question is whether all economic activity and expenditures should be associated with the presence of beaches or beach trips. We believe research efforts should be directed toward developing more appropriate data. Without knowing the distribution of economic activity within particular coastal counties, it is not possible to differentiate economic activity such as tourism spending associated with a "beach effect" from the remainder of the county-level activity, or to determine the magnitude of the upward bias in coastal county-level data if used for the purpose of representing coastal economic activity. This problem limits the usefulness of county-level data, and researchers should use caution in their application and interpretation. Below we provide evidence of the magnitude of the error that such county-level estimates can contain.

A number of points should be kept in mind when considering the role of coastal tourism in local or national economies. One concern is the purpose of the trip; a specific trip made for the purpose of beach recreation is a "distinct trip," while a trip made to visit friends or relatives and recreational attractions not located at the shore area coupled with some time spent at local beaches is a "multipurpose trip." One must be careful to account for the time or proportion of the trip that was spent at the beach, or that involved a specific trip to a beach (the time or proportion of the trip not spent at the beach is therefore irrelevant, as are associated expenses). Another major concern involves expenses, and the issue is to identify expenditures that are uniquely related to the presence of the beach (and/or beach trip) or related to the proximity of the beach. This problem becomes compounded for multipurpose trips. It can be argued that any expenses other than lodging, food, entertainment, transportation, and entrance fees (parking or beach) should not be deemed directly related to beach use activities. Two examples allude to potential problems. In one, a vacation is made to a beach resort community, but all the time is spent at the pool; should all, a portion, or none of the expenses be counted as beach related? Were any expenses beach-dependent? Could the same vacation have been spent at a resort away from the beach? Another example concerns purchases of durable and nondurable goods (small appliances, clothing other than beach apparel) made because of shopping convenience and leisure time in beach communities; should these be included as typical expenses of beach trips?

Studies discussed below that have examined the impact of tourism on local economies are all based on field survey data. The use of surveys and survey data and the issue of sample design introduce concerns. Before tourist expenses are projected, both in these studies and in future attempts, one needs to identify only relevant expenditure items made by beach tourists/users, and determine the appropriate trip expenses or proportion of total trip expenses due only to beach-related trips or visits. Such a process will avoid the problem of artificially inflating projections of tourist expenditures. For example, if only 50% of total expenses are relevant and pertain only to beach use from a particular survey, projected estimates of total expenses for a region or state would then contain an error of two-times; the sample and projected estimates should be reduced by half. Care must be used in developing any projections of sample-survey data, because the projected estimates, in turn, form the basis of arguments of the relative contribution of coastal tourism to local and regional economic activity. Similar remarks apply even more so in attempts to measure associations between beach nourishment and additional economic activity generated, if any.

Bell and Leeworthy (1985, 1986) were the first investigators to conduct a detailed analysis of the net economic value of beach use and its economic importance for an entire state. They focused on Florida, using field surveys of residents and tourists. Overall results indicated that an estimated average of $450/household per year was spent by residents and an average of $395/household per year was spent by nonresidents who visited Florida beaches in 1983. The estimated average travel expense was projected on the basis of the participation rate (that is, the inverse of the participation rate multiplied by the average travel expense), to yield total expenditures associated with beach use estimated at $2,276 million in 1983 ($1,123 million for residents and $1,153 million for nonresidents, 1983 dollars). These total expenditures are referred to as total sales impacts by Bell and Leeworthy; this term is slightly misleading, because this estimate reflects actual sales directly related to beach use (direct expenditures) and not impacts resulting from these sales, such as indirect effects. Economic impacts were then derived in terms of output (sales), employment (jobs), wages (income or earnings), and state tax revenues. However, the impacts are not derived in the typical manner; impacts for residents were projected based on ratios, and for nonresidents impacts were derived from an export-based theory (Bell and Leeworthy 1986: 8–10, 19–25). Total sales impacts were estimated at $4,581 million from total expenses of $2,276 million in 1983, an implied multiplier effect of 2.013.

Probably the most serious shortcoming of this study concerns whether the sample was representative of the general population of residents and tourists in Florida (a sample of 911 residents and 826 nonresidents). Can these sample statistics reasonably approximate the population statistics? Bell and Leeworthy then project total sales to $2.27 billion and sales impacts to $4.58 billion to the state of Florida in 1983 ($3.206 billion and $6.452 billion, respectively, in 1992 dollars; Bell and Leeworthy 1986: 30).

In another study, Stronge (1994) used survey data from the Florida Department of Tourism to advance the case for the economic importance of beach tourism. Beach tourists were identified on the basis of a response to a question about what particular facilities and programs they enjoyed during their visit or trip. Beaches were one of the options tourists could check off; those that checked off this category were classified as "beach tourists." Economic impacts were estimated on the basis of the average expenditure of beach tourists, the percentage of surveys that had the "beaches" response checked off (to establish and represent a participation rate), and statewide multipliers obtained from the U.S. Bureau of Economic Analysis's Regional Input-Output model. The economic impact from direct spending of beach tourists was estimated at $7.9 billion in 1992 in Florida. Other economic impacts estimated pertained to output (sales), earnings (income), and employment (jobs). Stronge (1992) argued that the contribution of beach tourism to Florida's GDP (referred to as "gross regional output" by Stronge) was estimated at $15.4 billion, developed from the product of the estimate of tourist spending and a statewide multiplier. But there are a number of potential problems with this estimate and claim.

The projected estimate of GDP was not based on the application of an input/output model. Rather, it can be thought of as a "back-of-the-envelope" economic impact assessment. Such an assessment may contain double counting. In addition, the portion of expenditures related only to beach use was not separated from overall expenses; rather, total expenditures of tourists who enjoyed beaches were treated as "beach expenses." Nor was any information gathered about the proportion of the trip and time spent at the beach (for example, the proportion of time spent on beach trips). These problems weaken the analysis and conclusions.

Houston (1995a,b) further extends the arguments on the impact of beach tourism following the approach of Stronge (1992). In both papers, Houston takes the Stronge result, compares it with overall tourism spending, estimates a ratio between beach tourism spending and overall tourism spending, and projects beach tourism spending for an estimate of its portion of state GDP, with a total estimate of $170 billion for all coastal states in the United States (Houston 1995a). This projected estimate is very loosely based on economic impact measures and probably contains double counting. A multiplier from previous research used to expand ratios of tourism spending is then claimed to represent and measure economic impacts. Although the projected estimate may overstate such impacts, further doubt is cast because of the statistical data taken from secondary, unofficial sources such as the *World Almanac* and press reports (*USA Today, National Geographic, Wall Street Journal*). Secondary statistical sources are usually not tested or examined for accuracy in the way that official statistics published by various branches of the government are, and can be misleading. The potential bias and error inherent in secondary-source statistics limits the usefulness of any research based on such data.

Notwithstanding that tourism in coastal areas can have an impact on spending (that is, effects over and above residents' spending) and possibly contribute to economic development if the tourism effect is large enough, several shortcomings of the Stronge and Houston papers weaken their results. For an economic impact analysis to accurately represent aggregate economic activity measures (such as GDP), one must be careful to avoid any potential double-counting problems; neither of these papers contains enough description to render a determination one way or the other. In both papers, expenses related to beach trips or use were not collected or separated from total expenditures. Rather, total expenditures were treated as beach related and representative of beach-use expenses, introducing an unknown upward bias into the projected estimates. Furthermore, to see the potential error contained within projected impacts, compare the Bell and Leeworthy (1985, 1986) estimate to that of Stronge's, both for the state of Florida. Stronge (1994) projected total sales of $7.9 billion related to beach tourists, and output (sales) impacts of $15.4 billion in 1992. The Bell and Leeworthy estimates differ remarkably from the Stronge estimates, being 2.5 times less regarding projected sales and 2.4 times less regarding sales impacts (all in 1992 dollars). This variation underlines our concern over the proper use of survey studies, and the need to isolate expenses just for the portion of the trip dedicated for beach use.

An additional issue about the effects of tourism on coastal communities has been raised by Manheim and Tyrrell (1986ab). These authors have argued that the influx of nonresident tourists during the summer season places an added, and previously ignored, burden on residents because of nonresidents' use of the local infrastructure, which in many cases has been developed on the basis of the needs of local residents and paid for by them through property taxes. The summer populations of beach communities literally explode, and the local infrastructure (roads, water, sewerage, waste hauling) either wear out or exceed their designed capacities more quickly. These costs are not internalized or borne by nonresident tourists, although they are for individuals who own summer homes in the communities. The issue mainly concerns the proportion of tourists who use hotel or motel accommodations and condo or beach cottage rentals in relation to the number of owner-occupied homes and apartments. Previous studies have not considered this aspect of tourism, and communities that advocate tourism need to take these hidden costs, similar to a negative externality, into consideration.

Cost-Benefit Analysis of Beach Nourishment Projects

A general overview of U.S. Army Corps of Engineers analyses and of economic analyses based on CBA will be useful here. A detailed review of U.S. ACOE studies of proposed projects in New Jersey was beyond the scope of this book, and the reader is referred to the individual reports (U.S. ACOE 1989a,b, 1991a,b, 1992, 1993c, 1994a,b, d, 1995).

In general, the U.S. ACOE analyses and economic analyses including CBA are thorough and well done. Development of the cost component is very detailed, usually accounting for all items involved with the project in question. Many of these cost items are based on detailed engineering studies. The benefit component is also very complete. All recent U.S. ACOE studies find economic benefits in up to five areas:

1. Storm reduction benefits
2. Benefits from the reduction in lost land
3. Benefits from intensification
4. Recreation benefits
5. Benefits from reduced maintenance and costs of shore stabilization at other sites

Storm reduction benefits. The first benefit item measures benefits as the reduction in storm-related damages prevented by the proposed project (that is, storm damage *without* the project less storm damage *with* the project). Although this definition is quite simplified, storm reduction benefits can have up to five distinct components and involve sophisticated computer models to develop estimates. These possible components are (1) reduction in the inundation of structures, (2) reduction in damage caused by wave attacks to structures, (3) reduction in damage associated with long-term erosion and storm events (shoreline recession), (4) reduction in maintenance costs associated with other shore-stabilization projects, and (5) reduction in public emergency costs that would arise from storm-flooding emergencies. To avoid double counting storm reduction benefits, the U.S. ACOE use a critical damage threshold, whereby only the maximum damage to any single structure pertaining to the first three components is used to measure storm-related damage and to estimate storm reduction benefits.

Reduction in lost land .The second benefit item is the reduction of the assessed value of real property that would be lost from erosion prevented by the project (that is, the assessed value of land lost *without* the project less the assessed value of land lost *with* the project) plus the value of the recreational component of the lost land when it is identified to be beach or recreational land (that is, derived from an estimate of the amount of beach users the lost land could support over a beach season valued as the sum of beach fees plus the economic value of beach use based on WTP without project conditions).

Intensification benefits. The third benefit item focuses on benefits due to increases in the assessed value of real property that are related to the presence of the proposed project.

Recreation benefits. The fourth item consists of the sum of the net increase in economic value (measured as willingness to pay) from a wider, more protected beach for current beach users, the net increase in the economic value of additional beach users due to a wider, more protected beach, and the economic value of preserving the beach (that is, its existence value) from a nonuser's perspective (that is, the sum of use and nonuse values). Most current U.S. ACOE studies use the contingent valuation method, although some studies use a unit-day-value method (a monetary value of the net increase in users as a result of the project, valued at either an average day-trip expense or an average entrance fee).

Benefits from reduced maintenance. The fifth item pertains to benefits from reduced shore-stabilization maintenance at other locations near the proposed project (derived as the maintenance costs of shore stabilization at nearby locations *without* the project less estimated maintenance costs of shore stabilization for these nearby locations *with* the project).

It should be pointed out that the benefit items used in U.S. ACOE analyses may be subject to debate and should not be taken as fact. For instance, the third benefit item, intensification benefits, implies that benefits arise from increases in the assessed value of real estate in close proximity of shoreline-stabilization projects. Not all coastal researchers may agree with this claim, nor have studies been conducted to rigorously examine and quantify this effect. Similar comments may apply to the other benefit items used in the U.S. ACOE analyses.

Limitations of U.S. ACOE studies include (1) inadequate sensitivity analysis of benefit items (where a variable is changed a little [increased or decreased] and the change in outcome is examined); (2) lack of a sensitivity analysis of cost items; (3) inadequate treatment of uncertainty in cost and benefit items, although some of this is conducted regarding the estimation of storm reduction benefits (through the use of storm damage–wave surge computer models to simulate storm damage); (4) insufficiently explicit treatment of risk and uncertainty; (5) no treatment of risk and variability involved with project lifespans and project outcomes, or exogenous effects from erosion (this is an area where there is much room for improvement); and (6) little or no research to support and validate the specific claims of benefits realized from shoreline stabilization.

Recommendations for future U.S. ACOE studies pertain to both cost and benefit components: (1) incorporation of uncertainty in cost and benefit items; (2) incorporation of risk and variability in project lifespans and project outcome (for example, more accurate estimates of lifespans and outcomes based on local experience); (3) greater coverage/application of sensitivity analysis to derivation of cost and benefit estimates; and (4) address the appropriateness of the benefit elements included in U.S. ACOE procedures.

The treatment of uncertainty, the subject of the first recommendation, involves two basic components. One of these concerns benefit and cost items; the other concerns the lifespan of shoreline-stabilization projects. Regarding the cost and benefit items, elements of uncertainty pertain to the frequency of their occurrence, and uncertainty over their future monetary value. Some elements of the costs and benefits of proposed shoreline-stabilization projects are stochastic in nature; this affects both the frequency of occurrence and the magnitude of the estimate for cost and benefit items. For example, over a ten-year period, some cost and benefit items may only occur one or two times in ten years because of the occurrence of significant coastal storms; the magnitudes of costs and benefits can also be highly variable (by orders of magnitude) at these times in comparison with nonstorm conditions.

In future studies, both of these elements of uncertainty (frequency of occurrence and variability of magnitude) should be incorporated into U.S. ACOE analyses rather than the use of average magnitudes and occurrences over a given time period. It is preferable to have some idea of the range of damage estimates and the range of storm damage reductions rather than a point estimate, such as an average value. This point needs to be emphasized. In CBA, care must be taken to have as realistic a situation as possible. If, for example, certain cost and benefit items only occur one or two times in a ten-year period, this needs to be reflected in the CBA. Also, when these cost and benefit items occur, their magnitudes will be highly variable and could differ by orders of magnitude compared with the remaining nine or eight years. This aspect should also be incorporated in CBA. All too often, the investigator uses the average value over the ten-year period as if it occurs in each period. This practice detracts from the realism a CBA should reflect. As an example, in the analysis of recreational benefits, ranges (upper and lower bounds and a mean) are developed in the analysis (U.S. ACOE 1989b: D-88, D-89). However, in the final analysis of benefit/cost ratios, this range is dropped and the ratios and CBA are based on the mean value.

Furthermore, uncertainty over the future monetary value refers to estimates of the value of cost and benefit items when their monetary value can vary over time because of scarcity, competing demands, inflation, and economies of scale. This element should also be incorporated into future U.S. ACOE analyses.

Incorporation of risk elements, the second recommendation, concern the expected project lifespan and the expected outcome. Neither factor is known with certainty, and there is a risk that projects may fail to achieve their expected lifespans and/or expected outcomes; for example, there may be a 10% probability that a project will fail (in terms of lifespan and outcome), based on expectations of the occurrence of significant coastal storms over the project planning period. These elements should be incorporated into future U.S. ACOE analyses. This recommendation must also include an effective monitoring program in which the relative effectiveness of the project can be measured, in terms of both achieving its expected outcome and achieving its expected lifespan. Variability in the expected lifespan can also exist across projects, and this element should be taken into account in future analyses and alternative approaches to CBA. The variability of expected lifespans is also affected by exogenous factors, such as erosion rates, which will differ spatially. Such effects should be accounted for.

The third major recommendation suggests an expanded application of sensitivity analysis in the derivation of cost and benefit estimates. The only sensitivity analysis in the U.S. ACOE studies pertains to the use of a range of discount rates in the present-value analysis of net benefits. Sensitivity analysis could be useful in evaluating project outcomes for different scenarios where both cost estimates and benefit estimates take on a range of values, for instance, a range of net benefits for different storm event scenarios (for example, net benefits for low storm activity, moderate storm activity, and high storm activity).

Other recommendations regarding future U.S. ACOE analyses concern: (1) the incorporation of some measure of the variability of the cost and benefit items; (2) the use of better damage data from more recent storm events as well as the commonly cited 1962 and 1984 "super" storms (here more recent data from FEMA and NFIP agencies are available); and (3) recreational benefits from increased use should not only incorporate costs of building additional parking facilities to accommodate new users but examine whether expanded parking facilities are even possible. In New Jersey, for example, there is little opportunity and space to expand parking facilities, and hence beach use is often limited by parking facilities and/or their absence. In addition, one needs to account for the cost of congestion at beach sites and for travel times that such increased use would impose (will the increased number of beach users on the proposed wider beach result in the same use density or a higher use density? and do we expect benefits from beach users to be constant as density and congestion increase?). Another recommendation is to use a sensitivity analysis of the distribution of increased beach users over the season to account for seasonal differences (differences in weather factors as well as water temperatures will affect beach use patterns; do we expect that all seasons will be the same over a ten-year period, an assumption presently made in U.S. ACOE analyses?). A CBA should be based on a scenario of the proportion of seasons with average, above-average, and below-average weather conditions.

CBAs of proposed shoreline-stabilization projects should take the following overall approach. CBA should be based on the present value of the expected value of: (1) costs, (2) benefits, (3) net benefits over a range of significant coastal storm conditions (for example, severe [>5 in 10 years], mild/average [3 in 10 years], light [1 in 10 years], based on past history when appropriate) and (4) net benefits should also be estimated over a range of probabilities of project failure, and pertain to a range of estimated seasonal conditions that would affect the benefit estimates over the proposed project period. In addition researchers should strive to account for the effects of erosion rates and spatial differences in these rates in CBAs and project analyses. Other decision-theoretic approaches should be explored and refined to better account for risk and uncertainty relative to CBA. Over the past three decades, for example, 5 storms of a 1-in-5-year recurrence interval or greater occurred during the 1961–1970 decade, 6 during the 1971–1980 decade, and 4 during the 1981–1990 decade. From the data for 1990–2000 (11 storms of comparable magnitude), the 1990s appeared to be a decade of much higher storm activity.

Expenditures and Impacts of Tourism on Coastal New Jersey

Studies regarding the effects of spending by tourists in shore communities are useful, because statistics on business sales and activity are difficult, if not impossible, to disaggregate into sales for shore and nonshore communities. These data are limited in that not all business sales and activity are measured: nontourist expenditures remain unknown. One also needs to be aware of the distinction between actual expenses and economic impact effects, and whether expenses represent direct or indirect sales.

A number of studies conducted for New Jersey's Division of Travel and Tourism have examined the tourism sector, tourism spending, and the economic impacts of tourism (R.L. Associates 1987, 1988; Opinion Research Corporation 1989; Longwoods International 1992, 1994a, 1995). These studies, which appeared after the *NJSPMP* was completed, have generated a great deal of interest from the public concerning the tourism sector, especially regarding the New Jersey coastal area, while coastal researchers have been rediscovering the importance of coastal tourism and are trying to establish some link between spending on it and spending on shore stabilization.

The first study funded by the state of New Jersey to determine the effects of tourism on coastal New Jersey was conducted in 1987 by R.L. Associates (1987). A random sample of all households located in non-coastal counties in New Jersey, and in local areas in New York, Delaware, Pennsylvania, Ohio, and Maryland, were interviewed via telephone. The same approach was used in R. L. Associates' 1988 study of coastal Jersey. Opinion Research Corporation (1989) also conducted a telephone interview of a random sample of households in the same areas, but used a different random sampling design from that of R.L. Associates. This sampling design, in turn, affected how the sample data were projected (that is, weighted) to generate state estimates of tourism expenditures. Because of these differences (other differences exist in relation to Longwoods' approach as well), one cannot compare the projected estimates across the studies; they can only be used as point estimates of tourism activity. A further problem concerning these three studies involves their research design. In any study of behavior where participation is a key factor, such as in recreational activity and in travel and tourism, the research design should include a segmented sample (that is, a two-part sample) like that used in the travel-tourism study conducted by Bell and Leeworthy (1986) in Florida, or estimates will be subject to sample selection bias. The first part serves to determine the participation rate, used as the projection or weighting factor; the second part obtains the sample data on use characteristics and expenditures.

Because of the problem of different designs across all three studies, Opinion Research Corporation in its 1989 study reprojected (reweighed) the sample data from the two previous R.L. Associates studies using techniques that were similar to those used in its 1989 survey study so that comparisons could reasonably be made across all three studies. Its results indicated that an estimated $6.2 billion was spent by tourists who traveled to the coastal region of New Jersey in 1987, an estimated $5.4 billion in 1988, and an estimated $7.4 billion in 1989 (see appendix B and Opinion Research Corporation 1989). As will be seen, the estimates from the Longwoods studies for their Barrier Island component are significantly smaller than these estimates, which is another reason why the studies cannot be directly compared.

The series of studies of the travel and tourism industry conducted by Longwoods International for the state of New Jersey began in 1991 (Longwoods International 1992, 1994a,b, 1995). The Longwoods studies used an entirely different design from that of most studies of tourism; researchers conducted a two-part survey, one of establishments and one of tourists. The first part was used to collect data on lodging expenditures from establishments in order to increase accuracy and to avoid recall error. The second part was used to collect sample data on tourism expenditures so as to determine the proportion of travel and tourism expenditures associated with specific types of accommodations used (including hotels, campgrounds, state parks, friends or relatives, day trips, and pass-throughs). Once the accommodation expenses were projected (from the first part of the survey), the remaining expenditure categories were derived on the basis of the proportion of all expenses they accounted for. For example, if hotel expenses were projected at $15 million, and if hotel expenses represented 38% of the total expenditure of tourists who stayed at hotels, projected total expenditures would be $39.47 million ($15 million/0.38). By knowing the proportion that the remaining expense categories represented of the total, estimates for these categories could be derived. If the category for food/restaurants represented 25% of total expenditures, its projected estimate would be $9.87 million ($39.47*0.25), and so on.

The Longwoods reports derived the impacts of tourism expenditures on the New Jersey travel and tourism industry by isolating and incorporating expenditures of tourists within the coastal fringe of the Jersey shore into a formal I/O model of the state of New Jersey (Longwoods International 1994b). It seems likely that this information was added to that used to represent a baseline I/O situation, comparable to measuring the impact of a proposed change or effect; that is, a baseline version of an I/O model is "run" and impact estimates derived, new information is added (tourists' expenditures), the model is "rerun," and changes in the impact estimates are obtained. This method involves some double counting, however, because the direct expenditures have already been accounted for in the baseline I/O model as reported by industry sector and have been grouped together with direct expenditures from residents. This results in overstated measures in comparison with the true measure of aggregate economic activity as represented by GDP for the coastal region. As a result some caution is advised in interpreting the expenditure impacts, and the estimates of direct expenditures are preferable to the estimates of expenditure impacts of tourism (the Longwoods reports develop both estimates).

Another difficulty with the Longwoods studies is that the projected estimates are on a county-level basis. County-level data cannot represent specific areas such as a coastal zone, a narrow area in close proximity to the coast. In order to isolate such an area from county-level data, one would need to know the distribution of retail establishments and the distribution of economic sales (expenditures) on a location basis within the entire county, that is, on a municipality basis; such information is either not available or not readily available. This method would still be subject to error. As a result of this difficulty, Longwoods included a separate survey component within its overall effort to isolate tourism spending activity in the coastal New Jersey area, that is, a Barrier Island component. Statements about the effect of tourism on coastal New Jersey can then be made, but only in reference to this Barrier Island component.

Longwoods International estimated that travel and tourism expenditures in the state of New Jersey represented $18.28 billion in 1990 ($18.83 billion in 1992 dollars), $17.84 billion in 1991 ($18.37 billion in 1992 dollars), $18.6 billion in 1992, $18.91 billion in 1993 ($18.36 billion in 1992 dollars), and $22.65 billion in 1994 ($21.44 billion in 1992 dollars) (table 8.2). Keep in mind that these estimates represent state totals. Concerning the coastal counties of Atlantic, Cape May, Monmouth, and Ocean, estimated totals were $9.1 billion in 1990 ($9.4 billion in 1992 dollars), $8.9 billion in 1991 ($9.1 billion in 1992 dollars), $9.6 billion in 1992, $9.7 billion in 1993 ($9.4 billion in 1992 dollars), and $12.56 billion in 1994 ($11.89 billion in 1992 dollars) or roughly half of the respective state estimates. If one treats gambling activity as being independent of the coastal area (that is, not dependent on coastal location), then the four coastal

Table 8.2 ∼ ESTIMATED EXPENDITURES OF TRAVEL AND TOURISM IN NEW JERSEY, 1990–1994

Type	Year	County	Cost	Cost92	Est. visitors
County level	1990	Atlantic	6220.26	6407.51	N.A.
County level	1990	Cape May	1458.93	1502.85	N.A.
County level	1990	Monmouth	758.56	781.39	N.A.
County level	1990	Ocean	677.76	698.16	N.A.
County level	1991	Atlantic	5910.94	6088.88	4115.3
County level	1991	Cape May	1510.71	1556.19	2583.9
County level	1991	Monmouth	753.56	776.24	1990.3
County level	1991	Ocean	692.66	713.51	1990.3
County level	1992	Atlantic	6704.51	6509.64	N.A.
County level	1992	Cape May	1094.00	1062.20	N.A.
County level	1992	Monmouth	1262.67	1225.97	N.A.
County level	1992	Ocean	892.53	866.59	N.A.
County level	1993	Atlantic	6453.49	6265.91	4427.9
County level	1993	Cape May	1390.23	1349.82	5348.9
County level	1993	Monmouth	1055.93	1025.24	2107.9
County level	1993	Ocean	769.02	746.67	2107.9
County level	1994	Atlantic	6865.98	6499.98	3767.4
County level	1994	Cape May	2527.51	2392.78	5461.3
County level	1994	Monmouth	1683.21	1593.48	2356.2
County level	1994	Ocean	1484.84	1405.69	2356.2
Barrier Is.–Jersey Shore	1992	Atlantic	30.75	29.86	117.80
Barrier Is.–Jersey Shore	1992	Cape May	509.24	494.44	2488.70
Barrier Is.–Jersey Shore	1992	Monmouth	28.91	28.06	90.9
Barrier Is.–Jersey Shore	1992	Ocean	193.78	188.14	579.9
Barrier Is.–Jersey Shore	1993	Atlantic	32.62	31.67	110.7
Barrier Is.–Jersey Shore	1993	Cape May	605.01	587.42	2954.1
Barrier Is.–Jersey Shore	1993	Monmouth	29.86	29.00	94.0
Barrier Is.–Jersey Shore	1993	Ocean	207.43	201.40	621.4
Barrier Is.–Jersey Shore	1994	Atlantic	44.78	45.22	117.5
Barrier Is.–Jersey Shore	1994	Cape May	554.58	527.85	2881.0
Barrier Is.–Jersey Shore	1994	Monmouth	39.15	39.89	140.2
Barrier Is.–Jersey Shore	1994	Ocean	178.78	172.09	781.7
Totals					
County level	1990		9115.51	9389.91	N.A.
County level	1991		8867.87	9134.82	8689.5
County level	1992		9953.71	9664.40	N.A.
County level	1993		9668.67	9387.64	11,884.7
County level	1994		12,561.54	11,891.93	11,584.9

(*continued*)

Table 8.2 ～ (*continued*)

Type	Year	County	Cost	Cost92	Est. visitors
Barrier Is.–Jersey Shore	1992		762.67	740.50	3277.3
Barrier Is.–Jersey Shore	1993		874.93	849.49	3780.2
Barrier Is.–Jersey Shore	1994		817.29	773.71	3920.4

SOURCE: 1990–91, Longwoods International (1992); 1992–93, Longwoods International (1994a); 1994, Longwoods International (1995).

NOTE: Cost is in millions of current dollars associated with the year of the study and refers to the projected expenditures on travel and tourism; Cost92 is in millions of 1992 dollars adjusted by the relevant consumer price index; Est. visitors refers to the estimated number of visitors in thousands. For the number of visitors, estimates were available only on a regional basis, and hence Ocean and Monmouth counties are considered as the Shore region (see Longwoods studies for details). *N.A.* = not available.

county totals excluding expenditures on gambling would come to $6.5 billion in 1990, $6.4 billion in 1991, $6.8 billion in 1992, $6.5 billion in 1993, and $9.3 billion in 1994 (table 8.3). These figures represent county-level data, which are not appropriate to represent coastal tourism activity, as mentioned above. Also, note that these estimates are within the same range as those produced by the earlier R.L. Associates and Opinion Research Corporation studies; hence the earlier research may represent county-level totals and therefore may also represent inflated estimates of coastal tourism economic activity.

Estimates developed for the barrier island, however, represent only one component of beach travel and tourism activity (that portion of tourists who rented accommodations along the shore communities) and underestimate the level of travel and tourism activity associated with beach travel (other components of beach travel consist of other overnight trips and day trips). In 1992, the first year data were collected, an estimated $740.5 million was spent by tourists and travelers who stayed at barrier island rental units, $874.9 million in 1993 ($849.5 million in 1992 dollars), and $817.3 million in 1994 ($773.3 million in 1992 dollars) (table 8.3). In 1994, the Barrier Island component represented 6.8% of total tourism expenditures of the four coastal counties, and 3.6% of the state tourism expenditure total. Expressed in terms of a three-year average (1992–1994), tourism expenditures of the Barrier Island component accounted for an estimated $787.9 million a year in 1992 dollars or 7.6% of a similar three-year average of the tourism expenditure total for the four coastal counties ($10,314.66 million/year) and 4.1% of the 1992–1994 average of the state tourism expenditure total ($19,289.24 million/year).

Table 8.3 ~ ESTIMATED EXPENDITURES OF TRAVEL AND TOURISM IN NEW JERSEY WITHOUT GAMBLING EXPENDITURES, 1990–1994

Type	Year	County	Cost	Cost92	Est. visitors
County level	1990	Atlantic	3622.23	3731.27	N.A.
County level	1990	Cape May	1458.93	1502.85	N.A.
County level	1990	Monmouth	758.56	781.39	N.A.
County level	1990	Ocean	677.76	698.16	N.A.
County level	1991	Atlantic	3421.15	3524.14	4115.3
County level	1991	Cape May	1510.71	1556.19	2583.9
County level	1991	Monmouth	753.56	776.24	1990.3
County level	1991	Ocean	692.66	713.51	1990.3
County level	1992	Atlantic	3574.75	3470.85	N.A.
County level	1992	Cape May	1094.00	1062.20	N.A.
County level	1992	Monmouth	1262.67	1225.97	N.A.
County level	1992	Ocean	892.53	866.59	N.A.
County level	1993	Atlantic	3285.75	3190.25	4427.9
County level	1993	Cape May	1390.23	1349.82	5348.9
County level	1993	Monmouth	1055.93	1025.24	2107.9
County level	1993	Ocean	769.02	746.67	2107.9
County level	1994	Atlantic	3663.14	3467.87	3767.4
County level	1994	Cape May	2527.51	2392.78	5461.3
County level	1994	Monmouth	1683.21	1593.48	2356.2
County level	1994	Ocean	1484.84	1405.69	2356.2
Barrier Is.–Jersey Shore	1992	Atlantic	29.80	28.93	117.80
Barrier Is.–Jersey Shore	1992	Cape May	509.24	494.44	2488.70
Barrier Is.–Jersey Shore	1992	Monmouth	28.91	28.06	90.9
Barrier Is.–Jersey Shore	1992	Ocean	193.78	188.14	579.9
Barrier Is.–Jersey Shore	1993	Atlantic	31.62	30.70	110.7
Barrier Is.–Jersey Shore	1993	Cape May	605.01	587.42	2954.1
Barrier Is.–Jersey Shore	1993	Monmouth	29.86	29.00	94.0
Barrier Is.–Jersey Shore	1993	Ocean	207.43	201.40	621.4
Barrier Is.–Jersey Shore	1994	Atlantic	34.56	32.72	117.5
Barrier Is.–Jersey Shore	1994	Cape May	554.58	527.85	2881.0
Barrier Is.–Jersey Shore	1994	Monmouth	39.15	39.89	140.2
Barrier Is.–Jersey Shore	1994	Ocean	178.78	172.09	781.7
Totals					
County level	1990		6517.48	6713.67	N.A.
County level	1991		6378.08	6570.08	8689.5
County level	1992		6823.95	6625.61	N.A.
County level	1993		6500.93	6311.98	11,884.7
County level	1994		9358.70	8859.82	11,584.9

(*continued*)

Table 8.3 ~ (*continued*)

Type	Year	County	Cost	Cost92	Est. visitors
Barrier Is.–Jersey Shore	1992		761.72	739.58	3277.3
Barrier Is.–Jersey Shore	1993		873.92	848.51	3780.2
Barrier Is.–Jersey Shore	1994		807.07	764.04	3920.4

Source: 1990–91, Longwoods International (1992); 1992–93, Longwoods International (1994a); 1994, Longwoods International (1995).

NOTE: Cost is in millions of current dollars associated with the year of the study and refers to the projected expenditures on travel and tourism; Cost92 is in millions of 1992 dollars adjusted by the relevant consumer price index; Est. visitors refers to the estimated number of visitors in thousands. For the number of visitors, estimates were available only on a regional basis, and hence Ocean and Monmouth counties are considered as the Shore region (see Longwoods studies for details). *N.A.* = not available.

Excluding gambling expenses, the three-year average for 1992–1994 for the Barrier Island component accounted for an estimated $786.9 million a year in 1992 dollars or 10.8% of the three-year average of the four coastal county tourism expenditure ($7,265.8 million/year) and 4.8% of the 1992–1994 average of state tourism spending ($16,392.93 million/year).

The usefulness of the Longwoods studies lies in their generation of projected *direct expenditures* and not their generation of economic *impact measures*. Direct expenditures represent the closest activity to aggregate GNP estimates, because they represent final goods and services sold and do not contain double counting. The Longwoods studies present only one aspect of beach travel in its Barrier Island component, and *underestimate* the importance and magnitude of tourism expenditure activity in the coastal region of New Jersey. To develop an estimate of expenditures associated with beach travel, similar estimates for day trips and overnight trips other than barrier island rentals for the four coastal counties are necessary. To gain some idea of the magnitude of beach-related expenses, an upper-bound estimate based on all three components of beach travel (barrier island rentals, other overnight travel, and day-trip travel) was developed. However, a word of caution. The estimates were developed for illustrative purposes rather than as a reliable point estimate. They are based on two separate Longwoods survey studies, and hence two different sampling bases, and there is some error from double counting (from overlap of the two different sampling bases). As a result, the estimated travel expense probably overstates beach-related travel expenses.

The Longwoods study for the 1993 season (Longwoods International 1994a) was the only season for which the New Jersey Division of Travel and Tourism supplied complete information to us (that is, all reports produced by Longwoods International for a particular year). The discussion that follows is based on the derivation in table 8.4. One component of the tourist survey conducted by Longwoods International (1994b) found that 12% of all overnight trips to New Jersey were beach trips, and 4% of all day trips were for beach trips; this allowed a derivation for beach trips of 7.62 million trips in total in 1993 (steps 1 and 2, table 8.4). An average trip expense was derived from projected total expenses and an estimate of the total number of trips by trip type (barrier island, other overnight, day) (step 3, table 8.4). On the basis of the estimated number of trips and the estimated average trip expense, an estimate for expenditures of all beach-related travel was developed at $2,095.88 million with gambling and $1,917.92 million without gambling (table 8.3). The Barrier Island component represented 41.74% ($874.92 million/ $2095.878 million) of the 1993 estimated tourist expenditures. If this proportion is representative across other years, the three-year (1992–1994) estimated average expense for beach trips would account for an estimated $1,887.64 million ($787.9 million/0.4174); similar estimates of tourism spending without gambling are 45.57% ($873.92 million/$1917.91 million) and $1,726.75 million.

The estimates illustrate the point that projected tourism expenses associated with beach trips based on the Barrier Island component are underestimates of such activity, whereas the county-level estimates of the four coastal counties are overestimates. The derived estimate, $1,887.64 million per year over the 1992–1994 period, represents 18% of the four coastal county three-year average, and 9.8% of the state three-year average (without gambling expenses, the comparable estimate is $1,726.75 million/year representing 23.8% of the four coastal county three-year average, and 10.5% of the three-year state average), illustrating the problem with conclusions that the majority of the state's travel and tourism industry is generated from shore-related expenditures. Efforts should be continued to develop expenses for all beach trips in the form of a range of estimates and not as a single-point estimate in future studies.

Shoreline Stabilization and Management Policy-Oriented Studies

This section discusses two studies, the *NJSPMP*, and a study conducted by ICF, Incorporated, regarding assessments of alternative shoreline-stabilization policies. Also included are a recent overall assessment of U.S. Army Corps of Engineers projects (self-study by the

Table 8.4 ~ DERIVATION OF ESTIMATED EXPENDITURES OF BEACH TRIPS IN NEW JERSEY IN 1993

1. Step 1	Trip type	No.	%
	Overnight	20.0M	13%
	Day	130.5M	87%
	Total	150.5M	100%

2. Step 2	Trip purpose	No.[a]	%
	Beach-overnight	2.4M	12%
	Beach-day	5.22M	4%
	Total-beach	7.62M	

a. Estimated based on % of trip purpose (from 2) times the No. in (1).

3. Step 3: Derivation of average expense (1993$)

Category	Overnight	Day	Barrier Is.
Total expense	$10,924.8M	$6,461.91M	$874.922M
Total trips	20.0M	130.5M	$637,991.56[b]
Average expense ($/trip)	$546.241	$49.517	$1371.368
Without gambling			
Total expense	$9,862.5M	$4,377.75M	$873.915M
Avg. exp. ($/trip)	$493.123	$33.546	$1369.792

b. 3,780,100 persons/(5.925 persons/trip) = 637,991.56 trips.

4. Step 4: Derivation of number of beach trips

Trip type	No.
Total-overnight	2.4M
less Barrier Is.	−0.638M
Other overnight[c]	1.762M

c. Overnight trips other than Barrier Islands.

5. Step 5: Estimated tourist beach expenses (1993$)

Barrier Is. trips	0.638M	@ $1371.368 = $874.922M
Other overnight	1.762M	@ $546.241 = $962.476M
Day trips	5.22M	@ $49.517 = $258.479M
Estimated 1993 beach-related expenses		$2,095.877M

Without gambling

Barrier Is. trips	0.638M	@ $1369.792 = $873.915M
Other overnight	1.762M	@ $493.123 = $868.883M
Day trips	5.22M	@ $33.546 = $175.11M
Estimated 1993 beach-related expenses		$1,917.91M

SOURCES: Expenses and Barrier Islands, Longwoods International (1994a); trip data, Longwoods International (1994b).
NOTE: M = millions.

U.S. ACOE) and an evaluation of national beach nourishment projects by the National Research Council.

The *NJSPMP*, prepared by the consulting engineering firm Dames and Moore, evaluated several alternative shoreline-stabilization plans for New Jersey using a cost-benefit analysis (NJDEP 1981). The alternative plans in the *NJSPMP* were classified as follows: (1) a Storm Erosion Protection alternative (nourishment equivalent to a 75-foot-wide berm with groins or a 100-foot-wide berm without groins); (2) a Recreation Development alternative (a berm width and beach width that would vary based on estimates of future recreational demand for beach use so as to provide a maximum of 100 square feet/person; either an increase or decrease in both berm and beach width compared with the Storm Erosion Protection alternative); (3) a Combination alternative (the maximum berm and beach width from the first two alternatives); (4) a Limited Restoration alternative (stabilization via nonstructural methods that would be greater than the level of protection from a Maintenance Program but smaller than the protection levels of the above three alternative plans); and (5) a Maintenance Program alternative (yielding the smallest level of protection of all alternatives by repairing and maintaining existing physical structures in place, and providing nourishment on an as-needed basis; a reaction effort rather than a preventative effort).

Parameters used in the CBA were a 50-year planning horizon (whereby the researchers developed a time plan for each alternative program if each were to be carried out over a 50-year period, including the maintenance, repair, and construction of any new hard structures needed and any periodic nourishment to maintain and/or increase beach width), and a discount rate of 9% (an interest rate used to express future value measures in present value or discounted terms). It should be noted that only in the case of the Recreation Development alternative was the beach and berm width estimated to increase with estimated recreation demand over time. In all other alternative plans, beach and berm width was essentially held fixed over time (that is, width was "stabilized" or controlled for processes of natural [long-term] erosion and storm [short-term] erosion).

Cost elements consisted of estimated engineering costs (those costs necessary to implement each alternative plan) plus estimated public service costs (estimated costs for increased infrastructure capacity from future estimates of the demand for beach use associated with each alternative plan). Engineering cost estimates were developed over a 50-year planning period based on projections of levels of engineering and labor effort needed to achieve each plan. Public service cost estimates were based on the product of the projected number of future beach users (projected demand) and an average cost of infrastructure use estimated at $1/beach user.

Benefit elements consisted of estimated recreational benefits (estimated benefits from the recreational use of the beach associated with each of the alternative plans) and property protection benefits (estimated benefits from protection of property associated with each of the alternative plans), both direct benefits attributable to beach stabilization. Recreational benefits were estimated from the product of the projected number of future beach users, or the future demand for beach use, and the opportunity cost of beach use estimated at $2/day per user (an average beach fee at the time). Property protection benefits were based on the estimated value of losses that would have occurred without the plan in place for each of the alternative plans (the difference between losses without the plan less losses with the plan for each of the alternative plans).

All estimated costs and benefits were then discounted over the planning period, summed, and a ratio of the discounted sum of benefits to that of costs (a B/C ratio), was calculated for each plan on a reach-by-reach basis.

Because the *NJSPMP* represents the basis from which all future shoreline-stabilization plans are developed, and because its emphasis was on the CBA of alternative plans, discussion of several possible limitations of the study is warranted. The research team were mostly engineers, and therefore it is not surprising that much emphasis was placed on the cost estimates at the expense of the benefit estimates. Limitations of the engineering cost estimates, if any (because the exact costs were not given over all 50 years prior to discounting in the *NJSPMP*), would exist if consideration was not given to changes in prices and costs over time due to inflation and/or deflation. It is reasonable to expect that over 50 years, the costs of fuel to operate machinery would fluctuate and increase along with labor costs, and so on. If these effects were not accounted for, the engineering cost estimates could be greatly understated.

Public service cost estimates could be limited for two basic reasons, one associated with the projected future recreational use estimates, the other with the infrastructure use estimate of $1/user. Here the rationale is that beach users use available public services to travel to and from the beach as well as at the beach, and from using these services they derive benefits. However, many users may not be residents of the beach community, and it is residents who pay for these local public services via property taxes. Thus these costs must be estimated. Limitations of the estimates follow. First, forecasts of any variable(s) into future time periods are very sensitive to the specific model and variables used. Second, for any forecast some sensitivity analysis or a range of estimates associated with the errors of forecast or levels of confidence of the estimates is necessary to provide perspective on the variability and reasonableness of the forecasts. Third, over time many variables regarding the public service cost estimate can change; the population and hence the number of future beach users, property values, and property tax rates can increase, and the cost of providing public services or infrastructure generally increases over time. None of these points, however, were discussed by the investigators in the *NJSPMP*. Moreover, the use of a constant figure of $1/user over all 50 years is not reasonable and seriously questions the validity of the estimates for public service costs. Again, the use of a constant figure over time will tend to understate these cost estimates, and misrepresent these costs.

More important, if both engineering costs and public service costs were underestimated and misrepresented, then any ratio of benefits to costs would favor the benefit side. Consider a simple ratio, a/b; to increase the value of the ratio one can either increase the numerator a (which represents benefits here), or decrease the denominator b (which represents costs here).

Now consider the benefit measures. The measure of recreational benefits was derived from the product of an estimated opportunity cost of beach use ($2/day per user) and an estimate of future use. As with public service costs, there are two components of this estimate, and hence two basic areas for limitations. The first is due to the estimates of future beach use, where the previous discussion concerning forecasts applies. The second basic reason concerns the use of opportunity costs as a benefit measure, the use of a constant figure across all beaches and stabilization projects in the state, and the use of a figure that does not change over time. The researchers ignore these issues. Opportunity costs are not an appropriate measure of economic welfare as measured by the area under a demand curve above price, and do not represent economic benefits, although they can represent a partial component of benefits. The use of a constant value across different beaches and stabilization projects means that either the researchers or the public perceive no differences between beaches located at Cape May and those at Asbury Park. Thus benefits from beach use do not vary across beaches or projects for the same reasons. From a research perspective, in order to compare one alternative to another, ideally one would like to have some variation in the variables across projects, for example, variation in the marginal benefits. Because there are physical differences across beaches along the New Jersey shoreline, and everyone has different tastes and preferences, one would expect that people will choose to go to specific beaches. In turn, differences in these preferences should be reflected in values placed on benefits from beach use as found in the Lindsay and Tupper (1989) study. Last, treatment of benefit estimates across time will tend to understate benefit estimates and, in turn, decrease the magnitude of a cost-benefit ratio. The researchers did not discuss any of these points.

Finally, consider the estimate of property protection benefits. This was derived from estimates of property value, not from actual property value or assessed property value available from tax assessor offices. The researchers estimated these benefits by first identifying what property would be lost or destroyed over a 50-year period if the stabilization plan were not implemented for each alternative plan. Then benefits were estimated from a product of the number of property structures that would be lost or destroyed by general type (business versus residences) estimated over the 50-year period and an average value for the specific type of property structure. This estimate depends on two parts, one part involving a forecast and the other involving the use of an estimate of the average value for the type of property structure. Limitations concern both of these parts. First, the part based on forecasted or future estimates can pose problems. The earlier discussion regarding difficulties of forecasts is appropriate, suggesting that the researchers should have provided some indication of the size of error or variability of the forecasted property lost, a level of confidence associated with the forecasts, and/or some type of sensitivity analysis. Second, the use of estimates of average value rather than actual values can introduce biases into these benefit estimates, since values differ across property structures and property.

In sum, the CBA performed in the *NJSPMP* is basically static, although some attempt was made to incorporate changes that occur over time, namely estimates of future beach use and estimates of future property lost or damaged. No attempt was made to incorporate any other dynamic elements or the risk associated with the expected outcome of the projects, where one could introduce uncertainty into the derivation of net benefits (benefits less costs). A dynamic analysis would compare and contrast the monetary value of a project's outcome if completely certain versus its outcome with the presence of uncertainty. In the case of beach stabilization, possible risk factors could involve effects such as erosion and storm damage, which could cause any project to fall short of 100% completion and effectiveness; uncertainty over available funds, which could threaten 100% completion of any project over the planning period; uncertainty over the estimated number of future beach users; and uncertainty over the value of estimated future property structures lost versus protected. In addition, the effect of sea-level rise in the future would increase the risk and magnitude of erosion and storm damage. Probably the most serious fault is the problem of downward bias in both the cost and the benefit estimates, which would tend to introduce either an upward bias or a downward bias in the magnitude of the B/C ratio, respectively, distorting it. The net effect is ambiguous, but does raise concerns regarding the validity and accuracy of the CBA in the *NJSPMP*.

The only study that examined policy options for areas within the coastal floodplain that our search located was conducted by ICF, Incorporated (1989) for the New England/New York Coastal Zone Task Force. This study of coastal floodplain management sought to determine: (1) the costs and revenues associated with governmental entities as a result of development in the coastal floodplain; (2) the costs and revenues of various policies targeted at coastal erosion and storm damage in the coastal floodplain; and (3) how these costs and revenues depended on sea-level rise. Regarding (1), revenues consisted of revenues from coastal development and from tourism and recreation; costs were due to damages from erosion and storm events and from stabilization efforts. Specifically, revenues consisted of the sum of property taxes, income taxes, beach-use fees, sales taxes, tolls, utility charges, accommodations taxes, and flood insurance premiums. Costs consisted of maintenance costs of local government services and infrastructure, maintenance costs of stabilization structures, fire protection, flood insurance claims, storm cleanup charges, costs of beach nourishment and dike building, and property acquisition.

Four policy options were evaluated: (1) no response; (2) beach nourishment only (to maintain the beach width, but not for protection of public and residential structures in close proximity to the beach); (3) dikes only (that is, revetments or seawalls; the report is not clear about this); and (4) property acquisition (when 50% or more of the value of a structure was damaged).

Two locations in New Jersey, Ocean City and Strathmere, were examined as case studies. Results of the study revealed: revenues to government entities from storm event damage and protection were estimated at $114.4 million for Ocean City and $9.9 million for Strathmere in 1987; costs due to damage and protective efforts were estimated at $130.6 million for Ocean City and at $3.2 million for Strathmere in 1987. The remainder of the study examined the effects of sea-level rise on the combined revenues lost and added costs for the above policy options at specific points in time, or "snapshots" (the years 2025 and 2050). Tables 8.5 and 8.6 contain these results for Ocean City and for Strathmere, respectively. ICF concluded that (1) the no-response option realized large losses and added costs in both "*storm*" and "*nonstorm*" conditions over time; (2) the option of beach nourishment prevented substantial losses and added costs in "*nonstorm*" conditions, but incurred large losses and costs in "*storm*" conditions because of erosion and overwash; (3) the dike project option incurred large costs and losses in "*nonstorm*" conditions reflecting one-time construction costs, and large losses and costs in "*storm*" conditions over time because of deterioration of dikes over time (their assumed lifespan appeared to be 25 years); and (4) the policy of property acquisition resulted in losses and costs on an order of magnitude higher in "*storm*" conditions, reflecting one-time property costs, although over time revenue losses and added costs were less than options (1), (2), and (3) under "*storm*" conditions. (It would be useful to know more about the effects of cumulative losses and added costs over time in these comparisons rather than the static approach taken here.) Similar remarks apply to the Strathmere application.

The overall study included four policy findings: (1) "new" development in coastal floodplains was found to be a net cost to governments, while "existing" development in many cases was worth protecting; (2) the "best" policy response was found to depend the existing level of development, the costs from damage, and the magnitude of revenues gained, and (2.1) in areas that were relatively less developed, beach nourishment was found to be a viable policy, while (2.2) in areas with high levels of development, protection via dikes was found to be a viable policy where large amounts of property were at risk of damage *and* where dike building could be coupled with a policy of halting further development; (3) optimal policies differed over time; and (4) the use of subsidies (for example, from NFIP) was found to have important consequences on the promotion of development. ICF policy recommendations are summarized below in the Policy Overview.

The ICF study's main limitation is that it contains only snapshot views, distinct years rather than cumulative effects over time; hence it is

Table 8.5 ~ COMBINED REVENUE LOSS AND INCREASED COSTS RELATIVE TO 1987 FOR OCEAN CITY, N.J. (1987 DOLLARS)

Year/ Sea-level rise	No response		Beach nourishment[a]		Dike[b]		Property acquisition[c]	
	Nonstorm	Storm	Nonstorm	Storm	Nonstorm	Storm	Nonstorm	Storm
2025								
Linear	18,498,578	87,841,610	1,108,412	88,286,990	41,284,079	299,051	18,498,578	667,338,706
Mid-low	23,045,581	93,358,730	2,463,138	97,053,793	42,062,207	299,051	23,045,581	772,692,035
Mid-high	26,060,507	111,688,471	3,263,657	122,239,897	45,077,133	299,051	26,060,507	838,690,128
2050								
Linear	22,094,151	99,328,566	1,108,412	101,344,744	22,094,187	99,328,566	25,947,729	44,340,276
Mid-low	28,547,967	115,419,783	2,929,374	128,788,189	28,547,967	115,419,783	31,217,619	53,758,775
Mid-high	31,746,387	155,047,215	3,958,614	181,381,048	31,746,215	155,746,215	33,679,370	63,526,340

SOURCE: ICF Incorporated (1989).

a. Beach nourishment is undertaken once every five years. Hence, to make the cost for nonstorm beach nourishment comparable to other yearly costs or revenues, they should be divided by 5.

b. Dike costs are listed in the nonstorm column. These are onetime costs, so they should be compared with other onetime costs or annualized before comparing with ongoing yearly costs or revenues.

c. Property acquisition costs are listed in the storm-related column. These are onetime costs, so they should be compared with other onetime costs or annualized before comparing with ongoing yearly costs or revenues.

Table 8.6 ~ COMBINED REVENUE LOSS AND INCREASED COSTS RELATIVE TO 1987 FOR STRATHMERE, N.J. (1987 DOLLARS)

Year/ Sea-level rise	No response		Beach nourishment[a]		Dike[b]		Property acquisition[c]	
	Nonstorm	Storm	Nonstorm	Storm	Nonstorm	Storm	Nonstorm	Storm
2025								
Linear	7,180,158	3,398,056	1,243,079	3,467,383	26,196,783	166,100	7,180,158	15,880,655
Mid-low	7,244,709	3,051,422	2,762,397	3,467,383	26,261,334	166,100	7,244,709	14,382,069
Mid-high	7,379,172	2,435,740	3,660,176	3,674,045	26,395,798	166,100	7,379,172	10,741,090
2050								
Linear	7,229,703	3,209,883	1,234,079	3,467,383	7,229,703	3,209,883	7,250,019	251,039
Mid-low	7,547,735	2,077,930	2,762,397	3,674,045	7,547,735	2,077,930	7,559,744	222,899
Mid-high	7,735,464	1,389,266	3,660,176	4,061,535	7,735,464	1,389,266	7,742,383	210,577

SOURCE: ICF Incorporated (1989).

a. Beach nourishment is undertaken once every five years. Hence, to make costs for nonstorm beach nourishment comparable to other yearly costs or revenues, they should be divided by 5.

b. Dike costs are listed in the nonstorm column. These are onetime costs, so they should be compared with other onetime costs or annualized before comparing with ongoing yearly costs or revenues.

c. Property acquisition costs are listed in the storm-related column. These are onetime costs, so they should be compared with other onetime costs or annualized before comparing with ongoing yearly costs or revenues.

somewhat static, whereas the evaluation of tradeoffs among policies should be conducted in an intertemporal context. Other limitations concern the derivations of added costs and lost revenues. Regarding lost revenues in property, sales taxes, and income taxes, it is not clear whether the analysis included all private households and residences located in the coastal communities in the study or only those residences in close proximity to the beach. Inclusion of income tax as a revenue item only makes sense for year-round residents and not for summer residents with a second home; again it is not clear why income taxes were used and whom they pertain to in the study.

In general, the ICF study is to be commended for its treatment of many complex issues involved with the provision of shoreline stabilization and for its examination of the tradeoffs among policy options. Investigators planning future policy-oriented studies have much to gain here.

Two other studies have examined overall projects of the Army Corps of Engineers and beach nourishment in general. One is a self-study conducted by the U.S. ACOE (1994e); the other is an evaluation of national beach nourishment projects conducted by the National Research Council (1995).

The U.S. ACOE (1994e) self-study was the first phase of a two-part process that examined cost and beachfill comparisons of U.S. ACOE beach nourishment projects over the 1950–1993 period, a total of 56 projects. The study sought to determine how well the U.S. ACOE staff was able to estimate beach nourishment costs and the amounts of beachfill actually needed. Results indicated that total costs of these 56 projects were $670.26 million with $403.26 million as the federal cost share (in current dollars) or $1,489.5 million with $881 million the federal share in 1993 dollars (ibid.: xv; it is noted that these estimates from the Executive Summary differ from those cited in the text as $1,459.31 million with $850.71 million as the federal share in 1993 dollars, p. 64). The amount of beachfill deposited was 167 million cubic yards (the sum of fill from 39 of 49 beach restoration projects and 33 of 40 beach nourishment projects).

Comparisons of estimated versus actual costs and fill quantities were developed for 80% of the 56 projects. These comparisons indicated that actual costs were 4.4% less than the estimated project costs (that is, $1,340.9 million versus $1,403 million, actual versus estimated, respectively, in 1993 dollars). Actual quantities of fill were 5.4% greater than estimates of fill (158.4 million cu. yd. versus 167 million cu. yd., estimated versus actual). The second phase of the U.S. ACOE self-study (Cordes and Yezer 1995; Hillyer 1996) concentrated on the role of shoreline-stabilization spending and its effect on coastal tourism activity. Results found that shore-stabilization projects had little or no effect on development, but that subsidized flood insurance encouraged development. Limitations of the 1994 U.S. ACOE study pertain to the selection of studies analyzed, and to the technique used to convert current dollar measures to 1993 dollar measures (constant dollars), in the use of a nonconventional price index rather than a price index developed by the statistical branch of the federal government (for example, a wholesale price index such as the Producer Price Index), although this seems minor given the small deviation between actual and estimated costs and quantities.

The National Research Council study convened a panel of experts to examine and evaluate beach nourishment projects in the United States (NRC 1995). They concluded that beach nourishment was a viable stabilization option for coastal communities and contributed to tourism. However, these conclusions depended on projects' being well designed, well built, and provided in areas that experience relatively minor levels of erosion. In the past, many projects had failed, the panel found. It recommended that projects be monitored and evaluated on a periodic basis and that future U.S. ACOE analyses incorporate risk and uncertainty into the economic and cost-benefit analyses as well as use the latest economic approaches (for example, CV techniques based on referendum-type formats) (Bockstael 1995; NRC 1995).

Assessment of Literature Reviewed

The basic issue one would like to address is whether depositing sand on a beach generates tourism and/or economic benefits. One can think of the coastal zone as a kind of "economic engine" in the sense that the coastal zone generates economic activity, such as income, sales, and jobs via tourism and businesses that are water dependent and/or must be located in close proximity to the coastal area. The studies and discussed above attempt to address different components of the beachfill–economic activity question. However, because these investigations are based on different research and sampling designs and have different objectives, the data and results are too fragmented for one to develop reliable and consistent estimates of economic activity. This means that the data from the literature are inadequate to develop point estimates of the magnitude of the economic activity associated with the coastal zone. Furthermore, studies that have tried to estimate the level of activity from coastal tourism have tended to ignore the effect of beach nourishment on coastal tourism activity. Data from the coastal tourism studies cited above are inappropriate to address the issue of whether beach nourishment projects, on their own, generate economic activity. In order to examine the issue, investigators must develop studies that incorporate research designs to isolate economic activity dependent on the coastal zone and/or on specific beach nourishment projects. Such studies may require data on economic activity and tourism expenditures that are location specific, in terms of their relative proximity to the shoreline and to beach nourishment projects, and that are collected on a seasonal basis. Such data are sensitive and generally hard to collect. Without them, however, one may not be able to advance beyond the current level of analysis and findings.

Policy Overview

Much of what follows regarding policy comes from the ICF study (1989) of coastal flood zones. Policy recommendations were offered by ICF in two categories, future development and existing development. Concerning future development, ICF recommended that (1) continued large-scale development would be a net cost to governments (that is, costs would exceed revenues); (2) NFIP should tighten the availability of flood insurance to discourage future development (such action would have an effect similar to what occurs when property owners are charged the full costs of flood insurance); (3) policies should be implemented whereby property owners are charged the full costs of cleanup and repairs; (4) policies should be designed to prohibit reconstruction of structures, and land should be rezoned following significant storm damage (when 50% or more of the value of a structure is damaged); and (5) governments should establish future policies on shoreline stabilization and announce these to the public. The idea behind this last recommendation is that if governments precommit to a policy of no provision of shoreline stabilization in areas facing new development, this will create disincentives to future construction and cause property owners to internalize and bear the full costs of damage and cleanup.

Regarding existing development, ICF admits that the choice of policy is not easy (ICF 1989: 60). Recommended policy options were found to depend on development levels: in areas with high levels of development it was recommended that policies protect existing structures, whereas in areas with low levels of development policies of stabilization were not recommended. To forestall further development, property acquisition, rezoning, tightening of insurance, and having owners assume the full costs of damage and cleanup and accept losses of capital investment in buildings (which would also mean losses to the tax base for the municipality) were recommended.

Suitability of Decision-Theoretic Models

During our assessment of shoreline-stabilization efforts in New Jersey we identified a number of limitations concerning economic analysis and models that are confined to cost-benefit analysis and expected-CBA given uncertainty (U.S. ACOE 1991c; Bockstael 1995; NRC 1995; Ofiara and Psuty 2001). Although CBA is a useful tool for shoreline-stabilization decisions on an individual project basis, it may not be the best tool to use for decision making at the aggregate or state level. Here we focus on its limitations in the treatment of risk and uncertainty, particularly the timing and abandonment of shoreline-stabilization investment decisions. A main concern is the treatment of the variability of returns where future returns are specified as a range of returns rather than as single-point estimates. In these cases the expected-CBA criteria ignores the variability of returns.

In order to demonstrate the flexibility of decision-theoretic models from finance we begin by examining cost-benefit analysis. In CBA one calculates the net discounted benefit, that is, discounted benefits less discounted costs, and selects those projects that yield a positive net discounted benefit. In the presence of many similar projects, the rule is to select the project(s) that yields the largest net discounted benefit.

Consider the following net returns (total returns less costs) in table 8.7, measured as the value of property damage prevented by four hypothetical shoreline-stabilization projects over time (where returns are expressed in thousands of dollars). On the basis of the CBA decision rule and a 3% discount rate, one would select Project B, followed by C. This represents the current basic method and decision criteria that are used by the U.S. ACOE in selecting among various shoreline-stabilization projects (U.S. ACOE 1991c, 1989a,b, 1994a,b).

Before the notion of uncertainty can be introduced, two different contexts in which uncertainty is often used should be distinguished. One context is from finance theory and is used to describe how risky a project is. This context is related to the variability of net returns, and is usually measured by the variance or standard deviation of returns over time (Levy

Table 8.7 ～ NET RETURNS OF FOUR HYPOTHETICAL
SHORELINE-STABILIZATION PROJECTS

			Year				
Project	*0*	*1*	*2*	*3*	*4*	*5*	*NPV*
A	−2,000	500	500	500	500	500	290
B	−1,000	200	1,200	1,000	500	0	1685
C	−2,000	1,000	333	1,000	1,000	0	1088
D	−5,000	3,000	2,000	250	0	0	27

NOTE: NPV = net present value.

and Sarnat 1994; Brigham and Houston 1998). To understand the variability or risk for the projects in table 8.7, observe the annual net returns
(nondiscounted) after the initial investment for periods 1–5. Project A has
a constant, steady return over periods 1–5, while Projects B, C, and D
exhibit more fluctuations and, hence, more deviations from a mean or
expected value over time, which is the definition of the variance (that is,
$E[X_i - \mu]^2$, where X_i refers to the individual values, μ the expected value
or mean, and $E[.]$ the expected value operator, and estimated from $[\sum_{i=1}^{n}$
$[X_i - \mu]^2]/[n - 1])$.

The mean and variance (sample variance) of net returns (nondiscounted) are calculated for each project in table 8.8. One can see that
Project A has the lowest variance and is the least risky, whereas Project D
has the highest variance and is the most risky. In general, projects with
fairly equal, steady returns over time, such as A and C, will be among the
least risky and have the lowest variance. Here, based on the principle of
the mean-variance rule, most would accept Project C overall, a project
with the second-highest net present value (NPV) of returns, in contrast to
the earlier decision to accept Project B using the CBA rule alone.

The other context of uncertainty related to investment projects is that
returns are uncertain in the real world, and the best one can do is to project a range of expected future returns based on some degree of probability
that they would be realized. In the above examples, it was assumed that the
net returns were certain, and hence only a single value for returns was used.
When net returns or net benefits are uncertain and are based on some
degree of probability that they will occur in the future, the method and
decision rule now become the expected-CBA method (or expected net present value rule). On the basis of this method, one would compute the
expected value of the return less the costs (where costs may also be expressed
as an expected value if future costs are also uncertain), discount the net

Table 8.8 ~ MEAN AND VARIABILITY OF NET RETURNS OF FOUR
HYPOTHETICAL SHORELINE-STABILIZATION PROJECTS

	Project A	*Project B*	*Project C*	*Project D*
Mean (μ)	500	580	666.60	1,050
Variance (σ^2)	0	262,000	222,278	1,887,499
Std. Dev. (σ)	0	511.86	471.5	1,373.9

expected returns, and sum over all periods. In general, the expected value of
net returns is equal to the product of the net return and its probability,
summed over all such returns in a given period, that is, $E(X_i) = \mu = X_1\pi_1$
$+ X_2\pi_2 + \ldots$ or $E(X_i) = \Sigma^n_{i=1} X_i\pi_i$, where X_i refers to the net return, π_i
its probability, and $E(.)$ the expected value operator. The variance is now
calculated as $\sigma^2 = (X_1 - \mu)^2 \pi_1 + (X_2 - \mu)^2 \pi_2 + \ldots = \Sigma^n_{i=1} (X_i - \mu)^2 \pi_i$.
When multiple time periods exist, these formulas must be based on dis-
counted net returns. The expected value of discounted net returns is
$E(NPV) = \alpha E(x_1) + \alpha^2 E(x_2) + \ldots + \alpha^n E(x_n) - C_o = \Sigma^n_{i=1} \alpha^i E(x_i) - C_o$,
where $\alpha = $ a discount factor $(1/1 - r)$. The variance of discounted expected
net returns is $\sigma^2 = \alpha^2\sigma_1^2 + \alpha^4\sigma_2^2 + \ldots + \alpha^{2n}\sigma_n^2 = \Sigma^n_{i=1} \alpha^{2i}\sigma_i^2$. The
decision rule would be to select the project(s) that yields net discounted
expected returns greater than 0 or, for similar, multiple projects, to select
the project that yields the maximum net discounted expected return. The
hypothetical shoreline-stabilization projects in table 8.7 are extended
below to introduce uncertainty regarding returns (that is, the value of
damage prevented).

In table 8.9, the net returns of the four projects are presented for two
periods. To simplify the discussion, assume costs are $0 or that costs are
equal for all projects in the initial period and that the discount rate is 3%.
Given this collection of shoreline-stabilization projects and using the
expected-CBA criteria, one would choose Project D over all projects, followed by Projects A and C. However, as one can observe, Project D is also
the most risky project, followed by Projects B and C. As mentioned earlier,
a fault with both CBA and expected-CBA rules is that they do not take into
account the variability of a project. As a result, two techniques were introduced in the field of finance that can account for the variability of returns
in the evaluation of investment decisions, the mean-variance rule
(Markowitz 1989; Levy and Sarnat 1994) and portfolio theory (Markowitz
1959; Brealey and Myers 1991; Levy and Sarnat 1994). Both of these meth-

Table 8.9 ~ NET RETURNS OF FOUR HYPOTHETICAL PROJECTS

	Project A		Project B		Project C		Project D	
	Return	Prob.	Return	Prob.	Return	Prob.	Return	Prob.
Period 1	500	0.5	200	0.8	0	0.6	1,000	0.8
	1,000	0.5	1,500	0.2	1,000	0.4	3,000	0.2
Period 2	500	0.7	1,000	0.8	0	0.5	1,000	0.8
	1,000	0.3	−1,000	0.2	2,000	0.5	10,000	0.2
Undiscounted returns								
Period 1								
Expected value (μ)	750		460		400		1,400	
Std. dev. (σ)	250		520		490		800	
Period 2								
Expected value (μ)	650		600		1,000		2,800	
Std. dev. (σ)	229		700		1,000		3,600	
Both periods								
Expected value (μ)	1,400		1,060		1,400		4,200	
Std. dev. (σ)	339.1		954.2		1,113.6		3,687.8	
Coeff. of var. (σ/μ)	0.24		0.90		0.79		0.88	
Discounted returns								
Both periods								
Expected value (μ)	1,341		1,012		1,331		3,998	
Std. dev. (σ)	324.8		830.8		1,055.8		3,481.1	
Coeff. of var. (σ/μ)	0.24		0.82		0.79		0.88	

NOTE: Prob. refers to probability.

ods consider the tradeoff involved between return (expected return, μ) and variability (σ^2 or σ) in determining a set of optimal investment projects.

The portfolio decision method, or portfolio theory, was advanced to consider the problem that arises when portions of multiple investments are to be undertaken simultaneously and when the decision maker seeks to reduce the overall level of risk by diversification, that is, by holding a collection or combination of investments, and at the same time maintaining some degree of stable returns. The standard result is that diversification can allow for fairly stable returns while reducing risk from a combination of investments. However, this method is based on the further assumption that investments are completely divisible (as are mutual funds and other financial assets). Although this assumption is not completely satisfied with lumpy (uneven, nondivisible) investment projects, such as public shoreline stabilization, in principle it may be possible to extend the decision rule based on the mean-variance model. To explore this issue, collections of hypothetical projects are examined below to determine if risk can be reduced from investing in a collection of projects (that is, a linear combination of shoreline-stabilization projects). Some may consider this to be a narrow application of portfolio-based models. However, our application is meant for exploratory and illustrative purposes. Ideally, the government should examine a variety of public projects (for example, Medicare programs, health care programs, welfare programs, education programs, shoreline stabilization) to maintain a portfolio of projects where tradeoffs between risk and returns are possible and projects are combined to reduce the overall risk. Our application demonstrates that there are some models from which CBA can be modified so it can be more flexible in decisions regarding shoreline-stabilization projects.

The mean-variance rule is based on decision criteria with which one selects projects as follows: (1) given two or more projects that yield similar returns, equal returns for simplicity, one would choose a project with the smallest variance (or standard deviation); (2) for two or more projects that have similar risk, one would choose that project that yields the largest expected return; and (3) where returns and variances differ between projects the decision criteria become complex and the decision maker must make a tradeoff between returns and risk. On the basis of this decision rule, Projects A and C have the same expected returns (undiscounted), but Project C is more risky, so Project A would be selected (see table 8.8). Regarding the case where returns and variances differ, an extension of the mean-variance rule is used where it is argued that the variance may be a misleading indicator of risk or where returns are not distributed normally (for example, log-normal distribution) (Levy and Sarnat 1994). The extended mean-variance rule is based on a measure of the relative dispersion of risk rather than the absolute dispersion of risk. The decision maker is assumed to be risk averse (we assume most public decision makers are in this category), and risk is based on the coefficient of variation, a measure of the *relative* dispersion of risk (that is, σ/μ). Under this new decision rule, Project D appears more favorable and has similar relative risk compared with Projects B and C (see table 8.6). Although this extension resolves some issues pertaining to risk, it does not cover all cases, and some cases can arise where the coefficient of variation between projects with different expected returns will differ. In such cases this extended decision rule breaks down.

A further development of decision-theoretic models in finance concerns option-pricing theory. Applied to the analysis of investment decisions, these models can address the issues of timing and abandonment. With regard to timing, the decision maker may be considering a situation in which future payoffs some three to four periods later may offset poor returns in early periods. To reap the future payoffs, however, one must start the project and accept poor or negative returns in the early periods; here the decision based solely on the early periods would be to reject the project, but on the basis of both the early and the later periods would satisfy expected-CBA criteria. Applying this model to shoreline stabilization, the planner may feel that stabilization must be in place (stabilization may take three years to complete, for example) *before* a period of high storm activity forecasted to occur some four years after the initial decision to provide stabilization is made. Here one cannot wait until the future to provide shoreline stabilization, it must be in place before the hazard occurs. In the situation of a high-risk, high-erosion area, the planner may feel that on the basis of a forecast of increased storm activity some three years later, if shoreline stabilization were provided now (year 0) there is a high probability that it would be neutralized (eroded or washed away) before the period of high storm activity occurs, and hence would not be effective in the future; that is, early returns would not be offset by future returns. Here it may be better to wait until just before the projected period of high storm activity and provide shoreline stabilization just before it occurs. In these cases, there is a value associated with the option to postpone the timing of investment projects.

Another use of option-price theory concerns the decision to abandon an investment project or to reverse the investment decision if a particular project begins to realize continued or sustained losses in future periods. In private industry, a decision to shut down a plant may be optimal in order to minimize losses. A similar example in shoreline stabilization is hard to realize when only the definition of benefits from shoreline stabilization as the value of property damage prevented is used. Because the federal government also provides flood insurance, applications to shoreline stabilization could be modeled in terms of a planner who must determine the provision of shoreline stabilization and the simultaneous decision of whether to offer flood insurance protection. Then benefits from shoreline stabilization would consist of the sum of two components, the value of property damage prevented and the difference between the payout for claims less insurance premiums received. In such a decision framework, it is possible for cases to exist where negative losses occur (that is, where the payout of claims exceeds the premiums received and value of damage prevented) and are sustained over time, as in areas of high risk where repeat claims are typical. In some cases it could be optimal to proceed with buyout decisions and actually retreat from the immediate shoreline zone. However, this outcome may in turn rest on decisions on how the buyout will proceed (at condemned/damaged levels versus full market value).

What do these approaches offer in terms of shoreline stabilization? A word of caution. Although in theory it is possible to combine projects in order to reduce risk in an aggregate sense, such an approach is naive; the key issue becomes the location of the projects. One needs to be careful to merge projects in a way that produces a linear combination of projects where two adjoining or contiguous projects are combined to form one continuous project, so that each component now depends on the other. For example, if three separate projects were being considered for a barrier island whose ends had various levels of erosion, all of the projects, one at each end and one in the middle, could be combined to form one large continuous project. Overall, the implication for shoreline-stabilization projects is to seek ways to combine projects in the same area in order to form larger projects. Such a practice has been followed from time to time and now there is further justification for it.

CHAPTER 9 ～

Hazard Management as a Component of Coastal Management

A 40-year period of relatively benign weather left southern Florida with a false sense of security regarding its ability to withstand hurricanes. This led to complacency about hurricane risk, leading to "helter-skelter" development, lackluster code enforcement, building code amendments, shortcuts in building practices, and violations that seriously undermined the integrity of the code and the quality of the building stock.

Insurance Research Council and the Insurance Institute for Property Loss Reduction (1995)

Foresight in the design and management of the built environment is key to effective coastal hazard mitigation. Traditional approaches to mitigation, . . ., are no longer enough. A community-based approach is required that integrates the principles and techniques of mitigation into the very manner in which communities are developed and redeveloped. An improved understanding of natural systems and appropriate siting, design, and construction of the built environment are essential to advances in hurricane and weather-related hazard mitigation.

H. John Heinz III Center for Science, Economics and the Environment (2000b)

Experience has shown that many shoreline-stabilization approaches once thought to be appropriate are not so anymore. Past management practices have attempted to maintain a static shoreline within an extremely dynamic coastal system, and these approaches have proved to be futile in some cases and detrimental in others. An emerging goal is to manage the coastline in a manner that is more compatible with the natural occurrence of high-energy events (the hazards associated with storms) that produce lasting changes. We refer to this approach as "hazard management."

Hazard management (or mitigation, as some refer to it) seeks to provide approaches that reduce the risk of damage and loss of life. Natural hazard management involves an understanding of potential natural hazards for specific areas and the use of specific concepts on which coastal land-use planning is based. In principle, hazard management offers flexible strategies that strive to work in conjunction with a dynamic coastal zone rather than against it.

A variety of approaches can be incorporated into coastal management efforts at the state and federal levels: (1) structural approaches; (2) nonstructural approaches; and (3) land-use management approaches. An important component of most of these is their ability to reduce the threat, or alleviate the effects, of coastal natural hazards before they occur. To work as intended, their application should be tailored to the characteristics of the site, the regional setting, and the objectives for the site and region.

Numerous reports (Platt et al. 1992; NRC 1995) have observed that the stabilization of eroding shorelines has shifted from "hard" engineered approaches to "soft" strategies. Earlier federal policies were found to be reactive to storm events rather than proactive to temper the conditions that exacerbate the magnitude of natural disasters. Most federal policies addressed only the short-term erosion that resulted from storms and paid little attention to long-term erosion due to the effects of relative sea-level rise and a decreasing sediment supply.

States such as North Carolina have moved decisively regarding eroding shorelines by adopting a retreat strategy in the form of minimum setback requirements for shoreline construction (Platt et al. 1992). Conversely New Jersey, with inadequate standards for rebuilding after a disaster, was limited in the extent of its intervention in coastal development. Overall there are few successful examples of the application of building-development setbacks to accommodate inland shifts of the shoreline, and in New Jersey as elsewhere the issue of retreat is a political firestorm that remains emotional, controversial, and largely unacceptable. However, Ocean City did remap building lots lost after the March 1962 storm, and the community of Bradley Beach did shift its boardwalk inland after a series of storms in 1991–92. Because of the enormous value of shorefront property, some have reasoned that arguments to prevent coastal development or to convert developed areas to natural environments hold little weight and will sway very few (Nordstrom 1994). Despite the actions of some states, retreat as a general management option in response to coastal erosion has not yet been widely applied.

In 1988 the National Research Council established a Committee on Coastal Erosion Zone Management in response to a request from FEMA and the Flood Insurance Administration (FIA) to provide advice on appropriate erosion management, support for data needs, and methodologies with which to administer these strategies through the National Flood Insurance Program as well as an evaluation of retreat via the Upton-Jones amendment. The 1987 amendment, discussed below, sought to aid owners whose structures faced collapse from erosion. In 1990 the NRC published *Managing Coastal Erosion*, which concluded that incentives were not in place to administer the amendment, and hence it was used very little. The report reviewed the current state of our understanding of the complex physical process of coastal erosion with its many natural and human-induced factors, and concluded that it was the human factors that were adding stress to the zone. This publication was followed by an assessment of the use of beach nourishment, *Beach Nourishment and Protection* (NRC 1995), five years later. In both reports the NRC sought to create a balance in approaches to erosion and provide an opportunity for science and engineering to be used effectively in the management of coastal processes. Studies like these both aid the understanding and provide verification of the hazardous nature of the coastal zone as they examine the effectiveness of current shoreline-stabilization approaches and management strategies designed to reduce exposure to natural hazards.

Hazard Management Applied to Coastal Policy

Because almost one-half of the U.S. population now lives within 80 km (50 miles) of the coastline, the coastal areas of the United States are under great pressure. The resulting development has disrupted coastal systems and is occurring alongside other fundamental, natural changes in their functioning (Culliton et al. 1990). As these cultural and natural processes reconfigure the coastal zone, the need for appropriate policies and management approaches increases.

Although past management approaches have addressed the coastal area as if it were static, the zone is a highly dynamic system requiring special attention and management considerations. In addition to the cultural forces that place coastal resources at risk, a combination of long-term sea-level rise and sediment loss, intertwined with coastal erosion, has made the shoreline an area of increasing exposure to the effects of storms. This process is believed to increase in a nonlinear manner over time. One approach to managing these risks with an emphasis on public protection is based on hazard management and land-use planning principles that seek to create safer communities as their primary goal. Hazard management involves recognizing and adapting to natural forces and is defined as any sustained action taken to reduce long-term risk to human life and property (FEMA 1995a,b); it is truly a management approach. On both a local and a national level, hazard management is becoming an accepted response both to existing hazards and to those that will arise in the future.

Although emphasis on pre-storm hazard management is important, the need for post-storm planning and recovery is especially relevant. Actions at the community level can be and have been successful in mitigating the effects of small storm events. Major storms, however, require advance preparations for both the pre-storm stage and the post-storm recovery process. Areas of high risk from natural hazards should become better identified and specific plans developed to reduce exposure in these locations. Both preparedness and post-storm recovery programs are basic to coastal hazard management.

Educating and increasing the awareness of the public to the threat of natural coastal hazards will assist with preparedness. Public involvement in both the planning and the development stages of hazard management reports encourage the subsequent implementation of acceptable management strategies for reducing risk along the shoreline.

Also important for management considerations are the many natural and cultural resources in coastal regions, such as in coastal New Jersey. The diversity of resources in New Jersey makes the state unique and attractive to many. Its coastal attributes include parks such as the Gateway National Recreation Area at Sandy Hook and Island Beach State Park; the wildlife refuges on the Atlantic Coast Flyway and their endangered and threatened bird species; highly developed recreational coastal towns such as Atlantic City that offer legalized gaming; the quaint Victorian style of Cape May; and facilities such as boardwalks, ocean piers for fishing, and the many attractive lighthouses. These and other valued resources warrant preservation by the citizens and officials of the state. Effective stewardship will require a delicate balancing of the system.

Another important component of the system is the continual need to update information concerning the coastal zone. This dynamic environment needs to be regularly monitored and the information collected must be shared appropriately. A strong database and the means to achieve it are paramount if we are to effectively manage the coast. Also needed is a regional instead of a piecemeal management approach to problems, as we argued earlier. All of these issues will require efforts that foster working relationships among all stakeholders (including the state, municipalities, counties, residents, and businesses).

New Directions and Objectives

Federal Policies and Directions

An important federal policy that continues to influence the nation's coast is the Coastal Zone Management Act of 1972. Until the late 1960s, decisions affecting coastal resources were made without coordination among federal, state, and local governments. Increased recreational, economic, and other uses of the coastal zone led to conflicts among the diverse groups involved. In response, Congress enacted the Coastal Zone Management Act (CZMA) in 1972, with the objective to preserve, protect, develop, and, where possible, restore and enhance the resources of the nation's coastal zone for current and succeeding generations.

The CZMA created a partnership among federal, state, and local governments to seek collective solutions to problems caused by competing coastal pressures. All activities within the coastal zone, and those activities outside this area that affect resources within it, are now subject to the multiple-use management regime established by the CZMA. Among its basic tenets are two mandates:

- To reduce the risk to life and property from coastal storms and erosion by directing coastal development away from hazardous areas
- To protect the dunes as a natural barrier to coastal storms

More than 95% of the nation's shoreline is managed under the CZMA through a network of 30 states, including New Jersey. To convince coastal states to join this voluntary program (the stick), the federal government provided them with two incentives (the carrot): (1) financial assistance to develop and implement state coastal management plans; and (2) federal consistency authority, a tool that enables states to address the adverse effects of federal activities on coastal resources within their jurisdiction.

Since completion of the *New Jersey Shore Protection Master Plan* (*NJSPMP*) in 1981, much progress has occurred within the federal government with regard to hazard mitigation, especially concerning floods and flood-related disasters. Important in this effort is the National Flood Insurance Program (NFIP), established by the National Flood Insurance Act of 1968 and amended in 1973 by the Flood Disaster Protection Act, collectively referred to as the Flood Insurance Act. The Flood Insurance Act made available flood insurance within communities willing to adopt floodplain management programs designed to mitigate future flood losses. A further requirement involved the identification of all floodplain areas within the United States and the establishment of flood-risk zones within these areas. Like those farther inland, coastal communities were qualified to participate in the program, because it was recognized that they face unique flood hazards resulting from storm surges and wave action generated in adjacent bodies of water. Flood Insurance Studies (FIS) and Flood Insurance Rate Maps (FIRMs) for flood-prone communities are essential documents that help achieve the goals of the NFIP. These studies and maps provide the technical information to communities that enable them to adopt floodplain management measures required for NFIP participation (FEMA 1995c).

In 1995 Congress mandated that there be a 30-day waiting period before coverage under a new contract for flood insurance or any modification to coverage under an existing flood insurance contract becomes effective, with two exceptions. The intent of Congress in requiring the waiting period was to prevent the purchase of flood insurance at times of imminent flood loss (FEMA 1995d). One exception to the rule involves the initial purchase of flood insurance in connection with the making, increasing, extension, or renewal of a mortgage loan. The second involves the initial purchase of flood insurance within one year of a map revision.

Another important development for the NFIP was passage of the Upton-Jones amendment to the Housing and Urban Development (HUD) Act of 1987. With this amendment, Congress could authorize payments from the National Flood Insurance Fund (a pool funded by NFIP premiums) for certain costs of demolishing or relocating insured structures imminently threatened with collapse from erosion (Platt et al. 1992). The intent of Upton-Jones was to encourage voluntary action by owners to remove their threatened structures. Prior to Upton-Jones, the NFIP only paid claims on insured buildings that had sustained physical damage as a result of flooding or flood-related erosion. The amendment allowed for payment of a claim prior to actual damage for the purpose of relocating or demolishing the structure (NRC 1990).

The Upton-Jones amendment thus sought to encourage the removal of structures in erosion-prone areas prior to their collapse, to avoid higher NFIP costs (which would result if claims on the damaged structures were honored and the structures rebuilt) and to reduce public safety hazards.

Although Upton-Jones stressed mitigation, its implementation was not very successful. Eligibility was too narrowly defined, and few claims were ever filed. Few property owners took advantage of the benefits, and when they did, they opted for demolition over relocation (Platt et al. 1992). The Upton-Jones amendment was eliminated from the NFIP in September of 1995 in the National Flood Insurance Reform Act of 1994, Title V of the Reigle Community Development and Regulatory Act of 1994 (P.L. 103-325).

In its place, a National Flood Mitigation Fund was created in 1994 that allowed funds to be used for mitigation activities such as relocation and governmental acquisition of structures subject to repetitive loss (P.L. 103-325). Money for this fund comes from surcharges placed on existing flood insurance policies. Selection of projects for support from the fund is based on those that reduce payments from the National Flood Insurance Fund. Although this fund is still in its early stages, it does represent a future mitigation opportunity.

An important mitigation development occurred when the Disaster Relief and Emergency Assistance Act of 1974 was revised by the Disaster Relief and Emergency Assistance amendments: the Great Lakes Planning Assistance Act of November 23, 1988. These amendments changed the measure's name to the Robert T. Stafford Disaster Relief and Emergency Assistance Act of 1988 (P.L. 93-288 as amended by P.L. 100-707), and added requirements for disaster preparedness plans and programs. Under the Stafford Act, the president must declare a disaster emergency prior to the authorization of any federal assistance. The Stafford Act provides up to 75% of the cost of hazard mitigation measures that the president has determined to be cost-effective and that substantially reduce the risk of further damage, hardship, loss, or suffering in any area affected by a major disaster (P.L. 93-288, as amended in 1988). The Stafford Act also created a Hazard Mitigation Program, which provides matching federal funds for state and local mitigation projects. These grant funds are also tied to disaster declarations.

Along with new opportunities for mitigation activities, there is a new direction proposed by FEMA. In 1995 FEMA announced its National Mitigation Strategy, "Partnerships for Building Safer Communities," discussed below (FEMA 1995a). The strategy is intended to generate a fundamental change in the public's perceptions of hazard risk, and to demonstrate that mitigation is often the most cost-effective and environmentally sound approach to reduce further losses. In accordance with this national initiative, states such as New Jersey can establish their own strategy for creating safer communities and in mitigating the effects of natural disasters that pose a threat to the resources in the coastal zone.

State Policies and Directions

COASTAL DEVELOPMENT, REDEVELOPMENT, AND PLANNING

Coastal storms often result in extensive damage to property and infrastructure. After such a disaster, reconstruction and repairs have typically restored damaged property to its predisaster condition without consideration of its continued exposure to natural hazards. Although this type of restoration returns a community to some form of normalcy rather quickly, the replication of predisaster conditions often results in a cycle of repetitive damage and reconstruction (FEMA 1990). Continued coastal development in this manner will undoubtedly result in future property damage and the increased need for costly shoreline-stabilization measures.

In an effort to break out of this cycle, communities and state governments have begun to adopt storm hazard mitigation strategies. In 1984, the New Jersey state Office of Emergency Management produced such a plan. The *State Hazard Mitigation Plan* was subsequently updated in 1993 as a result of a presidential disaster declaration for the January 1992 coastal storm (NJOEM 1993). Another revision to the plan was completed in May of 1994. In this version, the plan outlines a system of risk reduction efforts in New Jersey to aid state and local emergency management officials in developing a hazard management program. In addition to New Jersey's individual efforts, FEMA is developing hazard mitigation plans that can be adopted at the state, county, or municipality level.

Typically, storm hazard mitigation plans are developed for postdisaster situations; however, such mitigation plans are essentially a predisaster plan for the next disaster. Postdisaster relief and planning can be accomplished through strategies such as (1) identifying hazardous areas; (2) educating the public about hazard management; and (3) changes in land-use management practices, construction practices, and shoreline-stabilization practices. Improving land-use planning practices can be accomplished through steps such as acquiring land in hazardous areas (for example, the Coastal Blue Acres Program), transferring development rights, rezoning (for example, setback limits and dune protection ordinances), relocating public roads and other public necessities away from hazardous locations, and exchanging land in high-hazard areas for safer property locations. By requiring new construction and the repair of damaged property to meet stringent flood hazard area standards, future damage can be reduced in principle. This reduction hinges on the severity of the storm, because in the case of extreme storm events, such small incremental adjustments may be overwhelmed.

Building and reconstruction moratoriums can also be imposed in identified hazardous areas. Redefining shoreline-stabilization measures entails establishing and maintaining nonstructural approaches such as coastal dunes and beach nourishment projects in appropriate locations. Hazardous areas need to be thoroughly identified through a systematic program and then routinely monitored for changes in vulnerability. Further, educational programs can be developed to educate the public about natural coastal hazards and the ways in which hazard management can reduce associated coastal damages.

NEED FOR A REGIONAL PARTNERSHIP APPROACH

Public officials and coastal managers of states with extensive open beach exposure (for example, New Jersey, North Carolina, and Florida) are further challenged by the demands the population places on the shoreline. High population densities and the accompanying infrastructure are continually pitted against the natural processes of the coast. Traditionally, the problems this juxtaposition poses have often been dealt with on a piecemeal basis. Such an approach does not take into account the downstream or cascading effects of a shoreline-stabilization action on other parts of the coast. For example, the installation of a groin in one municipality may trap sand that had been naturally transported to an adjacent municipality to maintain its beach width.

As demand for use of the coast continues to grow, there will be a concomitant need for better information and more creative management strategies. If present and future managers are going to successfully balance these stresses on the coastal zone in a manner that fosters access, reduces hazards, and preserves the ecological integrity of the system, the development of an integrated coastal management approach should be addressed on a regional basis. Specifically, coastal managers must look at the overall currents and wave actions, sand transportation, existence of shoreline-stabilization structures, and erosion rates that affect the entire shoreline. However, for their efforts to be effective, consideration of coastal processes at a microlevel is also needed to address variations that a larger state-level (aggregate) approach would ignore or brush over. Therefore, delineation of the coastal zone into distinct subsets with similar coastal processes (for example, littoral cells, reaches) characterizes the preferred approach that leads to the creation of a nested hierarchy. Partnerships that transcend community boundaries are necessary to achieve this goal. Coastal management strategies should also be identified that function at the regional level, and active working committees should be formed to provide initiative and communication with other such committees in order to fit into larger program objectives.

CREATING AN INFORMED PUBLIC AND FOSTERING PUBLIC INVOLVEMENT

The process of developing an informed public goes beyond merely providing information to coastal communities. An informed public results from the development of communication goals and the establishment of a communication plan or process. Establishment of a mechanism for community involvement and participation is critical to instituting any successful partnership.

This education process will benefit from procedures followed in the field of conflict resolution and negotiation (Fisher and Brown 1989; Fisher, Ury, and Patton 1991). It must include all stakeholders affected and allow access to all efforts to be successful and achieve a consensus. Furthermore, all viewpoints from the community must be acknowledged and addressed. Decisions on how to encourage and incorporate the input from the community are an important component of a successful public participation process. Citizen meetings, questionnaires, and information hotlines are examples of how to involve the public. It is important for resource managers to clearly define the community's role in the project as well as to mutually agree with the members of community on the level of their input into the project. Involving communities from the outset is the best way to deal with both the technical aspects of any project and community concerns.

Involving an informed public in the implementation of a management strategy requires creating a productive dialogue between resource managers and the community. The development of an effective communication strategy includes setting communication goals, defining the audience, deciding how the goals will be implemented, and determining how the resource managers will respond to the community. The key principles of successful negotiation that have been developed by the Harvard Negotiation Project will be useful to any project that involves public policy and affects the public.

National Mitigation Strategy

The reorganization of FEMA on November 28, 1993, brought about the creation of a Mitigation Directorate. Mitigation, which involves reducing the impact of natural hazards, became one of the key elements in FEMA's reinvented organizational structure (FEMA 1995a). Director James Lee Witt raised the agency's mitigation efforts from a low-level office to one of its four main branches. FEMA announced its National Mitigation Strategy, "Partnerships for Building Safer Communities," which makes hazard risk reduction a national priority and sets mitigation as the cornerstone for creating safer communities (FEMA 1995a). Development of this strategy signifies a major redirection of national policy for managing high-risk, natural hazard areas. Another significant change is the elevation of FEMA to the level of a Cabinet department in February 1996, giving FEMA and its policies even more importance. FEMA's new focus involves retreat, removing people from hazard areas, providing support for public safety, reducing the costs of recovery following damage from natural hazards, and reducing payouts from the National Flood Insurance Program by 50% by the year 2010 (1995a).

The National Mitigation Strategy seeks to fundamentally alter the general public's perception of the risks of natural hazards and the mitigation of that risk, and to demonstrate that mitigation is often the most cost-effective and environmentally sound approach to reducing losses from natural hazards. The long-term goal of this strategy is to increase public awareness of natural hazard risk and, within 15 years, to reduce the risk of loss of life, injuries, economic costs, and disruption of families and communities caused by natural hazards. To meet its National Mitigation Goal, FEMA will

- Conduct studies to identify hazards and assess the risks associated with those hazards for communities throughout the nation
- Encourage applied research that will develop the latest technology in response to natural hazard risks and promote the transfer of that technology to users: state and local government, the private sector, and individual citizens

- Create a broad-based public awareness and understanding of natural hazard risks that lead to public support for actions to mitigate those risks
- Provide incentives and encourage mitigation activities and redirect resources from both the public and the private sector to support all elements of the strategy
- Provide national leadership in the achievement of the National Mitigation Goal, provide coordination among federal agencies to promote hazard mitigation throughout all federal programs and policies, and provide coordination with other levels of government and the private sector (FEMA 1995a)

Mitigation in New Jersey

The State Hazard Mitigation Plan

After the *NJSPMP* was completed in 1981, federal and state governments began to shift away from postdisaster assistance and move toward the use of mitigation measures to alleviate or avert damages prior to a disaster. Although the federal government had been discussing the use of such measures for a number of years, little was done to implement a national strategy until 1995. Prior to this time, the implementation of a successful mitigation strategy was the primary responsibility of state governments. In 1985 the NJDEP, in cooperation with the NJOEM and other state emergency management agencies, developed a hazard reduction mitigation plan for the state: the *New Jersey State Hazard Mitigation Plan: Section 406* (NJ-HMP) (NJDEP 1986). It was updated in 1992 (NJOEM 1993) and revised again in 1994 (NJOEM 1994).

The NJ-HMP strives to reduce or eliminate the threat of loss of life and property from natural disasters by providing recommendations and approaches that together constitute a system of mitigation measures for the state to administer. The NJ-HMP also serves as an aid to local and state emergency officials through its plan to mitigate hazards and its establishment of a framework to coordinate efforts between FEMA and the State Interagency Hazard Mitigation Team, or SHMT (NJOEM 1994). Although the NJ-HMP recommends mitigation activities for both coastal and riverine areas, only the measures appropriate to the coastal area are discussed here.

History of the New Jersey Hazard Mitigation Plan

1985 NEW JERSEY HAZARD MITIGATION PLAN: SECTION 406

During March 28–29, 1984, a severe northeaster tracked slowly along coastal New Jersey. The resulting storm surge caused flooding, damage to shoreline-stabilization structures, and severe beach and dune erosion. A second northeaster (April 4–5, 1984) followed that produced extensive riverine flooding throughout the Passaic River Basin (NJDEP 1985a; U.S. ACOE 1985). On April 12, President Ronald Reagan declared the four coastal counties disaster areas (FEMA-701-DR). Following the president's declaration, the state of New Jersey applied for disaster relief funds available under Section 409 of the Disaster Relief and Emergency Assistance Act of 1974. Under this act, as noted, the federal government may contribute up to 75% of the cost of hazard mitigation measures that are determined to be cost-effective and substantially reduce the risk of further damage, hardship, loss, or suffering. In order to receive disaster relief funding, the state had to evaluate and work to mitigate future natural hazards, which led to the development of the NJ-HMP.

The *New Jersey Hazard Mitigation Plan: Section 406* is a comprehensive document that describes damages from the riverine flooding and coastal storm of 1984. It also summarizes existing state mitigation measures and proposes mitigation measures to reduce future risks. The mitigation team recommended 23 short-term and long-term mitigation activities (that is, work elements) to assist affected coastal regions in their recovery. These work elements consisted of mitigation actions for specific sites as well as for the entire coastal area.

Short-term recommended measures included revisions to the computations of storm surges at the Atlantic City Steel Pier, technical assistance to communities to implement the NJ-HMP recommendations, structural fortification, a grant to study Long Beach Island's evacuation and warning systems, and the endorsement of changes to the building codes of the Building Officials and Code Administrators (BOCA). *Long-term* mitigation measures included dune enhancement and restoration projects, continuation of shoreline profiling to establish a time series of erosion rates, implementation of acquisition/relocation projects, and restriction of development seaward of Ocean Avenue in Sea Bright and Monmouth Beach until emergency beachfill could be provided.

On September 27, 1985, while the NJDEP and NJOEM were preparing the NJ-HMP, Hurricane Gloria passed along the New Jersey coast, which was already weakened from 1984's storm events. Although sustained damages were less than those inflicted by the previous year's storms, additional long- and short-term mitigation measures were developed as a result of the subsequent presidentially declared disaster (FEMA-749-DR). The short-term recommended measures included amendments to CAFRA, such as prohibiting development in V-zones (velocity zones, discussed below), instituting a 50-foot setback from shoreline-parallel structures, and supporting changes to BOCA for wind speed. Long-term mitigation measures included continuation of NJDEP regulatory jurisdiction of the coastal area, continuation of the New Jersey Hurricane Evacuation Study of the U.S. Army Corps of Engineers, establishment of dunes seaward of boardwalks in Monmouth County, and a proposed acquisition project for Whale Beach in Cape May County.

COASTAL STORM 1992 (FEMA-936-DR-NJ)
INTERAGENCY HAZARD MITIGATION TEAM REPORT

Between 1985 and 1992 the state handled approximately 18 storm-related flood emergencies (NJOEM 1993; U.S. ACOE 1993a,b). Each of these events was managed by local, county, or state funding and NJOEM's mitigation efforts. On March 3, 1992, all four coastal counties were declared major disaster areas by the president after a January 4, 1992, storm that devastated the coast. The northeaster produced coastal flooding and further erosion of beaches that were already weakened by a similar storm in October of 1991 (NJOEM 1993). Although the NJ-HMP had been updated as necessary during the 1985–1992 period, it was revised to include mitigation measures developed as a result of the 1992 storm in order for the state to qualify for disaster-assistance funding via the Stafford Act.

The SHMT made formal mitigation recommendations in its subsequent 1993 report (NJOEM 1993). These included the creation of a permanent group to oversee mitigation administration, along with public awareness/alert warnings, coastal issues, hazard identification, and social services. Each recommendation, when implemented, was to serve to prevent or reduce losses in the event of a similar future coastal disaster. The first recommendation of the NJ-HMP was the creation of a permanent State Interagency Hazard Mitigation Team. During the production of the 1985 NJ-HMP, a temporary State Interagency Hazard Mitigation Team had been created to assist NJDEP and NJOEM in identifying an interagency approach to hazard mitigation that supported the goals of mitigation (NJOEM 1993). Following the 1993 NJ-HMP, the team was formally established by executive order of the governor, with the NJOEM designated as the lead agency to coordinate it. Support agencies include NJDEP, the state Department of Transportation, state Department of Community Affairs, Office of Statewide Planning, U.S. ACOE, and others involved with the state's mitigation efforts.

STATE OF NEW JERSEY OFFICE OF EMERGENCY MANAGEMENT DR-973-NJ HAZARD MITIGATION PLAN

In December 1992 New Jersey experienced its second presidential disaster declaration since 1985. Between December 11 and 17, a severe northeaster caused widespread coastal damage from flooding and high-velocity winds. Following the storm, the state's *Hazard Mitigation Plan* was revised to reflect the new hazard mitigation needs and to reprioritize existing recommendations. The SHMT coastal area recommendations in the after math of the storm fell into six categories:

Beach and dune areas: Reemphasize dune restoration and maintenance; create a priority list of shoreline stabilization projects, replace bulkheads not in compliance with the federal flood insurance criteria for shoreline stabilization devices; revise ordinances prior to emergency work in communities with repeatedly damaged beaches; develop an acquisition program for postdisaster assistance; and update beach-dune topographic mapping.

Flood warnings: Continue both offshore data collection and the installation of backbay flooding remote sensors.

Regulations: Institute regulations for permits that address the installation and/or repair of retaining walls. The SHMT also recommended that NJOEM develop procedures and policies to regulate the collection and disposal of debris and implement a disaster debris management plan.

Specific coastal areas: Study the South Cape May Meadows area to assess the erosion problem and develop alternatives to emergency dune management.

Flood insurance: Adopt ordinances requiring landlords of rental properties within high hazard areas to provide flood insurance information for their tenants.

Planning: Continue the SHMT; HUD and the State Department of Community Affairs should advise municipalities of the availability of Community Block Grant and Small City Block Grant Programs for emergency response and risk-reduction funding.

Accomplishments of the New Jersey Hazard Mitigation Plan

Since the creation of the NJ-HMP, the NJOEM has been involved in the state's coastal hazard mitigation efforts. In Section VI of the plan, NJOEM (1994) listed the accomplishments, to date, of this process. Specific measures that have been implemented include:

1. Technical assistance from NJDEP to implement the NJ-HMP's recommendations
2. Implementation of the Shore Protection Fund
3. A study of Long Beach Island, completed in 1986 by the Division of Water Resources, Bureau of Flood Plain Management
4. An increase in regulatory jurisdiction in the coastal area for NJDEP through revision of CAFRA
5. Permanent restriction of development seaward of Ocean Avenue in the communities of Sea Bright and Monmouth Beach
6. Continued long-term monitoring of beaches and dunes by NJDEP
7. Continued pursuit by NJDEP of the proposed park plan for Whale Beach, Sea Isle City
8. In 1984, completion of the report *Assessment of Dune and Shorefront Protection Ordinances* (NJDEP 1984a)
9. Receipt by NJDEP of a $2 million grant for dune restoration for the 1984 storm damages
10. Revision of shoreline-stabilization rules to require adherence to coastal program regulations as a condition of state shoreline-stabilization expenditures to municipalities
11. Completion by NJDEP of storm vulnerability studies and site-specific storm hazard mitigation studies for the barrier islands in 1985
12. Achievement of compliance with the National Flood Insurance Program by all jurisdictions in the state due to aggressive efforts by state flood plain management officials
13. Completion of the *New Jersey Hurricane Evacuation Study, 1992*, used to increase the state's level of preparedness
14. Initiation by the state of the investigation of several mitigation measures, including studying the use of submerged concrete breakwaters as a means to stabilize oceanfront beaches
15. Support by NJDEP of further recommendations to BOCA

Recommendations to Improve the State's
Hazard Mitigation Efforts

Since the inception of the hazard mitigation plan, the SHMT has implemented many mitigation measures in the coastal area. Although there has been some success with the plan, there is still a need to redefine and improve the state's mitigation measures and incorporate new strategies into the NJ-HMP to continue to reduce the public's exposure to coastal hazards. Several mitigation measures can be executed through the NJ-HMP to further prevent storm damage:

- Promote the continuation of dune enhancement and maintenance programs
- Identify high-hazard areas along the coast and periodically update the list to reflect any changes
- Continue public awareness and education programs
- Once FEMA's national guidelines have been established, assist local municipalities in developing their mitigation plans
- Prohibit new development and redevelopment in high-hazard zones
- Promote the acquisition of structures (facing imminent collapse) in identified high-hazard areas
- Incorporate advances in wind-load reduction technologies into the state's building codes
- Provide technical assistance and incentives to communities that participate in mitigation efforts
- Design shoreline-stabilization efforts to be contingent on local efforts that implement mitigation strategies
- Utilize FEMA's Hazard Mitigation Grant Program funds to implement mitigation activities
- Continue to develop mitigation strategies using a regional approach instead of on a piecemeal basis

Consistency of State and Federal Efforts

The *NJSPMP* addressed hazard management (or mitigation) in a section titled "Disaster Mitigation and Recovery," under the Federal Programs and Policies portion of the Policy Review. The plan stated that there are federal disaster relief programs designed to provide assistance to states, local governments, individuals, and owners of selected nonprofit facilities to alleviate the suffering and damage that result from natural disasters (NJDEP 1981). It also noted that these programs assist with the reconstruction and rehabilitation of devastated areas, including those located in hazardous areas on barrier islands. The programs the *NJSPMP* specifically refers to are those of the U.S. Army Corps of Engineers and the Federal Disaster Assistance Administration (FDAA), now known as the Federal Emergency Management Agency, administering the federal Disaster Relief Act of 1974.

The *NJSPMP* further recommended that federal disaster mitigation and recovery programs be redirected to ensure that the programs do not invite past mistakes by encouraging reconstruction of storm-damaged structures in high-hazard areas (NJDEP 1981). Several options were listed for both moderate- and high-level protection. One moderate-level option was to strengthen the role of the U.S. ACOE in coastal stabilization by emphasizing the natural protective capabilities of beaches and dunes and the need to preserve them. It was also recommended that the U.S. ACOE shift from the increasingly expensive structural control of erosion and flooding to cooperative land management (NJDEP 1981). In recent times, the U.S. ACOE has shifted to softer approaches such as beach replenishment and nourishment projects, but these newer measures are not less expensive. Much controversy surrounds beach nourishment projects because of their high costs. Another recommendation related to moderate-level stabilization was that FEMA and the Small Business Administration (SBA) should consider developing regulations that would make receiving predisaster planning and postdisaster loans or grants conditional on the establishment of state disaster recovery plans that would (1) incorporate the state's disaster legislation and require its full implementation; (2) recognize that barrier islands are especially vulnerable to disaster; (3) promote regulations to adequately protect human life by discouraging development in high-hazard areas; and (4) foster contingency redevelopment plans by the state to encourage reconstruction away from barrier islands (NJDEP 1981).

Regarding shoreline stabilization, the *NJSPMP* recommended the establishment of mechanisms for identifying areas and types of facilities in coastal high-hazard locations that, when severely damaged by storms, would not be eligible for federal assistance for reconstruction purposes in the same location (NJDEP 1981). Relocation assistance would be available to aid individuals and businesses to move away from high-hazard areas. One specific recommendation concerned amending the Federal Flood Disaster Protection Act of 1973 (P.L. 93-234) to restrict disaster assistance from being used for reconstruction in high-hazard areas and instead to promote its use to provide relocation assistance for those who voluntarily elect to move to safer areas, similar to the Upton-Jones amendment.

The *NJSPMP* also recommended that the Disaster Relief Act of 1974 be amended to require disaster preparedness plans and programs, and that postdisaster recovery assistance should be contingent upon the adoption of such plans and programs. Furthermore, the act should authorize the establishment of "Recovery Planning Councils" prior to a major disaster to assist in developing and gaining approval of predisaster contingency plans for barrier islands (NJDEP 1981).

It should be emphasized that some of these philosophical directions were incorporated into the Stafford Act. These were provisions for revising and broadening the scope of existing relief programs, to encourage the development of comprehensive disaster preparedness and assistance programs by states and local governments, to encourage hazard mitigation measures to reduce losses from disasters, and to provide federal assistance programs for both public and private losses resulting from disasters (P.L. 93-288, as amended 1988).

Since the development of the 1981 *NJSPMP*, there has been much discussion regarding mitigation on the federal level as a means of alleviating or averting disasters, but little action leading to successful implementation of such means. Thus the implementation of successful mitigation strategies becomes an important task for the state.

The 1994 version of the NJ-HMP provides an outline for a system of risk reduction in New Jersey and serves to aid state and local emergency management officials in developing a hazard management program. The NJ-HMP also lays the groundwork for coordination between FEMA and the State Interagency Hazard Mitigation Team to review hazard management and mitigation plans and facilitate specific projects (NJOEM 1994).

The NJ-HMP also suggests that a positive relationship between government and the private sector be developed as a means of persuading the public of the viability of hazard management and mitigation (NJOEM 1994). In addition, various nonprofit groups could serve as valuable resources in seeking to influence the public. Overall the NJ-HMP provides for the means by which to carry out mitigation activities, but it is limited by the funds available at both the state and the federal level. An important task for NJOEM therefore involves obtaining funding from the state and federal government.

Formulating the State Plan in Terms of the
National Mitigation Strategy

The NJ-HMP incorporates directions similar to those enunciated in the National Mitigation Strategy and could readily be extended to make it more compatible with the national objectives. The following discussion establishes the compatibility of the two programs and demonstrates the similarity between the state and federal approaches. Emphasis is placed on the development of procedures that effectively translate the national strategy into useful strategies for state and local governments, the private sector, and the public. Where possible, greater emphasis should be placed on public safety and the mitigation strategies that are best suited for supporting this objective.

New Jersey is in a position to build upon existing NJ-HMP mitigation strategies to extend the state's mitigation efforts into FEMA's new initiatives. Ways to achieve this in relation to FEMA's five initiatives include:

1. Hazard Identification and Risk Assessment
   - A coordinated program (at state, federal, and local levels) of dune creation, restoration, maintenance and expansion for emergency recovery and long-term protection should be developed.
   - Beach and dune topographic maps should be produced that reflect post-storm conditions and can be further used by FEMA in Flood Insurance Administration reviews.
   - Alternative methods for obtaining additional offshore water and weather data to fill the void left by the removal of Large Navigational Buoys (LNB) should be explored by the National Weather Service.
   - Additional remote sensors in backbay areas should be installed to monitor bayside inundation.
2. Applied Research and Technology Transfer
   - A prioritized list of coastal municipalities that require revised wave-height analyses based on FEMA's current sand dune evaluation mapping criteria or flood insurance map revisions should be developed by NJDEP.
   - Appropriate long-term and short-term strategies for enhancing public safety should be examined.
   - Coastal dune development assistance should be provided and

standards developed, with specific dune dimensions, for protection levels available.

- Emergency work eligibility criteria for beaches with repetitive damage should be revised to include local adoption of beach and dune system management ordinances in conformity with the August 20, 1990, version of the state of New Jersey's Rules on Coastal Zone Management.
- The state or counties should be required to have an acquisition plan as a condition for future disaster assistance after coastal storms.
- Erosion problems associated with the South Cape May Meadows area should be examined along with alternatives to continued emergency dune management.
- A comprehensive plan utilizing flood proofing, relocation, zoning, and setback measures in the north end of Atlantic City should be conducted.
- State regulations requiring permits for installation and/or repair of retaining walls should be developed for waterfront bulkheads and related construction.
- Setback regulations should be developed that are directed toward the establishment of variable width buffers and setbacks that can shift in high-rate erosion zones.
- Standardized policies and procedures to regulate the collection and disposal of debris should be developed along with a disaster debris management plan.

3. Public Awareness, Training, and Education
   - The information generated from remote sensors in backbays should be used to forecast and warn communities of emergency situations.
   - Municipalities should be encouraged to develop ordinances that require landlords of rental properties within high-hazard areas to provide flood hazard information (and information on the availability of flood insurance protection) at the time of the rental.

4. Incentives and Resources
   - Dune creation programs should be established that create procedures and funding of emergency recovery and long-term and short-term buffering.
   - The state should establish a policy for the expenditure of public funds on beach nourishment.
   - The public assistance cost-sharing formula (under Section 406 of the Stafford Act) could be modified to place more responsibility at

the state level—for example, 40% federal and 60% nonfederal—
and make the difference available under Section 404 for property
acquisition.

- Municipalities should be advised by HUD and the state Division of
Community Affairs of the availability of Community Block Grants
and Small City Block Grant Programs for emergency response and
risk reduction funding.

5. Leadership and Coordination
- Policies should be created that establish long-term objectives for the
management of the coastal zone.
- Strategies that work toward achieving short-term and long-term
objectives should be developed.
- Communities will require assistance (finance and technical) in
achieving the steps required to be consistent with short- and long-
term objectives.
- Communities will require assistance to complete tasks/recommen-
dations of the NJ-HMP.
- The SHMT could act as a coordinator between the community,
FEMA, and state.
- Standards should be developed in accordance with the state's rules
on coastal zone management, which would apply to all development
within the coastal zone as defined by CAFRA
- A hazard mitigation executive order should establish a permanent
state hazard mitigation team.

Shoreline Mitigation Options

STRUCTURAL APPROACHES

As discussed earlier, previous attempts to confront shoreline erosion have sought to address the symptoms of sediment loss by
erecting barriers to the penetration of storms and high waves. These hard
structures were developed over decades, and their remains are seen in
many portions of the coast. There are structures erected parallel to the
coast that involve sand retention and stabilization features, such as revetments, bulkheads, and seawalls. There are also structures erected perpendicular to the coast to intercept shoreline transport, such as groins and
jetties.

It is now recognized that coastal structures are an effective means of
preventing the inland penetration of storm waters, but that they do not
work in conjunction with the natural system. Instead they form a barrier,

and anything seaward of these structures—as well as downdrift, in certain cases—is sacrificed. In the process, the beaches and dunes that represent the coastal area and attract visitors will eventually be lost. However, many have argued that in certain cases the use of hard structures, if supplemented with beachfill, can provide an adequate level of protection (NRC 1990, 1995).

NONSTRUCTURAL APPROACHES

Dune maintenance and nourishment. The establishment of a dune maintenance and nourishment program and the artificial creation of dunes are forms of hazard management and mitigation. As we have seen, dunes serve as a protective buffer that exchanges sand with the beach to reduce rates of shoreline displacement, and they act as a barrier to storm surges and high water levels. Therefore coastal dunes have been shown to be an acceptable means by which to reduce some of a storm's negative impacts. Dunes, artificial or natural, can be created and enhanced through vegetation, which is affordable and can be long-lasting. Dunes can be built or maintained to achieve certain dimensions of the beach profile. Such dimensions relate to a predetermined level of buffering desired from storm surge and erosion. Although they are not a panacea for shoreline stabilization and provide less than complete buffering, dunes can be well worth the effort as a hazard management and mitigation option. Many New Jersey communities have already established dunes and have dune ordinances in place; others are striving toward that goal.

Beach nourishment. Another nonstructural approach is beach nourishment. Beach nourishment, as described earlier, involves replacing sand on an eroded beach, usually with sand from an outside source. This process is time consuming and expensive. Beyond these constraints, several other factors need to be taken into consideration when undertaking a beach nourishment project, such as the rate of loss of beach in the region, the availability of beachfill to be used, methods used for the process, and the suitability of beachfill. It is important that beachfill material closely resemble that of the original beach. Furthermore, beach nourishment should be recognized as a short-term approach. It does not prevent erosion, but only prolongs shoreline displacement. This approach thus requires a continual financial commitment, which is expensive over a long time horizon. If undertaken, beach nourishment should be practiced regionally, and in a manner consistent with the objectives for the region.

LAND-USE MANAGEMENT

Zoning or land-use regulation. The application of zoning or land-use regulations constitutes local governments' primary means of regulating both land use (residential, industrial, or commercial) and land quality and characteristics (bulk, height, setbacks) (Beatley, Brower, and Schwab 1994). Although many coastal communities in New Jersey have adopted mitigation ordinances to protect their dunes and to regulate development, there are several additional mitigation strategies that can be incorporated into zoning ordinances to further enhance public safety. These measures include the designation of high-hazard zones and the establishment of maximum-density development thresholds.

If coastal communities were to clearly define and establish high-hazard zones, ordinances could be adopted to prohibit or restrict development in these areas, thereby reducing the public's exposure to hazards. High-hazard zones should include, but not be limited to, areas of dune fields and areas that experience frequent washover, rapid erosion, or island breaching. These zones could also include areas that, under normal conditions, would have dunes even if no dunes are present at the time (NJDEP 1984b). By identifying such high-hazard areas, communities will also be able to create buffer areas and promote beach-dune interaction. Ideally high-hazard zones should be periodically monitored to account for the dynamic nature of the coastline and should be shifted appropriately as the coastline changes.

Once these high-hazard areas have been identified, certain structures may become classified as "nonconforming uses" in order to achieve hazard management and mitigation objectives. Under the Municipal Land Use Law (N.J.S.A. 40: 55D-1), structures deemed "nonconforming uses" can only be restored or repaired after a disaster if the damages are not substantial (NJDEP 1984b). Although communities may otherwise define "substantial damage," it is generally taken to be less than a 50% loss (NJDEP 1984b). Thus nonconforming structures damaged by more than 50% would not be permitted to be rebuilt unless they complied with existing mitigation ordinances.

Another land-use strategy that municipalities can use to reduce hazard risks is to impose a maximum-density development threshold. By setting a maximum threshold, the community will lower its development density over time, and hence the extent of population and property exposed to coastal storm events in the future will greatly be reduced. Restricting density development also ensures that during times of emergencies, the population will not exceed the carrying capacity of the community (including its road capacity, water supply, medical assistance, sewerage, and land area).

In the *Coastal Storm Hazard Mitigation Handbook*, NJDEP recommends that municipal ordinances be adopted that prohibit construction of high-rise or multifamily structures in high-hazard areas (NJDEP 1984b). Preventing the construction of these types of structures in such areas is another way to substantially reduce the exposure of the population to future storms. An additional benefit to this type of zoning is that as the shoreline erodes, single-family dwellings will be easier to relocate compared with larger structures (NJDEP 1984b).

In sum, the application of zoning is the primary means for municipalities to regulate land-use development, and it can be used to enforce hazard management and mitigation strategies. By designating high-hazard coastal areas, communities may limit the type of development (single-family versus multifamily dwellings) and the density of the area. Ordinances can also be used to create buffer areas and preserve beach-dune systems. Zoning can be an effective means of reducing both human and structural exposure to coastal hazards.

Construction setbacks. Setback lines, an extension of zoning, are an extremely effective hazard management and mitigation strategy. Established by either state or local regulations, setback lines may prohibit any development other than specifically exempted water-dependent uses or shoreline-stabilization measures seaward of the line. In high-hazard coastal areas, setback lines can also be used to create buffer zones against the impact of coastal storms. These zones reduce the impact of development on beach and dune systems, reduce people's exposure to the effects of coastal storms, and provide an area for natural dune migration.

Setback lines may be established on the basis of a combination of factors, such as erosion rates, wave run-ups, V-zone (velocity-zone) boundaries, presence of dunes, vegetation line, shoreline-stabilization structures, distance from shoreline, and elevation (NJDEP 1984b). One of the strictest setback lines is employed by North Carolina. For small-scale development in beachfront areas, all new development involves a setback equal to a distance of 30 times the average annual erosion rate for that particular stretch of coastline, measured from the first stable vegetation line (Platt 1992). Development must also be landward of the crest of the "primary dune" and the toe of the "frontal dune." By comparison, New Jersey employs a "fixed line" to determine the state's coastal high-hazard setback line. The New Jersey Administrative Code (NJAC) 7:7E-3.18(d) (1995) requires that permanent structures be set back from oceanfront shoreline-stabilization structures, typically including bulkheads, revetments, and seawalls, by a minimum of 25 feet. This setback line is an important mitigation effort in preventing potential structural damage from storm wave run-up and penetration.

Although the setback line reduces potential damage to buildings behind shoreline-stabilization structures, New Jersey has not adopted setback lines that could protect coastal dunes. Rather, most coastal communities rely on dune ordinances that establish a static dune-protection area. In the 1984 assessment of dune ordinances, NJDEP found most dune ordinances only described a fixed and static line, such as a building line or a dune area, which did not account for future beach erosion or dune migration landward past the line. As consequence, the ordinances do not prevent construction in natural dune areas that are landward of the building line (NJDEP 1984a).

An opportunity exists to create dune protection zones and setback lines in the state's emerging hazard management and mitigation strategy. These zones could include a description of the dune's location and could be marked on a map with appropriate definitions provided. Landward of this zone, a setback line could be established that would provide adequate space for an extension of the natural beach-dune process. Such setback lines would allow dunes to extend landward naturally as wind and washover accumulates sand in the area.

In order for setback lines to be most effective, the location of the line should be flexible; that is, it should be able to shift as conditions change. Hence the boundaries of the zone must be periodically reviewed and adjusted to account for natural beach-dune profile changes over time. However, coastal researchers do not advise that such lines be moved seaward following extensive beachfill projects that widen beaches. Ordinances should be written to allow for review and redesignation of dune zones and setback lines every two to ten years or following significant storm events. These changes could be located on maps as well.

In summary, setback lines can be an effective method of addressing coastal hazard reduction. By requiring that new development and redevelopment following post-storm cleanup be located certain distances landward from coastal dunes, a buffer zone for hazard management and mitigation is created. This area would protect dunes as well as reduce the public's direct exposure to the effects of coastal storms. In order for dune protection zones or buffer zones to function well, the dynamics of dunes need to be accounted for. To be most effective, setback lines should not be static or fixed. For maximum effectiveness, setback lines and dune areas require periodic review to determine and incorporate changes as the coastline changes.

Elevation of structures. Another land-use option involves the elevation of structures above an established base flood elevation. This can be accomplished by placing a structure on pilings, allowing flood waters and storm surge to pass underneath without significant damage to the structure. This strategy will reduce the amount of structural damage associated with the event.

The elevation of structures is a damage reduction requirement for all communities that participate in the NFIP. Participating communities are required to adopt regulations that will protect any new construction from inundation by a 100-year flood (that is, a flood that has a 1% chance of occurring in a year, or a probability of once in 100 years). Base flood elevations (BFEs) and Flood Insurance Rate Maps have been developed by FEMA to assist communities to define and depict a 100-year coastal floodplain and the corresponding elevations of a 100-year flood. The 100-year coastal floodplain is then divided into a velocity-zone (V-zone) and an A-zone (involving bayside and other nonopen ocean shorelines).

The V-zone (velocity zone) is that portion of the 100-year floodplain that would be inundated by tidal surges with high-velocity wave action (also known as the coastal high-hazard area) where wave heights of at least a 0.9 m (3 ft) breaking wave would reach and where the still-water depth during a 100-year flood is less than 1.22 m (4 ft) (figure 9.1). Additionally, the FIRMs for coastal communities take into account BFEs that incorporate wave heights or wave run-up associated with a 100-year flood. Although erosion is taken into consideration when determining V-zones, it is often difficult to project erosion into the future, because the erosion rate is nonlinear and is also related to episodes of human manipulation of the sediment budget. The A-zone extends inland from where the V-zone ends and is defined as the portion of the 100-year floodplain not subject to significant wave action (that is, to breaking waves under 0.9 m (3 ft) high, see figure 9.1) (FEMA 1986a).

Although these zones are adjacent to each other, construction requirements in the V-zone and A-zone differ. In coastal V-zones (as opposed to riverine V-zones), any new construction and substantial improvements (greater than 50% of the value of a structure) to existing buildings or

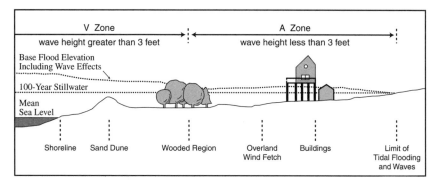

Base Flood Elevation
Including Wave Effects

100-Year Stillwater

Mean
Sea Level

| Shoreline | Sand Dune | Wooded Region | Overland Wind Fetch | Buildings | Limit of Tidal Flooding and Waves |

V Zone — wave height greater than 3 feet

A Zone — wave height less than 3 feet

Figure 9.1. Relationship of V-zone and A-zone to water depth and breaking waves. (Adapted from NRC 1990.)

structures must be elevated on anchored pilings or columns so that the bottom of the lowest horizontal structural members (excluding pilings and columns) is at or above the BFE to qualify for federal flood insurance (FEMA 1986b). A registered professional engineer or architect must certify that the structure is securely fastened to these anchored pilings or columns to withstand high-velocity flood waters and hurricane wave surge forces. In addition, it is required that no fill be used for structural support of new or substantially improved structures in V-zones, and that sand dunes not be altered so as to increase the potential for flood damage. In coastal A-zones, new construction or substantial improvements must be elevated so that the lowest floor, including basements, is at or above the BFE. However, the use of fill-raised foundations or piles and columns can be used to attain this elevation (FEMA 1986b). Figure 9.2 demonstrates the different construction requirements associated with V- and A-zones.

BUILDING AND HOUSING CODES

Building codes represent an effort to improve the overall quality of new structures through the requirement of minimum uniform levels of materials. In hazard areas, the effect of the codes is to strengthen new structures to enable them to better withstand hurricane-force winds and the pressure exerted by waves and storm surges, and thereby achieve a reduction in the amount of structural damage from coastal storms in principle. This involves specifying standards for design, materials, and building practices of all new construction (building codes) and construction that involves substantial improvements to existing structures (housing codes). Because housing codes apply to improvements of existing structures, their overall effectiveness is limited and dependent on the

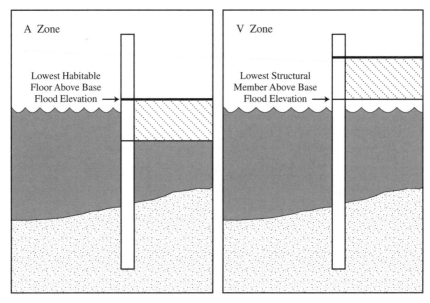

Figure 9.2. Comparison of construction requirements in A-zone and V-zone. (FEMA 1986b.)

proximity of existing structures to the waterfront, unlike the standards for new construction.

Building codes can be mandated at the national, state, or local level and can vary substantially in their requirements. Although some standards have been produced on a national level, most model building codes are developed on a regional basis, reflecting differences in building environments, and are enacted and administered at the state and local level (FEMA 1991a). Responding to different environmental factors, four major model code organizations have been created: the Building Officials and Code Administrators (BOCA), the Southern Building Code Congress International (SBCCI), the International Council of Building Officials (ICBO), and the Council of American Building Officials (CABO). The codes promulgated by these groups may take the form of either performance standards or construction specifications. New Jersey and most of the eastern and midwestern states have adopted BOCA's model codes, which for the most part are performance standards (NJDEP 1984a).

New Jersey and Building Officials and Code Administrators codes. In New Jersey, the Building Officials and Code Administrators National Building Code has been adopted as the Uniform Construction Code (N.J.S.A. 52:27D-1 et seq.) and must be used by all municipalities

(NJDEP 1984a). Municipalities must follow the standards as set by BOCA codes as their only construction standards and cannot supplement them with more stringent standards if so desired. The New Jersey Department of Community Affairs administers and enforces these codes at the local level. Since the passage of the 1981 Shore Master Plan, the Uniform Construction Code has continued to be updated to reflect BOCA's new recommendations to further reduce property damage from water-related effects.

In 1984 flood-proofing codes were adopted nationally for the first time and have subsequently been revised (NJDEP 1984a). In the 1993 version of BOCA codes, Section 3107.0 refers to flood-resistant construction. This particular code applies to all buildings and structures under construction in areas subject to flooding, and to all buildings undergoing alterations and repairs that exceed 50% of the value of the building (BOCA 1993). Flood-prone areas are determined from the 100-year flood elevation depicted on Flood Insurance Rate Maps, where this elevation, in turn, is used as the base flood level. Any building or structure located within the flood-prone area must have its lowest floor elevated at or above the base flood elevation. Structures located in the V-zone must have all structural members that support the lowest floor located at or above the base flood elevation. The flood-proof code also requires that the structural systems of all buildings and structures in flood-prone areas be designed and anchored to resist flotation, collapse, or permanent lateral movement due to structural loads and stresses from flooding and storm surges (BOCA 1993).

FEMA and model building codes. One aspect of FEMA's mitigation effort to reduce the public's risk from hazards is to promote the adoption of state and local loss reduction standards for all new construction and for renovations and substantial improvements on older structures (FEMA 1991b). The brochure *Disaster Mitigation: Reducing Losses of Life and Property through Model Codes* (FEMA 1991a) emphasizes the adoption of model building codes as a primary means for reducing the nation's exposure to risks from natural and man-made hazards. FEMA constantly strives to incorporate flood-proofing standards into model building codes by working closely with the four national building code organizations. Through the efforts of FEMA, most of the organizations, including BOCA, have adopted a majority of the National Flood Insurance Program building regulations (FEMA 1991a).

In order for communities to participate in the NFIP, communities must meet FEMA's minimum floodplain management standards. To help communities do so, FEMA has produced manuals and guidelines for the design and construction of buildings in areas subject to coastal storms and flooding. In 1986 FEMA distributed the *Coastal Construction Manual,* the second edition of these guidelines. The manual is meant to assist builders, designers, communities, and homeowners in complying with the performance standards of the National Flood Insurance Program. FEMA stresses that the manual is not intended to encourage construction in hazardous areas, but is meant rather to ensure that construction in these areas is designed and erected in a manner that minimizes potential wind and flood damage (FEMA 1986a). The *Coastal Construction Manual* includes new or expanded guidelines on construction materials, foundations, fasteners, anchorings, braces, shapes and configurations of houses, and ability to withstand wind and water loads. In the second edition changes included design guidance for breakaway wall enclosures, maintenance recommendations, and revisions of design procedures that reflect the American Society of Civil Engineers' standards and other design information. Overall the manual emphasized FEMA's goal to elevate, floodproof, and securely stabilize structures in high-hazard coastal areas to avoid or reduce potential damage.

Wind-load reduction standards. After a succession of severe coastal storms in the 1990s, emphasis on building codes now involves developing structures that will withstand hurricane-force winds as well as flooding. Although wind standards had been adopted earlier, Hurricane Andrew in south Florida in 1992 showed that homes that were constructed to survive 120 mph wind speeds still sustained considerable damage. One explanation for the severe damage was that enforcement of existing codes was inadequate (FEMA 1995b). An alternative is the enforcement of "prescriptive" codes rather than the present "performance" codes (FEMA 1991a). Whereas performance codes determine how a building should perform, a prescriptive code indicates specific design and construction criteria for new structures, such as size, dimensions, and material composition. Prescriptive codes that provide specific building designs against wind damage are beginning to be incorporated into model building codes.

Leadership at the state level. Building codes can be a proven method for reducing a coastal community's risks from coastal hazards, but to achieve a maximum level of effectiveness, they need to be amended and adopted at a national and state level. For the most part, coastal structures in New Jersey have withstood storm waves and winds fairly well because of the adoption of model building codes. However, most of these struc-

tures have not been tested by hurricane-force winds or water-surge veloc-
ities. Hurricane Andrew proved that the adoption of performance standards was not enough, and there is a need for better building code
enforcement on a national level. If the state of New Jersey is to avoid a
disaster similar to the one that occurred in Florida, appropriate prescriptive codes should be the absolute minimum for all new construction and
should be incorporated into the Uniform Construction Code. Because
coastal communities are unable to mandate new building codes, either
BOCA or the state should develop prescriptive codes with which to
enforce building compliance and adoption. Alternatively, new legislation
could be introduced that would enable communities to supplement
BOCA's codes with more stringent flood-proofing controls. Last, as the
insurance industry continues to develop wind-reduction incentive programs, grading systems, and other programs, the state of New Jersey
should support these efforts and possibly even work collectively.

RELOCATION FROM HAZARD ZONES

Relocation of existing structures from eroding and/or flood-
prone shorelines is another hazard management and mitigation option
the state and coastal communities have, although it is an extreme measure and one that is quite controversial. It is probably the most effective
means for reducing the potential for loss of life and structural damage,
however, since structures can only be damaged if they are located in high-
hazard areas.

There are a number of institutional and economic impediments associated with relocation. If space is not available (that is, adequate setback
space) on the same lot for the structure to be moved inland, an alternative site must be acquired and prepared (NRC 1990). Further problems
arise if the alternative site lacks the view and/or direct access that are often
the reasons for shoreline property ownership in the first place. If relocation is to become a realistic hazard management and mitigation option,
much more public education regarding the hazards of living near the
shoreline is necessary. The identification of high-hazard areas can assist in
these efforts and foster public awareness.

ACQUISITION OF PRIVATE PROPERTIES
IN HAZARD ZONES

The acquisition of private properties in high-hazard zones is still
another hazard management and mitigation option that coastal researchers,

land-use planners, and public policy makers have voiced in recent times. By acquiring high-hazard lands, coastal communities and the state can reduce the loss of life, personal injury, and damage to property caused by coastal storms. Acquisition of hazardous areas can also reduce the potential storm-related losses and vulnerability of coastal communities by creating the opportunity for natural coastal protection areas (buffer zones) and the development of coastal dunes. Beyond protecting the public's safety, there are also other associated benefits to coastal land acquisition such as an increase in public beaches, public access, recreational opportunities, and conservation areas.

Acquisition of coastal land has several drawbacks, primarily the high cost of acquiring developed property and the loss of property taxes (ratables) to a community. However, if the acquired property is to provide public access for recreational opportunities, the loss of property tax revenue may be partially offset by an increase in revenue derived from beach fees and other resulting services (NJDEP 1984b). A hidden benefit to community, state, and federal governments is that land acquisition programs can provide a cost savings to local, State, and federal agencies by decreasing the need for emergency expenditures and post-storm clean-up and repairs.

High-hazard coastal areas can be acquired through voluntary means (for example, willing sellers, land gifts to the state such as conservation easements) or condemnation for a "fee simple" or "less than fee simple" purchase following post-storm events. Acquisition of "fee simple" property is the most desirable type of acquisition, because the ownership and property rights are completely transferred (Clayton 1987). "Less than fee simple" purchases can also be a useful acquisition strategy. Examples of less than fee simple acquisitions include easements, leases, transfer of development rights, and donations. Although less than fee simple acquisitions cost less than do fee simple purchases, they may not provide total ownership and control of the acquired land.

The implementation of land acquisition programs can be as either pre- or post-storm programs. Pre-storm acquisitions are preferable to post-storm acquisitions, because the exposure of people to coastal hazards will be minimized prior to any disaster (NJDEP 1984b). However, many property owners in high-hazard areas are reluctant to move until their property has been substantially damaged by a storm. Thus if the state or community condemns the property, the cost of pre-storm acquisitions can become quite high. In the 1981 *NJSPMP,* post-storm acquisitions were preferred to pre-storm acquisitions under the rationale that expected damages to structures from a disaster could reduce the cost of post-storm buyouts (NJDEP 1981). However, this approach to post-storm acquisition has changed. In recent years, as an incentive to property owners, states with successful post-disaster acquisition programs have offered to purchase the property at the fair market value as it existed immediately prior to a disaster. The state of New Jersey's Coastal Blue Acres program uses this concept to determine the fair market value for post-storm acquisition.

Coastal Blue Acres Program. In 1995 the state legislature of New Jersey established the Coastal Blue Acres Program, known as "Blue Acres," a $15 million bond program to assist coastal communities in acquiring land most susceptible to storm damage and erosion from willing sellers (NJSGA 1995; P.L. 1995, c. 204). Hence acquisition of high-hazard coastal property has become a more viable hazard management and mitigation strategy for the state. Beyond serving as buffer areas within the coastal communities, all Blue Acres parcels will be utilized as conservation or recreational sites with public access. Administered by NJDEP's Green Acres Office, the Coastal Blue Acres Program is divided into two specific types of purchases: pre- and post-storm acquisitions. Post-storm acquisition involves the purchase of developed properties damaged by a storm, whereas pre-storm acquisitions involve the purchase of open, undeveloped, or largely undeveloped parcels that are low-lying or in close proximity to the water.

Under the post-storm acquisition program, $9 million is available for coastal communities to purchase lands in the coastal area that have been damaged, on up to a 50% grant–50% loan or matching funds basis. In order for storm-damaged property to qualify for Blue Acres, the value of improvements to the property must be reduced by at least 50% from damages caused by storm or storm-related flooding, and half of the cost must come from nonstate sources. In an effort to expedite post-storm buyouts and provide emergency relief, the state's portion of the post-storm purchase only needs the approval of the state legislature's Joint Budget Oversight Committee. Not only will a quick response provide the community with a unique opportunity to relocate citizens away from hazardous areas, but it may also help victims stabilize their lives as quickly as possible (Patton 1993).

The remaining $6 million is available to coastal communities for pre-storm acquisitions on up to a 75% grant–25% loan or matching funds basis, to assist them in acquiring lands in the coastal area. Acquisition criteria focus on lands that are prone to recurring damage from storms or storm-related flooding, that may buffer or protect other lands from such damage, or that support recreation and/or conservation. Because this type of purchase is planned rather than on an emergency basis, pre-storm acquisition projects must first be approved by the state legislature.

In administering the Coastal Blue Acres Program, priorities on the use of its limited funds must be followed, such as the purchase of property that in turn increases public safety and natural coastal protection. Before any Blue Acres parcels are acquired, NJDEP Green Acres staff must first determine an acquisition strategy for the Blue Acres program that ensures that parcels purchased promote the act's objectives. By creating and administering an acquisition plan, the Blue Acres program can reduce the fairly continuous stream of expenditures and investments by local communities and the state for post-storm cleanup and repair and shoreline-stabilization projects in known high-hazard areas. In order to achieve a reduction of the public's exposure and associated long-term risks from coastal hazards, these high-hazard areas need to be identified and characterized. Once they have been so designated, an advisory committee could assist the Green Acres staff in developing a list of specific parcels or areas to consider.

FUNDING SOURCES IN SUPPORT
OF ACQUISITION

As the coastal zone of New Jersey continues to be developed, property values will continue to increase, making land acquisition in the future an expensive option. The level of funding available for acquisition is therefore a critical aspect of the program's success. Several federal, state, and private sources of funding exist that can be potential sources of additional funding. Utilization of these sources could increase the amount of money available under the Blue Acres program.

Green Acres program. The state of New Jersey's successful Green Acres program provides funding for up to 50% of the total cost of an open-space acquisition project (P.L. 1995, c. 204). Because the Green Acres program grants funds for acquisition projects across the state, including coastal areas, it is another potential funding source for acquisition of coastal land in high-hazard areas. Under this program, land with unique natural features, water frontage or water resources, and/or other characteristics can be acquired for conservation and recreational purposes.

Unlike the Blue Acres program specifically for the coast, which only allocates funds to local governments, nonprofit and private groups can qualify for up to 50% of the cost of the project in matching funds under Green Acres.

Nonprofit sector. Because many nonprofit groups can qualify for matching Green Acres funds, several such groups have actively participated in acquisition projects. Besides grants from the Green Acres program, these groups also receive funding through gifts of personal and private land and monetary donations. Although most nonprofit groups acquire undeveloped lands to preserve and protect the biodiversity of unique habitats, some acquisitions are used to promote the expansion of public recreational areas as well. The Trust for Public Land, the Nature Conservancy, the New Jersey Conservation Foundation, and the New Jersey Audubon Society are examples of nonprofit groups that have participated in land acquisition projects in New Jersey. They could be potential partners, providing the funds that qualify as the nonstate share of the project, for coastal land acquisition via Green Acres.

Private sector. Private corporate or personal donations are also potential contributors to the nonstate's share of a project's cost. Large companies or businesses benefiting from the shore area and related tourism could make tax-deductible contributions toward the Coastal Blue Acres Program. Private donations of land or money could also be a large source of funding for the nonstate share of the program as well (NJDEP 1984b). Communities or groups acquiring property are encouraged by Blue Acres to make private donations part of the nonstate component.

New Jersey Shore Protection Fund. The Shore Protection Fund was created by the state legislature for the purpose of financing shoreline-stabilization projects (P.L. 1992, c. 148). The fund now receives $25 million annually from New Jersey's realty-transfer tax, to be allocated toward shoreline-stabilization projects. It is possible that this fund could be a funding source for coastal acquisition projects as part of an approach to shoreline management, where, for example, coastal acquisition could obviate the need for a $5 million shore-stabilization project and thus potentially provide up to $5 million in additional funds for land acquisition.

FEMA's Hazard Mitigation Grant Program. Under Section 406 of the 1993 Stafford Act amendments, FEMA may contribute up to 50% of the cost of funding for approved hazard management and mitigation measures. After a declared disaster, the state and local governments can apply through the Office of Emergency Management for funding for acquisition and relocation projects that result in the protection of public or private property (P.L. 100-707, 1993). Many states affected by the 1993 Midwest floods have used these funds to successfully relocate both people and buildings and to acquire substantially damaged property (FEMA 1993b). Such hazard management–mitigation grants can be a valuable source of funding for communities acquiring parcels through New Jersey's Blue Acres postdisaster grant/loan program.

HUD Community Development Block Grant Program. Each year, states and small cities (under 50,000 in population) apply for Community Development Block Grants (Non-Entitlement), available from the U.S. Department of Housing and Urban Development to assist in promoting sound community development. Under the program, these grants may be used toward projects such as the acquisition of real property, as long as no less than 70% of the funds are used for activities that benefit low- or moderate-income persons (U.S. HUD 1996). HUD also offers loans to communities to finance property acquisition under Section 108 of the Community Development Block Grants. These loans are granted for projects that either principally benefit low- or moderate-income people or assist community development to remedy problems (via acquisition or demolition) that present a serious and immediate threat to the health and welfare of the community, such as a coastal storm disaster (U.S. HUD 1996). Either of these Community Development Block Grants could be a source of funding for a community's share of the Blue Acres coastal acquisition program.

Small Business Administration Disaster Loan. The federal Small Business Administration (SBA) Disaster Loans, granted after declared disasters, are administered by FEMA. The SBA offers low-interest loans to repair or replace damaged property and personal belongings not covered by state or local programs or private insurance (FEMA 1994b). Under the SBA loan, homeowners are eligible for loans of up to $200,000 for repair or replacement of real estate; the loan may be increased an additional 20% for the use of mitigating devices for damaged real property. Businesses may receive loans worth up to 100% of their uninsured, SBA-verified disaster-related losses, and these monies similarly may be increased an extra 20% for the use of mitigation devices for damaged real property. It follows that SBA loans could be used in conjunction with other disaster benefits an owner has received to lower the overall cost of postdisaster acquisition projects.

National Flood Mitigation Fund. The National Flood Mitigation Fund was created in 1994, replacing the National Flood Insurance Program's Upton-Jones amendment. This fund is to be used for mitigation activities such as the relocation and acquisition of structures that experience repetitive losses (P.L. 103-325, 1994). Funding for this program comes from surcharges on flood insurance policies. The National Flood Mitigation Fund can also be a source of coastal land acquisition funding.

Individual and Family Grant Program. Homeowners who do not qualify for an SBA loan or who do not have other financial or insurance resources can apply for a loan under the federal Individual and Family Grant Program. Individual Grant Program loans are another supplementary recovery loan available to help individuals or families recover from a disaster. Under the Stafford Act, an individual or family may receive a grant for up to $10,000 for disaster assistance; 75% of which the federal government will fund. The remaining 25% of the cost is paid to the family or individual from funds made available by the state (FEMA 1995e). Similar to SBA loans, Individual Grant Program loans could also be used in conjunction with other disaster benefits an owner receives to lower the overall cost of postdisaster acquisition projects.

In summary, acquisition of property in high-hazard areas is a hazard management and mitigation option for local and state governments that reduces the public's exposure to coastal hazards. With the inception of the Blue Acres program, acquisition of coastal property in high-hazard areas has now become a viable mitigation strategy for the state. Because only $15 million is available under Blue Acres, the number of acquisition projects is limited. Therefore, state and coastal municipalities can take advantage of existing complementary acquisition funding sources to expand the application of this hazard management and mitigation strategy. Acquisition should also be considered as part of a local or regional program to further reduce risk and loss.

Other Mitigation Options

INSURANCE

New property owners that use financing for property situated in identified high-hazard areas must obtain flood insurance as a mortgage requirement. Because the entire coastal zone of New Jersey is considered a high-hazard area, insurance premiums along the shore have increased over the past few years. The rates for flood insurance premiums are set and mandated by the federal government and the state insurance commissioner, and increases in premiums are at the national level; thus every property owner located in a state affected by flood insurance sees an increase. Special areas or particular zones, such as the coastal zone, are not targeted for higher premiums associated with higher risk. It is possible that in the future, as post-storm cleanup costs escalate, insurance premiums may become associated with the level of risk, resulting in higher premiums for higher risk. Such a policy was instituted in 1991 to account for investment risk to help rebuild the banking industry following the 1980s savings and loan crisis (Benston and Kaufman 1997).

WIND-LOAD REDUCTION INCENTIVES

In the insurance industry, coastal areas are increasingly viewed as areas of unacceptable losses in terms of homeowners' insurance. Wind damage is one concern. FEMA in the past has provided technical guidance on the NFIP wind requirements, but it is the private insurance industry that has taken a more aggressive approach on reducing wind-load related losses to residential structures. Following Hurricane Andrew, which caused $16 million (1996 dollars) in private insurance losses, the insurance industry was unable to receive reinsurance and was forced to find ways to cut its losses (Moore 1996). In an attempt to avoid repeat disasters, the insurance industry has begun to implement building standard changes that include incentives for reducing wind-associated damages (FEMA 1995b). Even prior to Hurricane Andrew, the insurance industry had created "Wind-Rite," a credit/debit system that allows for a more precise and accurate rating system (FEMA 1995b). Depending on the type of construction practice to reduce wind hazards, premiums and deductibles will correspond to reductions in wind-related hazards. Hence premiums/deductibles will decrease if construction practices decrease the threat of wind hazards and increase if no hazard-prevention construction practice is used (FEMA 1995b). Also prior to Hurricane Andrew, the insurance industry began a program to improve construction practices by developing a grading system for rating construction and building code enforcements to be used by insurance inspectors-adjustors, training personnel, and making efforts to incorporate these changes into universal building codes (FEMA 1995b).

The insurance industry has also begun a research and development program for the design and construction of wind-resistant houses in the hope that this will create consumer demand for stronger, safer homes (Moore 1996). In North Carolina, for example, a pilot project named Blue Sky is in the process of developing houses that resist hurricane-force winds (Ross 1995). The aim of Blue Sky and other programs is to reduce property loss, create a safer community, offset insurance costs (that is, claim payouts and hence premium receipts), and provide other tax benefits to homeowners.

Mitigation Incentives

As part of its mitigation efforts, FEMA has initiated a Community Rating System (CRS). The CRS seeks to reward communities with a rebate for additional activities they undertake, beyond the minimum requirements of the NFIP, in which they practice good floodplain management and try to minimize flood damages. Participation in the CRS is voluntary, and it is the responsibility of local governments to submit documentation that shows implementation of the different creditable activities (Beatley, Brower, and Schwab 1994). Insurance premiums are reduced for those property owners within these communities, and in turn, communities receive a rebate on their insurance premiums.

There are four categories under the CRS program, with 18 mitigation activities that the CRS will credit: public information, mapping and regulation, flood damage reduction, and flood preparedness. Points are assigned for these activities, depending on the extent to which the community has successfully achieved the CRS objectives (table 9.1). The points accumulated from each individual measure are summed to determine the community's total points, from which the premium reduction is based and calculated. Premium reductions range from 5% to 45% for property within Special Flood Hazard Areas (SFHAs) (table 9.2). There is a maximum 5% reduction allowed for property that is outside of SFHAs, because premiums are already low in these areas and because means by which credits are given are based on the zones assigned to the 100-year flood level (Beatley, Brower, and Schwab 1994).

Although the CRS rating system was designed with good intentions, it is nearly impossible to achieve the highest rating. Presently, in order for a community to receive a high CRS rating, it would have to remove nearly everyone from the floodplain. As a result, communities that are credited with initiating many mitigation activities, such as Ocean City, are still only rated as Class 8.

Table 9.1 ~ EIGHTEEN MITIGATION ACTIVITIES IN THE COASTAL RATING
SYSTEM PROGRAM

Activity	Maximum points	Average points	Applicants (%)
300 Public Information			
310 Elevation Certificates	137	73	100
320 Map Determinations	140	140	92
330 Outreach Projects	175	59	53
340 Hazard Disclosure	81	39	40
350 Flood Protection Library	25	20	77
360 Flood Protection Assistance	66	51	45
400 Mapping and Regulatory			
410 Additional Flood Data	360	60	20
420 Open Space Preservation	450	115	42
430 Higher Regulatory Standards	785	101	59
440 Flood Data Maintenance	120	41	41
450 Stormwater Management	380	121	37
500 Flood Damage Reduction			
510 Repetitive Loss Projects	441	41	11
520 Acquisition and Relocation	1,600	97	13
530 Retrofitting	1,400	23	3
540 Drainage System Maintenance	330	226	82
600 Flood Preparedness			
610 Flood Warning Program	200	173	5
620 Safety	900	0	0
630 Dam Safety	120	64	45

SOURCE: Beatley, Brower, and Schwab (1994).

Table 9.2 ~ PREMIUM REDUCTIONS BASED ON THE EIGHTEEN
COASTAL RATING SYSTEM MITIGATION FACTORS

Community total points	Class	SFHA credit (%)	Non-SFHA credit (%)
4,500	1	45	5
4,000–4,499	2	40	5
3,500–3,999	3	35	5
3,000–3,499	4	30	5
2,500–2,999	5	25	5
2,000–2,499	6	20	5
1,500–1,999	7	15	5
1,000–1,499	8	10	5
500–999	9	5	5
0–499	10	0	0

SOURCE: Beatley, Brower, and Schwab (1994).
NOTE: SFHA = Special Flood Hazard Area.

Post-Storm Recovery Plans

Post-storm recovery plans should be developed and periodically revised to serve as an aid in reducing future losses in high-hazard areas. These plans can provide guidelines for a more idealized distribution of land use, avoiding development in problem areas and shepherding the environmental, economic, and cultural resources of the area. For most coastal areas, the post-storm period is the only "window of opportunity" in which to alter existing land uses and to move toward a long-term objective. The post-storm recovery period is recognized as an appropriate time to apply incentives aimed at reducing vulnerability and increasing public access (Burby 1998b). Although the idea is to develop post-storm plans, these plans can, in turn, evolve into hazard management approaches for the next storm and future conditions.

Post-storm plans should incorporate the interests of all levels of government. Objectives may largely coincide among the levels, but there may be differences. Some communities will have made progress in the creation of storm hazard mitigation and post-storm recovery plans; others may have developed general concepts with little long-term strategy or plans. To develop a unified set of plans, local plans will need to be blended into a regional plan, which, ultimately, is in concert with the long-term objectives of coastal management for the state. For this to happen, it is essential that leaders at higher levels establish and direct efforts toward long-term objectives and provide overall direction for coastal management policies. Local officials should be informed about the regional goals and how local plans serve as components of a larger system. Throughout this process, there should be assistance in the form of information about the long-term objectives of the state, coastal dynamics, the application of hazard management and mitigation strategies, and opportunities for financial support both in the pre-storm period and in the recovery period. An important product of the recovery plan should be the identification of approaches that communities can use to achieve long-term objectives and, in turn, qualify for federal and/or state funds. This information can serve as input that state and local decision makers will refer to in developing a uniform set of hazard management and mitigation strategies for the post-storm recovery period.

Future of Hazard Management in
Shoreline Management

As a result of FEMA's new direction and other institutional and economic policies, subsidies that once existed for coastal development and rebuilding may no longer be readily available. With this change comes an opportunity to use coastal planning ideas as a guide to begin to redevelop specific areas of the coastal zone. Instead of the previous focus, which was on "defending the line," the new focus could be directed toward "coastal hazard management." More emphasis could be placed on public safety issues that protect the public and minimize future damage rather than on protecting property. A further goal could be to integrate coastal natural hazard management and mitigation across various jurisdictional levels with available public funds. That is, as redevelopment or recovery plans parallel regional and state long-term objectives, there is more opportunity to qualify for public funds that support local hazard management programs. Conversely, those local activities that do not contribute toward the long-term objectives of enhancing public safety and reducing damage may not qualify for public funds beyond the community level. In principle, coastal hazard management plans should originate at the local level, then synthesized at the county level to develop a regional approach, and then integrated into an existing statewide coastal hazard management-mitigation program for funding and implementation. State and possibly federal leadership will be a key factor in developing long-term objectives and in obtaining the public funds necessary to institute regional management-mitigation programs.

Conclusion

The possible array of approaches to hazard management extends from constructing hard structures or barriers, to working within the natural system using beach nourishment and dune construction, to applying programmatic land-use controls and revised building codes, to the acquisition of properties in hazardous locations. The particular approach selected is, in turn, influenced by the severity of erosion, by the long-term objectives of the region or reach, by economic considerations, by political or community considerations, by a community's preferences regarding risk, and by available technology. Some approaches are short-term responses to a particular event; others are part of a long-term program. It is possible, of course, to select short-term options within a longer-range program. Indeed, whatever combination is selected will represent a pragmatic approach to decision making in the coastal zone.

As previously emphasized, the coast should be viewed as a large sand-sharing system that is integrated regionally with a large range of cultural and natural resources. The interdependence of natural and cultural resources begs for regional management along with special attention to particular local concerns. Each of the options and approaches discussed should be treated as serving not only to assist in local settings but also to help achieve the broader regional objectives of managing the coast.

CHAPTER 10

The Reassessment of Coastal Management in New Jersey

We need to integrate social systems, environmental systems, and the built environmental systems in ways that have not been done before. . . . It is a call for a new way of thinking about how we monitor, assess, and ultimately reduce our vulnerability to . . . [environmental hazards].

S. L. Cutter, *American Hazardscapes: The Regionalization of Hazards and Disasters* (2001)

It is time for an evolutionary nationwide shift in the approach now being used for coping with natural and technological hazards by universally adopting goals that are broader than local loss reduction. . . . Continuing along the same research and management path of the past several decades will bring increased frustration for researchers, practitioners, and policymakers. A broader perspective is needed so that far more complexity in both natural and human systems can be taken into account. . . . Any new approach must be compatible with those social, cultural, and economic forces that cannot readily be changed and must encourage modification of those that need alteration for the overall good. . . . A principal goal must be to foster truly long-term mitigation and loss reduction and to avoid burdening future generations with unnecessary hazards.

D. S. Mileti, *Disasters by Design* (1999)

Background: *The* New Jersey Shore Protection Master Plan

Historical Basis

When the Beaches and Harbors Bond Act was enacted in 1978, the New Jersey Legislature required that the State Department of Environmental Protection develop a comprehensive shoreline-stabilization plan, which was eventually released as the *New Jersey Shore Protection Master Plan* (NJDEP 1981). Part of this plan's purpose was to assist decision makers in the NJDEP with the allocation of shoreline-stabilization funds.

Prior to the act, funds were allocated by the state legislature for specific projects, or small amounts were committed to conduct repairs or to react to emergencies. With a substantial amount of funds, $10 million, it was possible to think beyond local needs and consider a more comprehensive approach, fostering certain overall objectives at the shore. As stated in the *NJSPMP*, the general objective was "to reduce the negative aspects of and conflicts between shoreline erosion management and coastal development, reduce hazard losses, and satisfy user demands in an equitable way" (NJDEP 1981, vol. I, p. I-2). The 1981 master plan was based on the following elements: physical processes causing erosion; geographical distribution of erosion; review of shoreline-stabilization approaches and plans; review of national and state policies related to shoreline stabilization and the coast; evaluation of alternative approaches to mitigating the effects of erosion; prioritization of stabilization projects; and a consideration of alternative land-use coastal management approaches. The result was a comprehensive plan consistent with state policies and objectives.

An important feature of the plan involved using the coastal reach as the basic planning-management unit. Each region or reach was in principle treated as a distinct geomorphological unit with distinct sediment compartments, and usually it was a division of the coastal zone delineated or separated by inlets. (The exception was the northern reach of Long Branch to Sandy Hook.) The implicit belief was that the management of sediment was best accomplished in terms of geomorphological entities rather than artificial political units. Management strategies could be specific to the individual reaches. This approach changed the scale of management by reducing the number of units to be considered, as the 44 political entities at the coast were combined into 13 reaches.

The general description of the conditions and processes of the New Jersey coastal zone contained in the 1981 assessment illustrates the high quality of this work, and it remains one of the standard treatments. However, the *NJSPMP* is over 20 years old, and there have been improvements in data and refinements in our knowledge about the coastal zone and in managing its hazards. A combination of programs at the national, state, and local level has also generated substantial data that did not previously exist or were not readily available earlier.

Philosophical Basis

The *NJSPMP* was an important step in the development of coastal management policies in New Jersey. It served to focus attention on broad issues and to demonstrate the concept of applying a regional approach to matters as basic as shoreline stabilization. It helped elevate the interest in management from local concerns for specific beaches to broader issues associated with entire reaches or barrier islands. Furthermore, it was the first to treat the coastal zone as a dynamic system subject to change, and to view these changes as part of the natural condition, to be recognized and accounted for in policy design rather than ignored or controlled as in past treatments. A predominant message throughout the *NJSPMP* is that the shoreline is continually being modified and that trying to reverse or control the changes is not only costly but futile. The first reports from the state's Engineering Advisory Board on Coastal Erosion in the 1920s contained a similar philosophy, recognizing that continued investment in the shore by the housing and recreation sectors was a major factor in continued financial commitments to shoreline stabilization (NJBEC 1930). Even though these earlier reports acknowledged that erosion and shoreline migration were inevitable, they suggested that short-term accommodations could be achieved. A half century later, until the *NJSPMP,* we appeared to be accepting the dynamics of the natural system but not incorporating them into the goal of managing coastal areas; a similar conclusion was reached by the second National Reassessment of Natural Hazards in the United States and by some experts in the field (Burby 1998a,b; Kunreuther and Roth 1998; Mileti 1999). The *NJSPMP* provides an indication of the philosophy regarding the state's policy for shoreline stabilization at the time. It was to be based on little or no disruption of coastal processes and sediment transport in the nearshore zone. Such a policy represented a movement away from "hard" structures along the beach or at inlets that significantly altered the process of sediment transport and delivery. It recommended that reach-level engineering programs be evaluated and implemented only if they passed a cost-benefit test, and that long-term projects requiring significant commitment of funds should not be implemented as emergency projects.

The *NJSPMP* recommendations reduced the reliance on short-term, stop-gap, local projects and replaced them with broader programs covering entire reaches. This approach allowed policy tools to work hand in hand with natural systems such as wide beaches and dunes in locations where these features existed or could exist. The *NJSPMP* also represented a shift in thinking from trying to protect everything to one that carefully considered only those projects where the greatest returns from expenditures of public funds existed. It further recognized that there might be opportunities to retreat from hazardous areas and to redefine land uses in exposed locations after damaging events. In general, the *NJSPMP* advocated support for those regions that required minimal investment to continue their economic growth and enhance their natural resources, but recommended that support be withheld from those areas located in sites of high rates of erosion and that required significant funds to maintain their existence.

Engineering Approach Applied to Reaches

The *NJSPMP* was primarily directed toward evaluating the characteristics of each reach and then recommending a course of action to maintain or enhance it using engineering techniques. The incentive for the plan came from the $10 million fund derived from the 1978 Beaches and Harbor Bond Act and a process allocating the funds among the highest-ranked applications for engineered coastal projects based on a cost-benefit ratio.

A portion of the *NJSPMP* was a cost-benefit analysis of five alternative engineering plans for each coastal reach: (1) storm erosion protection; (2) recreational development; (3) a combination of storm erosion protection and recreational development; (4) limited restoration; and (5) maintenance. On the basis of the property values protected, the costs of the engineering plans, the relative contributions of each reach to the recreational economy of the state, and additional infrastructure needs, it was concluded that four reaches justified support. Three of the reaches, Peck Beach, Absecon Island, and Seven Mile Beach, achieved relatively high values in terms of the recreational development alternative (Option 2), whereas the Sandy Hook to Long Branch reach was justified in terms of the maintenance alternative (Option 5). None of the other reaches or alternative engineering plans satisfied the 1:1 benefit/cost ratio, although the recreational alternative plan for Long Beach Island reach came close.

It is important to note that the *NJSPMP* incorporated considerable information beyond the narrow requirements for calculating a benefit/cost ration. A lengthy section on basic coastal processes explained the reason for some of the erosion problems. The issues of sediment deficits, society's interference with sediment transport, and the impact of sea-level rise are raised as important variables in dynamics of the system. The concept of hazard mitigation is proposed as an approach to management of the coastal zone. Generally, all of the concerns raised in the *NJSPMP* exist today, and most of the management approaches considered in it remain viable. Some limitations exist, however, because background data were missing in some instances, because information was in the development stage, or because an approach was introduced but procedures for implementation did not follow.

The Process of the Reassessment

A reassessment of the *NJSPMP* was undertaken in the mid-1990s, under the auspices of the Institute of Marine and Coastal Sciences at Rutgers University at the request of the NJDEP. The overall goal of the project was to update existing information, gather new information, and then to communicate these findings and develop strategies for coastal management at all levels (state, public, and stakeholders). The update and expansion of the *NJSPMP*, the *Coastal Hazard Management Report* (Psuty et al. 1996), was based on the incorporation of new data and insights from coastal management, including new information on basic coastal processes, directions in land-use management, options for development and redevelopment, creation of a land acquisition program, establishment of pre- and post-storm plans, and recommendations for procedures and criteria for implementation. Two essential components of this effort were the use of a public participation process and an evaluation of existing and potential approaches to coastal management. As a primary outcome, it developed a blueprint for action to deal with issues such as severe storms, erosion, and environmental quality. Such a process included economic as well as environmental concerns and permitted flexibility in policy formulation, as opposed to the narrower statewide policy approach in the existing *NJSPMP*. In part the reassessment worked to assist resource managers in making informed decisions on a range of strategies appropriate to the various reaches that constitute the shore region. Thus the effort was to update the *NJSPMP* as a basis for broadening the options within the realm of coastal management and to include procedures and criteria for exercising the options. Rather than a shoreline-stabilization plan, the new assessment serves as a resource and guide for future coastal and hazard management plans.

To address these tasks required expertise from the fields of the coastal sciences, coastal engineering, economics, management, and public policy. Accordingly, the Institute of Marine and Coastal Sciences drew upon the knowledge of individuals at a number of educational and research institutions in New Jersey. The core of the project was the Working Committee, consisting of research personnel and staff from the institute augmented by outside individuals (figure 10.1). The Working Committee was responsible for updating the coastal data and literature base; organizing the public participation, education, and outreach components of the project; and maintaining records. Specific duties included:

- Retrieval and collation of existing data sets
- Development and dissemination of education and outreach material
- Coordination and conduct of one-on-one meetings
- Organization and conduct of public meetings
- Establishment and management of Citizen Advisory Committees
- Documentation of information generated by the public and results of public interactions
- Production of interim and progress reports

Assisting the Working Committee were groups of scientists, technicians, government agency representatives, elected officials, environmentalist groups, town agency personnel, and concerned citizens. Through a process of seeking nominations, pursuing areas of expertise, and encouraging participation, a cadre of advisors was organized into an Advisory Committee, split into three groups (see figure 10.1). Each of the subadvisory groups was a significant asset to the process. The Scientific-Technical group was a major source of data identification. The Local Policy group included mayors, freeholders, and other elected officials and provided a direct means of communicating and interacting with local elected officials and town agencies. The Citizen Advisory group was late in forming, but became extremely valuable in coordinating the county meetings and searching out local ordinances. Consultants were also used to prepare and analyze information on issues requested by the public for which outside expertise was required.

Key Tasks of the Reassessment
PUBLIC PARTICIPATION AND
EDUCATIONAL OUTREACH

The reassessment's efforts to promote public participation and conduct educational outreach involved both obtaining new information to expand on the state of existing knowledge as summarized by the *NJSPMP* and communicating these findings. Key tasks in the outreach effort were conducted by the Working Committee, whose members worked with staff from the Center for Environmental Risk Communication

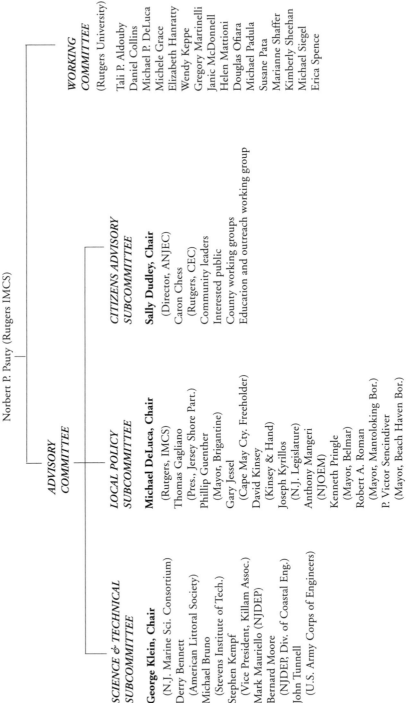

New Jersey's Shoreline Future: Preparing for Tomorrow

Project Leader
Norbert P. Psuty (Rutgers IMCS)

ADVISORY COMMITTEE

SCIENCE & TECHNICAL SUBCOMMITTEE

George Klein, Chair
(N.J. Marine Sci. Consortium)
Derry Bennett
(American Littoral Society)
Michael Bruno
(Stevens Institute of Tech.)
Stephen Kempf
(Vice President, Killam Assoc.)
Mark Mauriello (NJDEP)
Bernard Moore
(NJDEP, Div. of Coastal Eng.)
John Tunnell
(U.S. Army Corps of Engineers)

LOCAL POLICY SUBCOMMITTEE

Michael DeLuca, Chair
(Rutgers, IMCS)
Thomas Gagliano
(Pres., Jersey Shore Part.)
Phillip Guenther
(Mayor, Brigantine)
Gary Jessel
(Cape May Cty. Freeholder)
David Kinsey
(Kinsey & Hand)
Joseph Kyrillos
(N.J. Legislature)
Anthony Mangeri
(NJOEM)
Kenneth Pringle
(Mayor, Belmar)
Robert A. Roman
(Mayor, Mantoloking Bor.)
P. Victor Sencindiver
(Mayor, Beach Haven Bor.)

CITIZENS ADVISORY SUBCOMMITTEE

Sally Dudley, Chair
(Director, ANJEC)
Caron Chess
(Rutgers, CEC)
Community leaders
Interested public
County working groups
Education and outreach working group

WORKING COMMITTEE
(Rutgers University)

Tali P. Aldouby
Daniel Collins
Michael P. DeLuca
Michele Grace
Elizabeth Hanratty
Wendy Keppe
Gregory Martinelli
Janic McDonnell
Helen Mattioni
Douglas Ofiara
Michael Padula
Susane Pata
Marianne Shaffer
Kimberly Sheehan
Michael Siegel
Erica Spence

Figure 10.1. Organizational chart for conducting 1996 reassessment

at Rutgers University to design a public participation process to ensure public access and input to the Project Team, to emphasize the dissemination and discussion of information on coastal management, and to draw upon working relationships between individuals and institutions in order to achieve the project goals. Part of this effort involved contact with the individuals who had prepared the *NJSPMP* in order to determine how they designed their outreach process and to learn from their experiences. Broad-based public participation has been shown to contribute to positive negotiations on a variety of public issues, including the siting of hazardous waste facilities, NIMBY (Not In My Backyard) issues, nuclear power plants, and even penitentiaries (Fisher and Brown 1989; Fisher, Ury, and Patton 1991). The process used in the reassessment was based on findings from conflict resolution and negotiation efforts such as the Harvard Negotiation Project, and its main principles include:

- Separating the people from the problem
- Focusing on interests, not positions
- Inventing options for mutual gain
- Insisting on using objective criteria (Fisher, Ury, and Patton 1991)

The Working Committee also developed a series of short handouts for distribution to the general public concerning the *NJSPMP* revision effort, fact sheets on coastal management issues, short reports that characterized the New Jersey shoreline, and educational activities to be infused with existing public school science curricula to raise awareness of coastal management issues among students.

Educational material was also disseminated at each of the one-on-one and public meetings used to gather input from the public and organized groups. Press packages were distributed in advance of each public meeting. To facilitate distribution, handouts were mailed to all key players and other interested parties compiled on a mailing list. The dissemination of educational material to school districts was accomplished with the assistance of the New Jersey Marine Educators Association, New Jersey Science Teachers Association, and the New Jersey Council on Elementary Science. The material was offered as in-service training for teachers in school districts with which the institute was collaborating, and it was also made available to existing reform efforts for math and science education, such as the Statewide Systemic Initiative sponsored by the National Science Foundation, and Hunterdon 2001, a countywide effort to disseminate "hands-on" science and technology into the classroom.

In addition, the Working Committee conducted one-on-one meetings with key players in New Jersey. These individuals represented interest groups and constituencies with strong interests in coastal management issues, such as local officials of coastal communities (mayors, freeholders, county engineers, environmental commissioners), state and federal officials, public interest groups, and marine trade groups. Members of the Project Team met informally with these key players to describe the project's goals and tasks, discuss local issues, and solicit support to disseminate information on the project to their representative constituency or group members. The one-on-one meetings were followed by more formal presentations at local meetings such as town committee meetings.

After this initial round of information exchange, the Working Committee conducted a series of public meetings that focused on the reassessment effort. The structure of these meetings was determined from the design of the public participation process, and included presentations by disciplinary experts as well as opportunities for dialogue between the Project Team and the general public. Public meetings were held in each of the four coastal counties (Monmouth, Ocean, Atlantic, and Cape May). The Working Committee also developed an exhibit that was displayed at organized events throughout the year such as the Barnegat Decoy Show and the Belmar Seafood Festival. The exhibit was designed to capitalize on organized events to further disseminate information about the project and to discuss it with event patrons.

NEEDS ASSESSMENT STUDY

A needs assessment study was conducted to identify data gaps in the coast's physical processes and the constraints driving the shoreline response. Such an assessment involves bringing forth new concepts related to managing coastal resources. It seeks to identify a broader range of options regarding the natural and cultural components of the area and to find strategies to incorporate these alternatives in a new mechanism for managing an evolving coastal zone. As part of this discovery process, the key task was to identify and evaluate the major techniques used in New Jersey for coastal management. The evaluation was made in terms of three different contexts:

1. Basic information on coastal processes, shoreline change, and approaches to shoreline stabilization
   - Wave and current data

- Sea-level rise
- Magnitude and direction of shoreline change
- History and evaluation of beachfills
- Distribution and performance of structures

2. Systematic dune-building and other hazard mitigation programs
   - Dune-building programs
   - Hazard management plans
   - Emergency beach and dune restoration grants
   - Local ordinances

3. Revisions to existing land-use practices
   - Evaluation of existing land-use regulations at all levels of government concerning coastal hazards
   - Review of case studies in land-use management that have successfully addressed coastal hazards
   - Development of strategies for coastal land acquisition, post-storm plans, and implementation mechanisms

It was recognized from the start that the third area was the most contentious element of this reassessment process and required special attention. For example, the option of acquiring presently developed land damaged by a storm as part of a post-storm effort fueled emotions and provoked intense debate. Nevertheless, it was determined that the issue warranted evaluation as one of many possible strategies. The public meetings also considered this component.

In the context of coastal hazard management approaches, a public dialogue was initiated to discuss the merits of formulating a post-storm plan. Such an option was also discussed at the one-on-one meetings, town committee meetings, and public meetings informally. The dialogue sought to consider the establishment of objectives, strategies, and procedures associated with post-storm evaluation of land use, and guidance for rebuilding portions of the coastal area to decrease exposure of people, property, and infrastructure to hazards in the future.

REASSESSMENT REPORT

From the basic information gathered in the form of data sheets and summaries of public comments, a report was prepared to (1) cover the needs assessment described above; (2) outline a variety of possible planning options to achieve the above goals; (3) discuss the outcomes expected from each option; and (4) address practical procedures for implementing each of the options. Again it was recognized that a number of the possible options would result in much debate, and that the public participation process was the best approach to encourage this discussion, in order to share concerns for coastal hazard management, to allow all sides to voice opinions on policy options and outcomes, and possibly to reach a consensus. Without such a process, proposed public policies fail in their intentions. The most difficult tasks facing such an exercise are to educate the public, address conflicts that arise, and prepare a package of policy options contributed by the group that represent an acceptable vision of coastal management. However, the reassessment's purpose was not to produce a consensus but, rather, to bring forward information on new data, new approaches, new policies and options, and the increasing concern for the safety of the public in coastal hazard areas. Furthermore, the report sought to identify measurable economic data to add to the set of available information to assist officials and agencies in decision making and policy development.

PUBLIC PARTICIPATION

An essential component of the reassessment effort was the design of a process that ensured the involvement of the public and stake-holders in the evaluation of existing and potential coastal and hazard management alternatives. The importance of this component stems in part from the shortcomings associated with the timing and structure of the public participation process used to prepare the *NJSPMP*. In that effort, the process was not initiated until the report had largely been drafted. This placed the public in a reactive position during the latter stages of the project, requiring it to respond to a report that was already prepared. Consequently, there was little opportunity for substantive change emanating from key stakeholders on the issues involved. In addition, the formal structure of the 1981 process to solicit input did not promote the incorporation of local concerns into the project.

As a result of these shortcomings, much attention during our reassessment effort was devoted to the design of a process that ensured public access and input to the Project Team. Given that the policy-making authority in New Jersey is vested in local governments and the alternatives presented by this study would rely on local governments for implementation, it was critical that the process involving reassessment be designed for grassroots participation. Consequently, the process was structured to ensure that citizens in coastal municipalities became stakeholders in the project and had a well-defined role in assisting with the direction of the effort. With this approach, results of the study were more likely to be accepted and used. In essence, the state government was not the only voice dictating how change would occur in coastal management. Rather, local governments would receive assistance in defining their long-term goals for coastal hazard management as well as help in evaluating the most appropriate strategies to meet their specific shoreline needs.

After initial efforts aimed at designing an accessible and involved public component, a Kickoff Workshop was held. It began to address key questions identified by the public and served as a key vehicle for the overall effort. Specifically, the workshop was organized to discuss and evaluate potential alternatives for coastal management in three major areas of public concern: shoreline management strategies, socioeconomics, and policy. Each thematic discussion was led by a chairperson who served as the facilitator for a work group. These chairs were asked to elicit a list of priority coastal management issues from the participants and to pose a series of questions to their groups (see table 10.1).

Work Group Areas

SHORELINE MANAGEMENT STRATEGIES

The work group for shoreline management strategies focused on four areas: strategies and associated needs in coastal areas, white paper topics, data sources, and contact groups for outreach efforts. The group discussed a range of strategy issues, and several were identified as priority areas requiring further attention:

- Incorporation of local needs into shoreline-stabilization strategies
- Advance planning to support a rapid response to emergencies
- Consistency among building codes and enforcement
- Strategies should reflect the inherent differences between postdisaster planning and long-term planning

The group recommended that scenarios or case studies should address specific issues, such as the effects of mitigation strategies on neighboring communities (that is, a regional approach) and the compatibility of various coastal management strategies deployed within an area—especially the compatibility of "hard" and "soft" engineering approaches.

Table 10.1 ~ QUESTIONS ADDRESSED AT THE JULY 12, 1994, WORKSHOP
TO DEVELOP POTENTIAL ALTERNATIVES FOR COASTAL
HAZARD MANAGEMENT

Shore Protection Strategies
- What are the most important shore management issues that must be addressed by the Project Team? Which of these issues should be addressed by white papers?
- What coastal research and engineering factors/properties must be considered in order to select the most appropriate shore management strategies?
- How would you characterize an area or coastal reach that is best suited to a "hard" management strategy? A "soft" management strategy?

Legal/Policy Issues
- What are the most important legal/policy issues associated with shore management that must be addressed by the Project Team? Which of these issues should be addressed by white papers?
- Which regulatory impediments (state and federal) are burdensome to local and county authorities responsible for shore management?
- What measures should the Project Team investigate to mitigate these impediments?
- What shore management policies have proved successful for other coastal states, especially those with high population density?

Socioeconomic Issues
- What are the most important socioeconomic issues associated with shore management that must be addressed by the Project Team? Which of these issues should be addressed by white papers?
- Should the costs associated with shore management be allocated any differently from the present method?
- What methods are best suited to ensuring public participation in the project to reassess shore management?

The group identified a broad range of existing data sources and compiled a list of outreach contacts for the Project Team. Finally, this group proposed five topics for white papers:

- Use of GIS (Geographical Information System) in the planning process
- Compilation of existing data
- Review of management protection priorities for individual coastal communities
- Postdisaster planning and long-term mitigation strategies
- Historical review of coastal planning and responses to disasters, including experience of other coastal states

SOCIOECONOMICS

From the questions posed for the socioeconomics work group, four major emerged. The first consisted of proposed white paper topics. These were:

- To assess the *NJSPMP* as a resource for data and methods
- To quantify the magnitude and distribution of benefits from shoreline stabilization and coastal hazard mitigation
- To specify the spatial and temporal scale by identifying stakeholders
- To identify linkages between management alternatives and environmental or natural resource indicators
- To specify how multiple impacts will be incorporated by cost-benefit analysis
- To clarify hazard mitigation options by linking socioeconomics with varying approaches to shoreline stabilization

The remaining considerations centered on the allocation of shoreline-stabilization costs, public involvement, and improving public access. The last issue concerned whether public funding for shoreline-stabilization projects should be linked to public access.

POLICY

The policy work group identified and discussed a variety of policy issues:

- Beach-ocean access
- Coastal hazard and resource protection area maps
- Economics of shoreline stabilization
- Public perceptions
- Use of flood insurance claims
- Adequacy of coastal flood insurance
- Regulatory versus nonregulatory approaches to shoreline stabilization
- Cost-sharing approaches

In addition, the group recommended that shoreline-stabilization policies in other coastal states be examined. Although New Jersey's shoreline is somewhat unique, the group noted that other states possess similar regions of coastline characterized by high population and use. North Carolina was identified as one such state that possesses barrier islands and employs a land acquisition program that appears to work. Along with the other two work groups, this group agreed that public input would be a useful resource for the reassessment.

Each chair summarized the results of his or her group deliberations during a final plenary session, and submitted written reports following the workshop. Priority issues were then to be addressed in white papers prepared by disciplinary experts, with drafts distributed to and discussed with the public to ensure that local concerns were indeed addressed.

Other Information Transfer

The Project Team also participated in a variety of local, national, and international meetings to keep abreast of reform efforts associated with coastal hazard management. As noted, a traveling exhibit was also developed and used for display at the project meetings as well as at organized events throughout the duration of the reassessment. Members of the Project Team capitalized on events such as seafood festivals and other shore-related activities to discuss the project with event patrons as well as to hand out project literature.

At these meetings and events, questionnaires were distributed to help gather public concerns. Flyers were also distributed in each of the coastal communities to foster more discussion of the reassessment. Committees and interested citizens were updated on the reassessment through the circulation of progress reports. To assist with the dissemination of information about the project, members of the Rutgers Cooperative Extension Service were briefed and asked to inform their constituencies. In addition, these individuals helped to publicize and arrange public meetings, and distributed written material such as fact sheets. Results of these interactions were fed directly into a variety of communication mechanisms designed to reach the public.

Committee Activities

One of the first actions taken by the Working Committee was to anticipate concerns of the public on issues associated with coastal hazard management. Several documents were prepared to address these concerns, including a one-page description of the project, a question-and-answer document, and a "top ten" list of characteristics developed to describe what the project was and what it was not. Once the priority issues had been developed via this process, fact sheets were prepared on specific topics such as sea-level rise, dune management, and coastal and shoreline management strategies.

Initially, key stakeholders were targeted for one-on-one discussions with project personnel to discuss the project goals and tasks, local concerns, and to solicit support to disseminate information on the project to their respective constituents, peers, or group members. Stakeholders included local elected officials (such as mayors and freeholders), county engineers, state and federal officials, public interest groups, and marine trades groups. These local meetings were followed by more formal presentations at "town meetings" in each of the four coastal counties included in the reassessment. The town meetings were held to initiate the outreach process and in turn to heighten public awareness of the project and coastal management issues. These meetings included presentations by experts from specific disciplines, fostered dialogue between the Project Team and the general public, and led to the creation of the Citizen Advisory Committee.

A kickoff meeting was held to organize the Citizen Advisory Committee. At this meeting, advisory committees were organized for each of the coastal counties (Atlantic and Cape May County committees were combined). This structure ensured that different priorities held by the public across the diverse regions of the New Jersey coastline would be present. Consensus was reached on the charge for the advisory committees and the process to be used to meet their objectives. Specifically, their charge was to:

- Identify local concerns (by community) related to coastal hazard management as input to the white papers
- Prepare a history of shoreline stabilization for each community, including any information on the date and extent of past beach nourishment projects, construction of engineered structures, and so on
- Collect copies of all local ordinances related to coastal hazard management, especially those that address dune management
- Become informed about the reassessment effort and assist with disseminating information on shoreline stabilization, shoreline management, and coastal hazards to local municipalities, via seminars, distribution of handouts, exhibits, school projects, and so on
- Establish a timetable for advisory committee meetings to complete the tasks identified above

Figure 10.2. Monmouth County chair David Grant and Citizen Advisory Committee members examine Sandy Hook shoreline documents. (Photo by Susane Pata.)

Chairs were elected for each coastal Citizen Advisory Committee, priorities were discussed, and plans were made to arrange a timetable and agree on a format for addressing the priorities. Project staff members were assigned as a liaison to each of these committees and helped them obtain any resources they required. Several Citizen Advisory Committee meetings were subsequently held, at which interested citizens gathered with committee chairs and Project Team members to raise and discuss issues of concern (figure 10.2).

Among the most effective means of soliciting input on local concerns were "beach walks" with local elected officials (especially mayors), emergency services personnel, and citizens. Most of the communities participated in this part of the process (figure 10.3). Held up and down the coastline, the walks provided the Project Team with examples of coastal management and mitigation measures that performed well and those that presented problems for local decision makers (figure 10.4). In some cases, the Project Team was able to deliver on-site advice and guidance to address problem areas. The intent of the walks was to ensure that local problems would be addressed by the reassessment, and to select suitable areas for case studies or scenarios to help guide coastal communities in their selection of coastal hazard management strategies to meet specific needs.

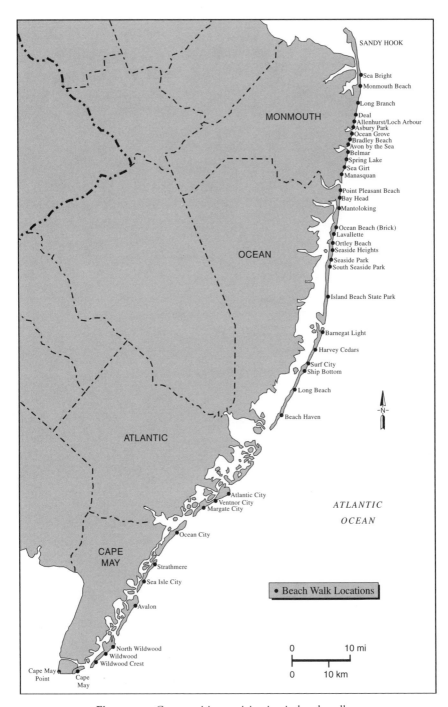

SANDY HOOK

MONMOUTH

● Sea Bright
● Monmouth Beach

● Long Branch

● Deal
● Allenhurst/Loch Arbour
● Asbury Park
● Ocean Grove
● Bradley Beach
● Avon by the Sea
● Belmar
● Spring Lake
● Sea Girt
● Manasquan

● Point Pleasant Beach
● Bay Head
● Mantoloking

OCEAN

● Ocean Beach (Brick)
● Lavallette
● Ortley Beach
● Seaside Heights
● Seaside Park
● South Seaside Park

● Island Beach State Park

● Barnegat Light

● Harvey Cedars

● Surf City
● Ship Bottom

● Long Beach

● Beach Haven

-N-

ATLANTIC

ATLANTIC

OCEAN

● Atlantic City
● Ventnor City
● Margate City

● Ocean City

CAPE
MAY

● Strathmere

● Sea Isle City

● Avalon

● Beach Walk Locations

● North Wildwood
● Wildwood
● Wildwood Crest
Cape May ●
Point
● Cape
May

0   10 mi

0   10 km

Figure 10.3. Communities participating in beach walks.

Figure 10.4. Members of the coastal community of Seaside Park discuss coastal issues with project leader Norbert Psuty. (Photo by Susane Pata.)

Television news interviews were another means of communicating information about the reassessment. Reporters from television stations interviewed Project Team members about the most pertinent coastal issues of the day and gave the team opportunities to discuss the reassessment effort (figure 10.5). The educational and outreach effort contributed more exhibits to enhance public awareness of the project, the issues of coastal hazards, and what citizens could do to become involved in the process. These were displayed in a variety of formal and informal venues ranging from legislative events at the statehouse to regularly scheduled meetings of public interest groups.

Education, Outreach, and Interpretive
Programs for the Precollege Community

The *NJSPMP* identified public education and training as a means to raise awareness of coastal management programs and policies. In addition, it stated that funding for these programs should be provided by the state and used to support public participation workshops, meetings, and hearings. Although these activities are necessary for informing the general public, they typically do not reach the precollege or K–12 community. Yet this audience represents the next generation of decision mak-

Figure 10.5. Norbert Psuty is interviewed by "New Jersey News" regarding the project and other shoreline issues. (Photo by Susane Pata.)

ers and was not overlooked. The involvement of children and young people in focusing on the significance of the environment and of recycling demonstrate the contributions that the precollege members of the community can make.

A broad range of topics and pedagogical techniques have been introduced and used to develop an informed public via the precollege community. Through the existing school system, information about public policy issues can be incorporated into existing curricula and designed to enrich the teaching of basic skills, problem solving, and the development of critical thinking skills. This approach reaches not only students, the decision makers and educators of the future, but their parents, the present community of decision makers. The Project Team made use of Project Tomorrow, an existing collection of science-based learning programs and approaches aimed at the precollege community based at the Rutgers's Institute of Marine and Coastal Sciences, to develop education and outreach activities on the topic of shoreline stabilization and coastal hazard management. The core of this effort involved activities that recognized the long-term nature of increasing public awareness to reduce the loss of life, injuries, economic losses, and the disruption of families and communities caused by natural hazards.

In a joint effort with Project Tomorrow staff, the Project Team developed additional educational activities for application in formal settings (public and private schools) and informal settings (nature centers, aquariums, and museums). These efforts focused on the development of educational materials through the enrichment of existing curricula, teacher workshops, and the establishment of library information systems and a home page on the Internet (site address accessed via *http://marine.rutgers.edu*).

Enrichment of Curricula

Efforts to enrich curricula were based on the guiding principle of the Project Tomorrow program, which stresses the paramount importance of a hands-on and minds-on approach that links real-time research to classroom science education for the development of problem solving and critical thinking skills among our students. Development of these skills, along with the support of basic skills training spurred on by the public's fascination with the ocean, is felt to be essential to prepare the next generation of informed decision makers and to raise environmental awareness generally among the public.

The Project Team organized a group of twenty K–12 educators to assist with the development of classroom and field activities focusing on issues associated with shoreline stabilization and coastal hazard management. Two supplementary curricula served as the basis for this effort— *Marine Activities Resources and Education (MARE)*, developed by the Lawrence Hall of Science at the University of California at Berkeley, and *Event-Based Science*, produced at the University of Maryland. Each of these curricula was used as a model to design and test a 14-lesson classroom and field activity guide for K–12 application. Activities were organized under four major themes: coastal geology, coastal biology, sustainable development (environmental planning and management), and global influences (sea-level rise and storm frequencies). The first version of this guide was produced and underwent pilot testing in several schools in the state during the 1996–97 academic year. The final guide was incorporated as a component of the *MARE* curriculum for the state of New Jersey. Teacher Enrichment Workshops were also scheduled once the pilot testing was completed. In addition, as part of this exercise, the Project Team developed an Internet activity that employs real data to demonstrate issues associated with sea-level rise in an interactive manner (*http:// marine.rutgers.edu*).

Teacher Workshops

Two enrichment workshops were conducted for precollege educators on the topic of sea-level rise and its effects on the coastal zone.

One workshop was designed for formal educators and the other for informal educators. These workshops produced supplementary curriculum materials to illustrate the importance of a healthy coastal ecosystem that can enable development to occur in a manner compatible with environmental concerns. The curricula addressed the 1996 science core standards for New Jersey students with a focus on development of critical thinking and problem-solving skills.

The Project Team also presented information on sea-level rise and coastal management strategies to precollege educators participating in the 1996 Earth Science Teachers Conference. This meeting provided an ideal opportunity to demonstrate classroom-based activities for incorporation into existing curricula. More informal workshops were held as well to continue the process of incorporating coastal hazard information into the school system. These included the Teach at the Beach Conference, the Global Change and Sustainable Development Teacher Workshop, and the annual meeting of the New Jersey Marine Educators Association.

Library Information Systems and Internet Home Page

During the reassessment, the Project Team collected a wealth of reports, articles, and other literature, ranging from technical reports on beach erosion and digital data of coastal structures to literature on the coastal economics and coastal history in New Jersey. This led to the creation of a comprehensive bibliography and an effort to incorporate this information into a library information system for broad dissemination and easy access. The complete bibliography was placed on the Internet and is accessed via *http://marine.rutgers.edu*.

The Project Team also constructed a home page for the project on the Internet (accessed via *http://marine.rutgers.edu*). The page became an effective means for sharing information on the project and enabled the Project Team to maintain a dialogue with the general public on coastal hazard issues, coastal erosion and stabilization, and coastal management. Requests for project information were also received and handled electronically.

Results of Public Interactions

From the participation process and the education-outreach effort, the Project Team was able to develop priorities that were respon-

Table 10.2 ～ WHITE PAPER TOPICS

Coastal processes
   Erosion and accretion
  *Dune management
   Sand transport

*Beach structures and engineering
   Hard and soft stabilization measures
   Artificial reefs
   New technology

*Socioeconomics
   Costs and benefits
   Beach use
   Tourism

*Sea-level rise
   Rate of increase
   Relationship of storm magnitude

*Storm frequencies

Education and outreach
   Formal and informal education
   Public awareness
   Informed decision making

*Coastal/state comparisons
   Federal coastal zone program
   New Jersey coastal program
   Coastal approaches used by other states
   Regional strategies

---

*Topic addressed in a stand-alone document as well as in the final report.

sive to the public, focus discussions on specific problem areas such as the need for recent information on dune management practices, and incorporate public concerns into the documents and reports generated by the reassessment. A summary report of the public meetings and interactions associated with the project and a summary report of the general meeting held on January 18, 1995, which served as a follow-up to a July 12, 1994, workshop, were distributed to all committee members. A list of the white paper topics resulting from these interactions is provided in table 10.2.

Many of the discussions centered on the need for new approaches to reduce the risk associated with maintaining the current infrastructure along the New Jersey coast in the face of the rising sea level, erosion, and frequent storms. It is clear that the highly formal, rigid approach of the past does not promote the dialogue necessary to develop coastal and hazard management strategies that respond to local concerns. Because the means to enhance the protection of life and property along the coast relies on local action by the citizens most directly affected by these concerns, the development of any new coastal hazard management policy must involve these stakeholders in the preparation of the plan and in its implementation.

Continued Public Involvement

One element of the stakeholder process was to engage coastal residents and decision makers in the collection and dissemination of information on coastal hazard issues and coastal erosion and stabilization. Beyond this, a further goal involved continued public involvement. This type of participation should be designed to give the public greater access to the ongoing decision-making process. Increased involvement can lead to a more informed public, whose decision making is separated from emotions and based on interests rather than on positions. The policies that emerge are then better understood and more likely to be accepted and heeded.

Because the manner with which communities approach coastal management depends on the existence of an informed public, one of the most significant accomplishments of the reassessment is the continuing work of the Citizen Advisory Committees. These groups played an important role in shaping the project and are the logical individuals to use in encouraging and facilitating implementation of new coastal hazard management policies. The experience of other coastal states and nations can help determine the appropriate structure and scope of activities for such committees. One such example exists in the United Kingdom (Oakes 1994).

Regional coastal groups have been used in England and Wales to prepare regional coastal management plans (Oakes 1994). Although these are voluntary groups, they are empowered by the government to improve management of coastal hazards. This includes:

- Furthering cooperation between agencies with responsibility for coastal management
- Sharing data and experience
- Identifying best practices
- Identifying research needs
- Promoting strategic planning for coastal hazard management

- Identifying impediments to the implementation of alternatives
- Maintaining awareness of policy developments, the results of research, and other initiatives

The groups meet quarterly and consist primarily of representatives from relevant government agencies, technical experts, and citizens. Periodically the group chairs meet to discuss common concerns and to develop a national approach to coastal hazard management.

This approach in England and Wales has fostered national mitigation strategies in which the effort is organized by littoral cells or reaches that recognize the importance of compatibility among measures deployed within a reach. Coastal processes do not respect political boundaries, and therefore regional approaches that respond to the broad scale on which physical forces operate along the coast are warranted. This situation requires greater coordination among coastal municipalities, especially to develop "reach-wide" hazard management programs. New Jersey's Citizen Advisory Committees, organized by county, are well suited to address coastal hazard management on such a regional scale or reach-by-reach basis.

It is hoped that the state of New Jersey, particularly its Department of Environmental Protection, will reestablish the Citizen Advisory Committees as a formal mechanism to assist with development and implementation of a new coastal management strategy, one designed to reduce the risk associated with coastal hazards as discussed in chapter 9. Each committee should consist of representatives of relevant government agencies (U.S. ACOE, FEMA, U.S. Geological Survey, NJDEP, N.J. Office of Emergency Management and Preparedness), technical experts from academia and the private sector, local elected officials and emergency management personnel, and representatives drawn from coastal municipalities and interest groups. These committees could interact directly with NJDEP on coastal management issues (figure 10.6).

The primary goal of these committees would be to assist the state with the development and implementation of a long-term, coastal hazard management program. Specific duties could include:

- Assisting in the preparation of regional coastal management plans addressing hazard management and coastal erosion
- Identifying and incorporating local concerns into the regional plans
- Preparing model coastal ordinances for local implementation
- Collecting, maintaining, and disseminating current information on

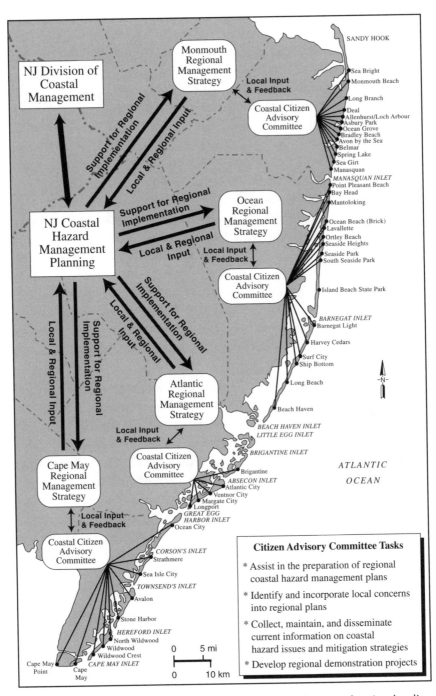

Figure 10.6. Possible organization appropriate to regional approaches in shoreline management.

coastal hazard issues, erosion and stabilization, and management strategies
- Developing regional demonstration projects

These duties should be carried out with state funding as an incentive to enable the committees to hire technical experts when necessary, establish regional information repositories at county libraries, conduct public outreach efforts, and establish volunteer mitigation and monitoring efforts. An example of the last initiative is the organization of community groups and schools to conduct remedial action on dunes damaged by storms or erosion and to monitor the recovery process. In addition, these groups could be mobilized and equipped to collect beach and sediment information that could be used by the advisory committees to track coastal change over time, similar to the water-quality monitoring efforts under way in Barnegat Bay and other in water systems throughout the state. This information could then be factored into the decision-making process.

Through construction of the bibliography, home page, and background information generated by the reassessment, electronic means are available to store, manage, and make available a great deal of information on coastal hazards and coastal management. This system should be maintained and continually updated as a support service for the advisory committees. The system also could be linked with county-based repositories and used to help coordinate activities between the regions or other advisory committees. Furthermore, paralleling the regional coastal groups employed in England, representatives from each of the advisory committees should meet periodically to foster interactions and transfer information among the county-based committees. These committees also could be used by the state to solicit public input on many coastal issues, not just those related to coastal hazard management.

Continued Education and Outreach

FEMA's National Mitigation Strategy focuses on how public education and training can effect changes in the way coastal communities manage their shorelines. Successful change is grounded in education. As we have seen, change can be accomplished through two principal target audiences, the formal precollege and the informal education communities.

Many coastal education efforts and programs exist to reach the formal and informal education communities. Obviously, partnerships with existing programs can make efficient use of resources and allow one to capitalize on the participants already engaged with these groups. Partnerships should be used wherever possible to get the most out of resources and expertise and make use of existing networks to disseminate information.

Formal Precollege (K–12) Education

Efforts are under way to expand the *MARE* initiative throughout New Jersey by Project Tomorrow staff. This, and other science-based supplementary curricula, can be used as an effective means to transfer information to the precollege community on issues associated with coastal hazards and coastal management. Hands-on activities, Internet activities, and a field guide produced by the Project Team and tested in New Jersey schools are being incorporated into the New Jersey *MARE* program. Supplementary materials are being made available to New Jersey educators through the Marine-Environmental Science Curriculum Repository located at Rutgers's Institute of Marine and Coastal Sciences.

Informal Education

INTERPRETIVE MEDIA, PLACARDS, AND SIGNS

Interpretive displays and public placards are recommended as a general means for creating a better understanding of coastal processes and society's vulnerability to storms and the longer-term implications of sea-level rise. Actions to consider include a pole or some other visible marker to depict the high-water marks and storm surges from past coastal storms as well as to display projected cumulative increases in water level due to sea-level rise over the next 100 years. Many New England communities used such a device after the 1938 hurricane. Interpretive displays can be located in popular shoreline locations such as boardwalks, nature centers, public meeting places, and the Coastal Heritage Trail that winds through Cape May, Atlantic, Ocean, and Monmouth counties.

SHORT VIDEOS

Brief educational videos have provided an effective medium for interpreting scientific information and represent an excellent tool for communicating the findings of the reassessment and for understanding coastal hazards and coastal management. Other states have had great success with this approach. The Louisiana Department of Natural Resources, for example, has developed a short video called *Reversing the*

Tide that highlights the importance of wetland restoration and the dynamic nature of the coastal zone. This video has been shown on public television and is available to all elementary, middle, and high schools at no charge.

PUBLIC SERVICE ANNOUNCEMENTS

The nation's Coastal Zone Management Program sponsors annual public events such as Coast Week and Estuaries Day. These activities feature canoe trips, interpretive marsh walks, beach cleanups, and bird watches, all designed to heighten public awareness of estuaries and the importance of their preservation. As part of these programs, radio and television stations air one- to two-minute public service announcements to increase public awareness of these events and their importance. Information on coastal hazards and coastal management could be incorporated in these events.

PRINT MEDIA AND NEWSLETTERS

Education and awareness efforts can benefit from the widespread distribution of local papers and popular press. Press releases to local papers, editorials, and short articles are effective means of increasing awareness of coastal hazard issues and management options. Short articles and editorials in environmental and civic group newsletters are also an effective way of keeping natural coastal hazards and management options in the minds of individuals.

SPEAKERS BUREAU

Throughout the development of the reassessment, a "speakers bureau" was used to conduct outreach activities with state and local government agencies, the precollege community, and the general public. The bureau consisted of an informal group of disciplinary experts and project staff who responded to requests for detailed project information or information on a specific coastal hazard issue and management options. This proved to be another effective means of raising public awareness and soliciting local input on the project.

USE OF EXISTING EDUCATION PROGRAMS

As noted, formal precollege and informal education programs should be created in partnership with existing programs in the coastal zone in order to take advantage of existing resources and audiences. This approach will provide a focus so that groups and alliances already in existence can synthesize existing information and concentrate actions.

The designation of the Mullica River–Great Bay as the Jacques Cousteau National Estuarine Research Reserve (NERR) site is a primary example of a fruitful educational partnership that has raised awareness of estuarine and coastal issues. As part of the 1972 Coastal Zone Management Act, the NERRs program officially recognizes the resources of the coastal zone and their national significance and works with federal and state authorities and academic institutions to establish, manage, and maintain reserves; to provide for long-term stewardship and to design educational programs for students and the general public. The Coastal Heritage Trail is an additional venue for the outreach and education program. The trail is visited by millions of New Jersey shoreline tourists and could serve as an excellent site for informal education programs.

Conclusion

Similar to most studies of public policy issues and concerns, the reassessment of the *NJSPMP* made a thorough and systematic effort to review and analyze existing information, to collect and evaluate new information, and to ensure broad-based participation from the general public and stakeholders throughout the process. In addition to the Project Team (composed of investigators from the disciplines of coastal sciences, coastal engineering, economics, management, and public policy), the process involved participants from many federal, state, local, and private organizations with a broad range of expertise. This inclusiveness, combined with public participation, ensured that the reassessment was based on the best available information on coastal hazard management. The previous chapters summarize the contribution from this process.

CHAPTER II

Coastal and Hazard Management: Trends, Options, and Future Directions

Finally, if homeowners and businesses believe that they will receive substantial disaster assistance in the form of low-interest loans or grants from the federal government, then they will have less incentive to invest in mitigation measures. . . . In developing a strategy for reducing losses from coastal hazards, a community needs to look at a package of mitigation and preparedness measures that will involve a number of different interested parties, each of which has a stake in the outcome.

> H. John Heinz Center for Science, Economics, and the Environment (2000b)

Federal and state programs affecting natural hazards have created a legacy of missed opportunities that complicates efforts to find a path to reforms based on appropriate decisions about land use.

> R. J. Burby, *Cooperating with Nature* (1998)

Hazard management is central to all the issues facing coastal New Jersey and coastal areas in general. Many of the questions of shoreline stabilization and of coastal management more broadly defined can be recast and approached in the context of the management of coastal hazards. Understanding the coastal environment and its processes ultimately provides insight into the nature of coastal hazards and possible ways in which to respond to them. To this end we have discussed engineered approaches to stabilize the coastline, coastal dunes to aid in buffering storm effects, the effects of sea-level rise and coastal storms on shoreline change, economic methods and applications in assessing shoreline stabilization and the economic importance of the coastal region, approaches from the field of hazard management to reduce the threat of loss and damage, coastal erosion and historical shoreline change—all necessary components in understanding hazard management and coastal management methods. In this

chapter recent trends in public policies and the private insurance industry regarding natural catastrophes are summarized, hazard mitigation
measures are evaluated, and a synthesis of options and future directions is
provided.

Public Policy, the Insurance Industry, and Natural Catastrophes: Recent Trends

Public policy initiatives have begun to readdress the issue
of natural catastrophes and natural hazards. Although these events are
highly unpredictable and traditionally have not been a major part of the
public policy agenda, fairly recent and damaging natural catastrophes
have brought the subject to center stage. Between 1989 and 1999 four
significant hurricanes struck the United States, causing damage that
exceeded a billion dollars in insured losses per event and totaling $23.4
billion, or an average of $5.85 billion per storm (ISO, Inc. 1999; H. J.
Heinz Center 2000a). In addition, the West Coast has experienced significant earthquake damage since 1989 with the Northridge earthquake
that struck San Francisco in 1994, causing over $12.5 billion in insured
losses and $50 billion in total losses; the Loma Prieta earthquake in 1989,
with property damage of about $6 billion, insured and uninsured losses;
and the Landers/Big Bear earthquake in 1992, with damages of over $90
million, insured and uninsured (Palm 1998).

Overall, the level of damages inflicted by catastrophes and natural
hazards as measured by insured losses has dramatically increased in the
United States in recent times (table 11.1, figure 11.1). The data demonstrate
that this change is nonlinear in nature over a longer time period. Before
1989 the insurance industry had not experienced single-event disasters
that exceeded $1 billion per disaster, but since 1989 up to 10 disasters
exceeded $1 billion per disaster when measured in 1997 dollars (Kun-
reuther 1998). Estimates of total losses, insured and uninsured, from catastrophic events in the United States for the 1989–1999 period of $286
billion[1] approach the losses experienced in the savings and loan industry
failures during the 1980s of $90–$130 billion (Thomas 1990), or
$140–$150 billion in discounted terms and $350–$500 billion in undiscounted terms (White 1991). Indications of conditions facing the coastal
United States point to increasing threat and rising damage.

A similar pattern exists for the state of New Jersey. Damages measured
by insured losses from natural hazards and catastrophes over the 1950–

Table 11.1 ∼ PRESIDENTIAL MAJOR DISASTERS REQUESTED
AND DECLARED, FISCAL YEARS 1984–1997

Fiscal Year	Number requested	Number declared	Percentage declared
1984	48	35	72
1985	32	19	59
1986	38	30	79
1987	32	24	75
1988	25	17	68
1989	43	29	67
1990	43	35	81
1991	52	39	75
1992	56	46	82
1993	51	39	76
1994	51	36	71
1995	45	29	64
1996	85	72	85
1997	66	49	74
1998	NA	62	NA

SOURCE: H. J. Heinz Center (2000b); U.S. General Accounting Office (1995).

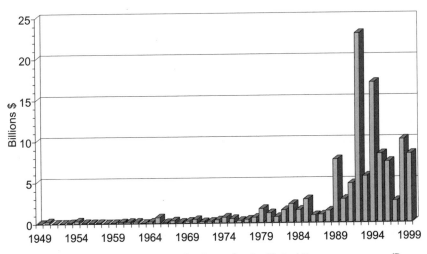

Figure 11.1. Annual insured catastrophic losses for the United States, 1949–1999. (Property Claim Services, Insurance Service Offices.)

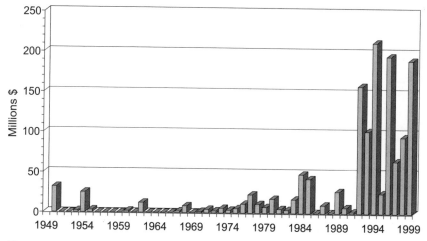

Figure 11.2. Annual insured catastrophic losses for New Jersey, 1950–1999. (Property Claim Services, Insurance Service Offices.)

1999 period show evidence of a nonlinear increase (figure 11.2). The increase in insured losses (in current dollars) from the 1950–1988 period to the 1989–1999 period was almost 254%. New Jersey exhibits more vulnerability to northeasters than to hurricanes, as shown in tables 11.2 and 11.3, where 16 of the most significant 22 hazard events were associated with northeasters.

Evidence based on insured catastrophic losses is not the only illustration of the seriousness of natural hazards and what is at stake. In recent times, population figures along the East and Gulf coasts of the United States have been on the rise, with an overall increase of 15% in coastal counties compared with 12% in noncoastal counties from 1980 to 1993 for Gulf and Atlantic states (table 11.4). In New Jersey the population increased 17% in coastal counties compared with 6% in noncoastal counties over the 1980–1993 period. This rising trend has increased the demand for housing and real estate, further escalating the value of limited coastal property. Considering all Gulf and Atlantic states, the increase in the value of insured coastal property from 1980 to 1993 for both residential and commercial properties was 178%, and for all coastal states in the United States the increase was 161%. Florida represented the largest value of insured coastal property with $871.7 million in 1993, followed by New York with $595.6 million and Massachusetts with $321.6 million (table 11.5). New Jersey ranked sixth with $152.8 million in 1993. Of concern to both the insurance industry and the emergency and hazard management

Table 11.2 ∽ TWENTY-TWO LARGEST INSURED CATASTROPHIC LOSS EVENTS FOR NEW JERSEY, 1950–1999

Year	Date and type of catastrophe	Estimated loss payments ($)
1992	Dec. 10–13: Wind, flooding, snow	150,000,000
1999	Sept. 14–17: Hurricane Floyd	95,000,000
1994	Jan. 17–20: Wind, snow, ice, freezing	65,000,000
1993	March 11–14: Wind, tornadoes, snow, etc.	55,000,000
1998	Sept. 6–8: Wind, tornadoes, flooding	55,000,000
1996	Jan. 12–13: Wind, snow, ice, flooding	50,000,000
1999	Jan. 1–4: Wind, snow, ice, flooding, etc.	40,000,000
1999	Jan. 17–18: Wind, tornadoes, flooding, etc.	40,000,000
1996	Oct. 18–21: Wind, flooding	35,000,000
1994	Feb. 10–12: Wind, flooding, snow, ice, etc.	35,000,000
1993	March 4–5: Wind, flooding, snow, ice	30,000,000
1994	Jan. 27–29: Wind, ice, flooding	30,000,000
1985	Sept. 26: Hurricane Gloria, wind, flooding	27,500,000
1950	Nov. 24–27: Wind	25,000,000
1984	March 27–30: Wind, tornadoes, snow, etc.	20,000,000
1984	April 3–7: Wind, tornadoes, flooding	20,000,000
1996	Jan. 17–20: Wind, snow, tornadoes, etc.	20,000,000
1996	March 16–21: Wind, tornadoes, flooding	20,000,000
1997	July 18: Wind, tornadoes, flooding	20,000,000
1997	Aug. 15–17: Wind, tornadoes, flooding	20,000,000
1998	May 30/June 1: Wind, tornadoes, hail	20,000,000
1977	Nov. 6–9: Wind, flooding	19,500,000
	Total	$892,000,000

SOURCE: Property Claim Services, Insurance Services Office (2000).

Table 11.3 ∽ TEN LARGEST INSURED LOSSES FROM HURRICANES FOR NEW JERSEY, 1950–1999

Year	Date and type of catastrophe	Estimated loss payments ($)
1999	Sept. 14–17: Hurricane Floyd	95,000,000
1985	Sept. 26: Hurricane Gloria, wind, flooding	27,500,000
1954	Oct. 15–16: Hurricane Hazel	10,000,000
1954	Aug. 30–31: Hurricane Carol	7,000,000
1976	Aug. 8–10: Hurricane Belle	5,125,670
1979	Aug. 30/Sept. 6: Hurricane David	4,800,000
1971	Aug. 27–28: Tropical Storm Doria	4,230,000
1955	Aug. 11–13: Hurricane Connie	2,000,000
1960	Sept. 9–11: Hurricane Donna	2,000,000
1975	Sept. 16–26: Hurricane Eloise	476,000
	Total	$158,131,670

SOURCE: Property Claim Services, Insurance Services Office (2000).

Table 11.4 ~ COASTAL POPULATION CHANGE IN GULF AND ATLANTIC STATES, 1980–1993

State	1980 Population		1993 Population		Percentage change, 1980–1993		1993 Coastal population as percentage of total population
	Coast	Total	Coast	Total	Coastal	Total	
Texas	1,287,865	14,229,186	1,411,245	17,568,472	10	23	8
Louisiana	1,461,593	4,205,909	1,401,493	4,258,216	−4	1	33
Mississippi	297,871	2,520,650	313,626	2,600,012	5	3	12
Alabama	443,442	3,893,874	488,144	4,109,593	10	6	12
Florida	7,659,364	9,746,320	10,501,222	13,527,968	37	39	78
Georgia	327,020	5,463,108	401,731	6,710,896	23	23	6
South Carolina	546,540	3,121,818	663,375	3,596,623	21	15	18
North Carolina	514,854	5,881,747	645,183	6,805,129	25	16	10
Virginia	822,706	5,346,838	1,031,896	6,358,402	25	19	16
Maryland	2,328,834	4,216,965	2,527,443	4,913,350	9	17	51
Delaware	594,335	594,335	687,214	687,214	16	16	100
New Jersey	1,328,691	7,364,847	1,554,286	7,796,867	17	6	19
New York	7,080,247	17,558,086	7,288,604	18,117,594	3	3	40
Connecticut	1,935,639	3,107,578	2,038,022	3,306,319	5	6	62
Rhode Island	947,156	947,156	1,008,938	1,008,938	7	7	100
Massachusetts	2,941,588	5,737,042	3,109,281	6,017,970	6	5	52
New Hampshire	274,983	920,605	354,169	1,117,261	29	21	32
Maine	548,082	1,124,648	635,635	1,244,147	16	11	51
Total	31,340,808	226,546,368	36,061,500	254,293,104	15	12	14

SOURCE: Insurance Research Council (1995).

Table 11.5 ~ VALUE OF INSURED COASTAL PROPERTY EXPOSURES BY STATE FOR GULF AND ATLANTIC STATES, 1980 AND 1993

State	Residential		Commercial		Total		Percentage change (1980–1993)		
	1980 (1980$)	1993 (1993$)	1980 (1980$)	1993 (1993$)	1980 (1980$)	1993 (1993$)	Res.	Comm.	Total
Texas	20,355,288	43,441,516	28,271,910	85,197,376	48,627,198	128,638,896	113	201	164
Louisiana	26,358,420	43,571,616	38,957,144	79,892,216	65,315,564	123,463,832	65	105	89
Mississippi	4,644,828	10,110,020	3,531,034	15,343,068	8,175,862	25,453,088	117	334	211
Alabama	7,506,048	16,946,840	7,689,400	19,961,476	15,195,448	36,908,316	125	159	142
Florida	177,709,426	418,392,736	155,213,287	453,288,896	332,922,713	871,681,664	135	192	161
Georgia	4,262,578	12,423,317	5,150,122	20,085,646	9,412,700	32,508,964	191	290	245
South Carolina	9,399,383	27,177,172	7,608,740	27,562,476	17,008,123	54,739,648	189	262	221
North Carolina	7,441,513	23,081,062	5,713,113	21,890,876	13,154,626	44,971,936	210	283	241
Virginia	13,643,584	32,872,384	10,912,406	34,895,616	24,555,990	67,768,000	140	219	175
Maryland	40,969,895	103,350,240	36,101,614	99,239,744	77,071,509	202,589,984	152	174	162
Delaware	12,453,050	31,458,320	9,059,654	36,272,740	21,512,704	67,731,056	152	300	214
New Jersey	35,614,156	95,903,160	15,477,539	56,866,296	51,091,695	152,769,456	169	267	199
New York	111,997,023	363,934,048	76,278,532	231,668,512	188,275,555	595,602,560	224	203	216
Connecticut	54,071,915	143,672,144	36,443,194	104,391,352	90,515,109	248,063,488	165	186	174
Rhode Island	17,296,893	46,156,076	15,460,444	36,918,872	32,757,337	83,074,944	166	138	153
Massachusetts	54,109,154	177,115,248	50,354,124	144,503,136	104,463,278	321,618,368	227	186	207
New Hampshire	6,072,027	18,233,048	3,943,331	16,691,802	10,015,358	34,924,848	200	323	248
Maine	11,692,564	31,902,472	7,901,399	22,634,296	19,593,963	54,536,768	172	186	178
Coastal total	615,597,745	1,639,741,419	514,066,987	1,507,304,396	1,129,664,732	3,147,045,815	166	193	178
U.S. total	4,240,948,850	10,378,875,018	3,807,860,750	11,043,124,377	8,048,809,600	21,421,999,395	135	192	161

SOURCE: Applied Insurance Research and Insurance Institute for Property Loss Reduction and Insurance Research Council (1995).

authorities is that the mid-Atlantic area (New York, New Jersey, Delaware, and Maryland) represents 32% of the total value, or $1,018,693,056, of coastal property in the Gulf and Atlantic states, implying a tremendous concentration of coastal property in harm's way. New York alone accounts for 19% of the total. An event similar to a Hurricane Andrew occurring in the mid-Atlantic area would prove devastating.

Although formulation of a federal policy response to natural hazards began in 1968 with the passage of the National Flood Insurance Act, only since the 1990s has the federal government attempted to readdress the increasing threat of damage due to increased population in the coastal zone. Policy initiatives involve the provision of the National Flood Insurance Program, in which a community initially prepares a hazard management plan that is evaluated by the Federal Emergency Management Agency. Continued participation in the NFIP is dependent on the community's adoption of hazard management approaches and/or contingency arrangements should a natural hazard occur. The program also outlines steps to be taken in post-storm recoveries that are meant to reduce the future threat of damage and personal loss. As an indication of trends in NFIP policies, the number of policies (both coastal *and* noncoastal counties) in effect by state and U.S. possession is contained in table 11.6 (up to January 31, 1998) and only for coastal counties in table 11.7 (up to June 30, 1998). In terms of the number of NFIP policies, Florida is ranked first, and the remaining top five states include California, Louisiana, Texas, and New Jersey. Four of these five states experience mainly coastal storm hazards. In addition, for coastal counties only, South Carolina, which experienced Hurricane Hugo in 1989, is ranked sixth; North Carolina, which is extremely vulnerable to hurricanes and northeasters, is ranked seventh; and New York is eighth (considering only the contiguous United States). Information on claims activities also identifies the same states as vulnerable, having historically experienced devastating coastal storms. Louisiana is ranked first, followed by Florida and Texas. New Jersey and California follow in terms of the number of claim payments. In terms of the dollar value of claim payments, South Carolina is ranked fourth, and New Jersey fifth. In repetitive loss payments, the five top states are Louisiana, Florida, Texas, New Jersey, and New York; all have experienced frequent significant coastal storms.

The inundation of the coastal zone by slow but accelerating increases in sea level further heightens its vulnerability (Psuty 1986; Najjar et al. 2000). Coastal scientists agree that this effect will heighten the damage of

State	Number of policies	Number of contracts	Coverage ($)	Premiums ($)
Alabama	31,267	23,693	3,040,828,800	12,590,206
Alaska	2,269	2,124	253,792,900	894,127
Arizona	25,844	25,061	2,787,814,000	9,370,586
Arkansas	13,132	12,994	830,143,900	4,804,696
California	379,227 (2)	361,986	54,938,258,200	148,117,394
Colorado	14,601	13,010	1,604,555,200	6,393,360
Connecticut	26,607	22,093	3,422,757,700	15,770,825
Delaware	14,216	10,543	1,766,024,500	6,384,508
District of Columbia	335	86	19,704,400	83,424
Florida	1,635,721 (1)	1,102,863	190,579,653,100	526,508,110
Georgia	51,823	49,362	6,774,150,700	22,160,783
Hawaii	45,536	14,117	5,334,955,100	13,713,026
Idaho	8,128	8,016	1,228,966,300	2,787,577
Illinois	46,254	42,595	3,995,726,500	19,576,667
Indiana	25,216	25,178	1,731,097,400	10,674,440
Iowa	9,763	9,749	721,136,400	4,616,649
Kansas	10,069	10,032	710,527,200	4,158,461
Kentucky	22,264	21,979	1,350,636,100	8,598,709
Louisiana	334,045 (3)	327,369	32,888,330,600	119,883,650
Maine	6,443	6,029	652,649,200	3,524,444
Maryland	45,678	27,843	4,516,394,900	14,800,547
Massachusetts	35,098	31,007	4,162,194,200	21,558,946
Michigan	26,293	25,019	2,241,978,500	11,341,981
Minnesota	13,184	12,889	1,137,582,800	4,572,150
Mississippi	41,378	40,606	3,491,664,200	15,302,304
Missouri	22,754	22,636	1,897,445,800	11,454,638
Montana	9,570	9,397	976,254,500	2,919,701
Nebraska	11,941	11,888	854,533,700	4,776,598
Nevada	11,877	11,791	1,575,761,300	4,993,559
New Hampshire	4,211	3,673	393,918,500	2,097,915
New Jersey	157,534 (5)	133,893	19,332,025,400	81,685,889
New Mexico	10,373	10,266	812,552,600	3,824,770
New York	87,543 (7)	82,833	10,876,776,400	49,791,682
North Carolina	73,660 (8)	65,141	8,843,212,300	31,423,193
North Dakota	12,686	12,497	1,257,150,200	3,949,441
Ohio	33,144	31,864	2,386,721,400	14,329,736
Oklahoma	14,182	14,022	1,076,584,300	5,532,267
Oregon	21,159	20,313	2,675,936,700	9,423,163
Pennsylvania	62,885 (10)	61,536	5,590,188,100	29,312,187
Rhode Island	10,579	9,176	1,260,573,200	6,668,231
South Carolina	105,847 (6)	74,823	14,796,686,500	43,794,852
South Dakota	4,528	4,528	456,029,100	1,705,060
Tennessee	13,104	12,725	1,262,243,700	5,472,716

(*continued*)

Table 11.6 ～ (*continued*)

State	Number of policies	Number of contracts	Coverage ($)	Premiums ($)
Texas	280,215 (4)	265,240	32,721,302,200	102,971,224
Utah	2,369	2,015	273,127,800	909,735
Vermont	2,422	2,375	202,402,600	1,247,577
Virginia	63,200 (9)	55,054	7,364,603,500	23,198,088
Washington	25,757	25,093	2,883,529,900	11,054,833
West Virginia	18,072	18,033	1,034,512,200	7,729,741
Wisconsin	11,707	11,584	865,058,800	5,028,062
Wyoming	2,818	2,812	388,351,800	1,092,419
Total	3,938,528	3,201,451	452,239,005,300	1,474,574,847

SOURCE: National Flood Insurance Program; H. J. Heinz Center (2000b).
NOTE: Numbers in parentheses indicate top ten states.

future coastal storms, so that low to moderate storms may approach the effects of moderate to severe storms, and severe storms could approach a "storm of the century" standard. Sea-level rise will continue to magnify the already nonlinear trend in damages and catastrophic losses.

From a public policy perspective, the change in the capacity of the property insurance industry resulting from recent catastrophes is a critical concern. This problem affects the states of Florida and California (much of the following discussion is from Kunreuther 1998, Roth 1998, and Lecomte and Gahagan 1998). Following the claims from Hurricane Andrew in Florida in 1992 and the Northridge earthquake in California in 1994, private insurance companies saw the profits they had gained over the past 10–15 years vanish in just one hazard event. Many private insurers were unable to meet the claims submitted and were forced into bankruptcy. Those that survived, did so barely or only because of arrangements with reinsurance companies. However, these survivors were left in a very weakened financial position, forcing them to withdraw from the property protection area. Hence, in both Florida and California, the state government had to intervene and provide flood insurance protection to affected homeowners in Florida and earthquake-related damage in California (Kunreuther 1998).

Given the prospects of a rising sea level, any significant coastal storm could have the potential to repeat the effects of a Hurricane Andrew, a Hurricane Hugo, or a similar event such as the Ash Wednesday storm of 1962 in New Jersey, in which over 80% of the coastline experienced extensive damage and erosion. In New Jersey and other historically vulnerable

Table 11.7 ~ NFIP ACTIVITY FOR COASTAL COUNTIES BY STATE, VARIOUS YEARS

State	Number of policies[a]	premiums[b] (thousands of dollars)	Total coverage[a] (millons of dollars)	Number of payments[a]	Total claim payments[a] (thousand of dollars)	Repetitive loss payments[c] (thousands of dollars)
Alabama	21,690	8,704	2,258	10,938	108,847	26,811
California	181,742 (3)	74,125	28,392	23,345 (5)	199,331 (7)	84,382 (6)
Connecticut	23,698	14,693	3,157	11,166	83,606	39,276
Delaware	15,247	6,708	1,848	2,589	15,305	6,021
Florida	1,624,220 (1)	532,534	190,850	92,456 (2)	1,069,313 (2)	178,681 (2)
Georgia	35,846	15,834	5,133	2,061	17,496	6,072
Hawaii	47,687	14,111	5,433	3,175	53,155	6,920
Louisiana	265,977 (2)	102,389	27,812	115,517 (1)	1,113,016 (1)	468,897 (1)
Maine	4,457	2,617	528	1,943	11,927	2,537
Maryland	31,470	7,972	2,706	1,890	10,899	505
Massachusetts	29,720	18,818	4,549	18,991 (9)	188,689 (8)	80,473 (7)
Mississippi	19,172	7,268	1,911	5,931	33,108	17,414
New Jersey	121,128 (5)	61,447	15,165	41,095 (4)	296,846 (5)	130,630 (4)
New York	45,103 (8)	30,022	7,056	23,147 (6)	186,224 (9)	112,751 (5)
North Carolina	57,035 (7)	24,503	7,213	19,644 (8)	218,359 (6)	51,985 (9)
Oregon	9,038	4,194	1,120	1,059	9,050	2,349
Rhode Island	6,128	4,227	764	1,229	8,661	3,283
South Carolina	105,311 (6)	43,506	14,662	21,742 (7)	384,120 (4)	55,389 (8)
Texas	123,826 (4)	45,091	14,069	50,015 (3)	406,115 (3)	151,516 (3)
Virginia	9,913	4,408	1,139	1,096	3,695	1,110
Washington	7,632	3,302	819	1,355	16,489	3,579

SOURCE: H. J. Heinz Center (2000b); FEMA unpublished data.

NOTE: Numbers in parentheses indicate ranking.

a. January 1, 1978–June 30, 1998.

b. As of September 16, 1998.

c. January 1, 1978–November 30, 1997.

states such as Louisiana, Texas, South and North Carolina, and New York, such events could have effects on the property insurance industry similar to those in Florida and California. Accordingly, agencies at both the state and the federal level may have to assume partial or complete financial responsibility in insurance coverage and devise or improve programs to reduce the threat of natural hazards and address shoreline management. One important aspect of this process will be increased scrutiny of the use of public funds for shoreline stabilization. Policy makers will require better information and better decision models, capable of handling tradeoffs involving risk and returns and the effects of risk outside the control of shoreline-stabilization projects (exogenous risk) such as coastal erosion and sea-level rise. Such models will offer the policy maker a set of tools with which to make more informed decisions regarding shoreline management under realistic scenarios, and can help all researchers and public officials better understand the complex, dynamic problem of hazard management and the provision of shoreline stabilization. Some of these elements include how much coastal management to provide, the identification of a set of efficient shoreline-stabilization projects, how risk and uncertainty influence outcomes, the timing and abandonment of shoreline-stabilization investment decisions, and possibly decisions about reoccupying sites after a storm and reinvesting in community infrastructure.

However, mixed messages have been sent to communities located in hazard areas. One of these has been the availability of national flood insurance protection, which has been offered at less than actuarially fair rates (that is, at subsidized or discounted rates) to encourage participation in the NFIP program. Another has been the availability of disaster assistance and relief from FEMA and other federal agencies when disaster strikes. For the United States as a whole, over the years 1989–2000 a total of $26.6 billion was obligated for disaster relief (table 11.8). Of this amount, public assistance, individual assistance, and hazard mitigation constitute the largest categories with a combined 87% of the overall total relief. The three largest states that received disaster relief were California with 35% of the total, Florida with 9%, and North Carolina with 4%. Both the availability of discounted rates for flood insurance protection and the almost automatic availability of disaster relief send a message that bailouts will continue for communities located in hazard-prone areas. A different message would be transmitted if flood insurance protection were priced at self-sufficient rates and if disaster relief were cut back a third or only made available for non–repetitive loss areas or for hazard mitigation

Table 11.8 ~ FEMA PROGRAM OBLIGATIONS BY STATE FOR FISCAL YEARS 1989–2000 (AS OF FEBRUARY 29, 2000), IN DOLLARS

State	Public assistance	Individual assistance	Hazard mitigation	Mission assignments[a]	FEMA adminis.	Total FEMA obligations
Alabama	182,358,351	64,483,081	27,484,099	4,818,790	21,873,560	301,017,882
Alaska	92,619,513	4,955,539	14,440,972	155,836	4,451,365	116,623,225
Arkansas	46,122,233	16,785,805	7,960,131	7,607,749	10,280,235	88,756,153
Arizona	128,324,192	2,933,629	6,588,779	3,852	3,265,140	141,115,592
California	6,304,738,711	1,771,333,027	949,716,284	46,859,616	344,059,991	9,416,707,629
Colorado	11,997,980	8,988,504	3,435,691	59,400	5,620,213	30,101,788
Connecticut	35,814,757	4,751,687	550,794	57,239	3,342,435	44,516,912
District of Columbia	6,989,778	–	410,653	803	123,116	7,524,350
Delaware	17,386,262	2,621,197	1,572,774	25,746	2,421,113	24,027,092
Florida	1,296,908,349	503,533,418	72,902,124	496,805,403	148,591,036	2,518,740,330
Georgia	431,872,269	97,183,082	69,521,611	19,287,012	53,265,888	671,129,861
Hawaii	162,452,380	40,914,341	5,803,887	22,741,823	47,883,879	279,796,311
Iowa	182,191,225	126,212,999	48,807,525	12,864,986	22,819,655	392,896,390
Idaho	37,347,076	9,279,731	8,299,450	738,574	7,608,723	63,273,554
Illinois	257,825,828	311,316,731	78,017,619	13,542,609	28,996,313	689,699,099
Indiana	70,451,910	18,088,622	6,328,736	758,571	7,614,559	103,242,398
Kansas	105,989,875	69,290,559	25,194,908	7,678,928	27,140,174	235,294,444
Louisiana	146,664,857	219,641,150	25,981,092	8,379,523	24,505,193	425,171,814
Massachusetts	114,276,692	42,952,840	16,080,325	406,357	11,468,685	185,184,900
Maryland	43,830,869	6,105,260	2,976,872	168,994	5,250,691	58,332,686
Maine	72,777,108	11,128,728	9,906,997	1,687,549	11,291,368	106,791,749
Michigan	72,246,557	26,811,623	8,437,581	–	4,904,731	112,400,492
Minnesota	362,130,245	75,817,900	55,694,911	1,010,168	28,257,522	522,910,745
Missouri	169,992,731	105,462,604	45,481,252	7,291,892	26,503,067	354,731,546
Mississippi	122,571,224	42,090,954	18,712,993	5,743,847	16,973,465	206,092,484

Montana	12,605,769	–	1,423,314	29,900	1,734,720	15,793,703
North Carolina	485,567,376	280,186,417	135,214,782	162,720,736	85,045,963	1,148,735,274
North Dakota	319,953,469	153,547,423	59,891,937	29,953,564	43,703,085	607,049,477
Nebraska	138,914,846	6,262,296	23,456,012	364,034	4,354,300	173,351,488
New Hampshire	22,472,739	3,295,519	2,250,099	94,306	5,087,671	33,200,333
New Jersey	102,809,866	53,606,088	3,542,032	734,582	26,771,469	187,464,038
New Mexico	8,659,358	748,061	480,229	235	857,114	10,744,997
Nevada	23,612,605	3,960,801	4,117,231	621,510	6,057,358	38,369,505
New York	373,704,616	74,842,877	32,096,199	2,762,744	43,372,272	526,778,708
Ohio	80,537,639	50,485,403	20,986,561	2,551,419	15,541,419	170,012,441
Oklahoma	81,079,924	23,207,280	15,343,980	36,489,227	34,171,990	190,292,401
Oregon	92,817,819	26,954,588	15,472,717	1,729,957	12,661,167	149,636,248
Pennsylvania	218,827,177	111,167,581	39,800,941	3,039,466	39,086,300	411,921,466
Rhode Island	16,242,003	25,965	471,329	22,811	986,410	17,748,518
South Carolina	277,271,601	107,271,040	12,661,378	37,429,369	20,891,515	455,524,903
South Dakota	97,109,927	48,220,636	14,643,457	5,485,923	24,189,350	189,649,293
Tennessee	174,099,425	15,741,571	16,074,352	212,033	15,331,937	221,459,319
Texas	168,305,188	309,568,692	51,541,922	4,912,762	53,286,641	587,615,206
Utah	1,845,728	118,147	95,048	16,000	1,213,214	3,288,137
Virginia	114,495,034	31,723,331	11,479,466	6,847,174	24,012,307	188,557,312
Vermont	36,589,678	5,082,102	4,989,134	199,726	9,217,514	56,078,155
Washington	291,812,340	60,626,806	41,938,098	1,359,020	18,805,748	414,542,013
Wisconsin	55,162,473	72,311,099	22,798,149	335,219	7,980,823	158,587,763
West Virginia	56,933,534	42,261,897	14,107,815	110,483	15,209,610	128,623,339
Wyoming	816,513	–	44,889	–	298,187	1,159,589
Total	13,728,127,619	6,927,367,684	2,413,091,252	1,374,917,086	2,175,240,808	26,618,744,449

SOURCE: FEMA, unpublished data.

a. Tasks delegated to other agencies.

projects. This message would then be that individual property owners must assume a larger share of the responsibility of occupying hazardous areas and that development in hazard-prone areas is to be curtailed.

As an example, relief efforts associated with the recovery from Hurricane Floyd in September 1999 for North Carolina amount to a total of $1 billion when all pending and obligated relief is counted (table 11.9). Of this amount, 24% was associated with hazard mitigation efforts. Such a level of support is to be commended, but consider what the message would be if at least half ($500 million) were to be committed to hazard mitigation. Furthermore, considering the scarcity of public resources and society needs as a whole, what is the best use of public funds? How often will the Hurricane Floyd scenario be repeated in the future? If its $1 billion were used for property buyouts and relocations and the transferred property turned into public lands for all to enjoy, such repetitive losses would in fact become cost savings and could realize tremendous public benefits. To further demonstrate the potential benefits from hazard mitigation efforts, consider several case studies.

Insights from Recent Hazard
Mitigation Efforts

Over 1982–1991, the state of Florida established and then adopted a series of building code requirements referred to as the Coastal Construction Control Line, or CCCL (FEMA 1997). These land-use and building requirements (pertaining to construction and elevation) were more stringent than those set forth in NFIP coastal requirements for property in high-velocity zones, and the wind-load requirements were more stringent than those specified in standard building codes. Hurricane Opal struck Florida in 1995, and the Florida Department of Environmental Protection reported that none of the 576 major habitable structures with CCCL permits sustained substantial damage (that is, damage that exceeded 50% of the market value of the structure) (FEMA 1996, 1997). In contrast, almost 770 nonpermitted, habitable structures seaward of the CCCL zone experienced substantial damage. Hazard mitigation in this case was accomplished through effective land-use and building code requirements.

Another hazard mitigation effort involved the acquisition, elevation, and relocation of residential structures in response to the 1993 Midwest floods in the city of Arnold, Missouri (FEMA 1997). Overall estimates of

Table 11.9 ~ HURRICANE FLOYD RECOVERY IN NORTH CAROLINA, 2000

Category	Individual and family grant program	Disaster housing	Small Business Administration loans	Public assistance	Hazard mitigation applications approved	Hazard mitigation applications pending	Disaster unemployment relief	Temporary housing relief[a]
Amount approved ($)	81,760,425	78,156,588	437,183,600	79,548,358	44,823,366	194,392,239	5,817,947	43,931,509
Applications approved (#)	21,762	36,188	11,247	537	768	3,062	5,813	
Avg. amount approved ($)	3,757	113,109	38,871	148,135	58,364	63,485	1,001	
Cases pending (#)	1,390	3,240	1,289					
Ineligible cases (#)	21,050	36,188						
Total applications (#)	47,099	66,015		221,295,717[b]				

SOURCE: North Carolina Emergency Management, Information and Planning Section, unpublished data on Hurricane Floyd recovery.

a. State temporary housing obligated a total of $46,000,000.

b. Total current losses eligible for public assistance; of this amount FEMA has obligated $126,402,255.

total damages of these floods ranged from $12 billion to $16 billion, and a total of $203 million was set aside in Hazard Mitigation Grant Program funds (including the 25% nonfederal match). After an extensive examination of communities that had suffered repetitive losses from riverine flooding, the city of Arnold was one site selected for a buyout program by FEMA in which damaged or destroyed property was purchased and assistance given for the relocation of residential structures to create a greenway to supplement the city's floodplain. The 1993 floods caused damage to about 250 habitable structures, 528 households applied for relief, and disaster relief amounted to over $2 million. In 1995, after the acquisition and relocation program, the fourth-largest flood in the Arnold's history occurred; only 26 households applied for relief, with damages less than $40,000. According to the city manager, the damage was less severe mainly because "most of the areas affected had been bought out, so the people weren't there" (FEMA 1995f). Similar programs were instituted in the city of Valmeyer, Illinois, which totally relocated, and in Wakenda, Missouri, which abandoned its site and disincorporated its municipality (FEMA 1997).

These two examples demonstrate that with planning and foresight, the use of hazard mitigation measures can achieve reductions in the losses and damages inflicted by hazardous events and catastrophes. One of the simplest approaches, to relocate out of harm's way, can be one of the most effective long-term tactics.

Policy Outlook

Two national policy developments vital for any projections of a state's future role in shoreline planning and management occurred in the 1990s. One development is the increasingly negative attitude at the federal level toward subsidies for beach nourishment. The federal government has zero-budgeted the beach nourishment program for several years. Local representatives have been able to find support for their projects as special appropriations, but the signs are that federal funding will become more and more difficult to secure and that the subsidy will continue to decrease. Consequently, states will have to make plans that involve much larger expenditures for beach nourishment or find some alternative to placing large quantities of sand (at great expense) on the beach. Second, a growing federal initiative involves a National Mitigation Strategy to reduce the losses from natural hazards, including coastal erosion and flooding. The national initiative supports moving people and structures out of hazardous areas. Funds in support of pre- and post-storm removal and relocation are fueling this initiative.

Because of the extent of development in coastal areas, mitigation is expensive. It will require a dedicated pool of public funds to accomplish the reduction of people and structures at risk. But it is also true that all the existing construction and infrastructure in coastal areas will be impossible to maintain without continuous substantial expenditures. Likewise, it will not be possible to attain the goal of reducing the exposure to natural hazards at the coast without large expenditures of public monies.

Any piecemeal approach will not be successful in facing the issue of the long-term erosion and drowning of coastal resources. Shoreline management and planning efforts must function at regional levels to be most effective. Experts agree that opportunities to exercise decisions will be in the immediate post-storm periods, but decisions and plans involving both post-storm cleanup and hazard management must be made before that time. Such long-range plans should be mindful of the availability of funds to execute those decisions. However, flood insurance protection and disaster relief have to be designed to support hazard management efforts. These programs, furthermore, must be tempered by an awareness of the actions taken by the private insurance industry when faced with recent catastrophic events, as in Florida and California. These new developments should be incorporated into national efforts and directives concerning national mitigation strategies.

Research that involves comparison and analysis of various policy options can assist in this effort. Only from such an exploration of possible actions, outcomes, and long-term implications can we begin to identify those steps that benefit society as a whole. In many cases decisions will reflect communities' visions of themselves, but ever increasingly these decisions will be influenced by state and federal policy directives and the availability of state and federal funds. We are either approaching or are at a turning point for many areas that have suffered repetitive losses, and policy directions may be forced to deviate from those of the past.

It is important to realize that much of the coastal barrier island development can be compatible with the changing natural system and the increased political opportunity to apply a hazard mitigation program. An effort to identify those areas that are prone to damage and the resulting economic hardship is required. The procedure to identify these areas must encourage full public participation in the process. There are a number of useful variables to aid in identification that have been described previously, such as repetitive insurance claims, elevation, exposure, rates of erosion, and proximity to water. There are a number of criteria presented in chapter 3 to describe critically eroding areas that could also be of use in this important process.

Options Addressing Hazards in the United States

Several independent reports of the late 1990s addressed concerns about the high cost of occupying hazardous areas, especially coastal zones. The Second National Assessment of Natural Hazards in the United States convened many of the country's leading experts on natural hazards. It pointed the way to viewing, studying, and managing hazards in the United States through the pursuit of "disaster-resilient" communities, higher environmental quality, economic sustainability, and improved quality of life for those residing in hazard-prone areas (Burby 1998a; Kunreuther and Roth 1998; Mileti 1999; Cutter 2001; Tierney, Lindell, and Perry 2001). In the major study (Mileti 1999) and series of reports, that came out of the assessment, the authors investigated why progress has not been made in reducing dollar losses from natural hazard events. A central problem uncovered was that people residing in hazard areas believe that technology can be used to essentially tame nature and to make them safe, and hence views of the dangers are based on this shortsighted and narrow conception of the relation between humans and the natural environment (Mileti 1999). The suggested approach is to recognize this perception, reorient thinking to develop communities that work with the natural environment, and shift toward a policy of "sustainable hazard mitigation" (Mileti 1999). To achieve this end a number of options and policy changes were addressed.

The national panel recommended that the federal government improve its collaboration with state and local governments to form partnerships and, at the same time, adopt a consistent policy that reduces subsidies (Mileti 1999). It was felt that as long as risk is subsidized (or people believe it is), it will be difficult to convince the public to take responsibility for decisions that put them and their property at risk. Hence all forms of federal subsidies and disaster relief need to be examined. Analyses should consider policies' possible effect on reducing or eliminating subsidization of risk before they are implemented. The costs of hazard control structures should be covered by state and local government and by those who will benefit from them, so that federal appropriations can be phased out. The federal government should also make better use of incentives and disencentives to achieve its desired outcomes. For example, all federal aid for infrastructure and development should be dependent on community participation in land-use and other planning. New federal policy could provide consistent regulation to ensure that the issue of natural hazards is considered in the process of planning for all development and redevelopment. And a limit should be established on recovery with repetitive claims (Mileti 1999).

The Second National Assessment notes that more has to be done to build public-private partnerships so that all parties can participate in improving building codes. More complete cost and safety information will help develop codes that are both understood and enforced. The insurance industry should continue to link insurance with the adoption of mitigation measures that reduce damage to existing structures and that limit new development. Minimum levels of damage resistance should be set, incorporated into building codes, and verified by inspectors so as to provide desired incentives for mitigation and produce more equitable insurance premiums (Mileti 1999: 284). There is also a need to monitor past and future efforts. For example, the NFIP has never undergone a thorough examination of its effectiveness. It is one of many programs that all in the national assessment felt was proceeding without a clear understanding of its costs and benefits. Databases should also be constructed and made available to all concerning policy program information, costs, benefits, losses and damages, and redevelopment costs. Other databases could contain information on localities' tolerances of future damage, so that simulations could be run to compare likely future losses with local tolerances to determine any differences or possible thresholds. Furthermore, all knowledge needs to be shared among the community of researchers, specialists, policy makers, and public officials within the United States and overseas. Doing so will expand knowledge of policy and mitigation measures that have worked elsewhere. And last, another benefit would come from sharing information concerning the assessment of hazards' risks, both in the United States and in other countries, to help assemble appropriate assessment procedures and determine the spatial dimension of changes in natural hazards.

The H. John Heinz III Center for Science, Economics and the Environment has been instrumental in bringing together experts to identify, evaluate, and develop strategies to cope with issues associated with coastal development. In a publication on the hidden costs of coastal hazards, center researchers draw attention to the large range of economic burdens encountered as a product of coastal damage that extend beyond the direct destruction of property and infrastructure (H. J. Heinz Center 2000a). They identify losses associated with interruptions of the business community on site and in adjacent areas, social and family disruptions, health costs, costs to public and private institutions, and the damages to natural resources and environmental systems. Elaborating on a recent NRC study (NRC 1999a), the authors seek to improve the methods of accounting for the losses associated with coastal hazards and to find alternatives to the policies that support or promote an increase in the vulnerability of coastal communities.

A second report on erosion hazards by the center stated that coastal erosion is an underappreciated hazard that may claim one in four houses currently located within 500 feet of the shoreline by the year 2060 (H. J. Heinz Center 2000b). The report recommends that FEMA begin to create erosion hazard maps to complement its flood hazard maps and make them available to the public and the banking and insurance industries. It is also recommended that the cost of erosion losses be incorporated into flood insurance rates at the coast.

A common theme in these reports and in a recent publication by the NRC (1999b) is that present hazard management efforts and the institutions presently undertaking them are ill suited to solve the problems that are being created by an evolving coastal zone. The challenge is to create new strategies to work with boundaries and attributes that will not be present for several decades while fostering a transition from the guidelines of the past to those of the future. A problem as well as an asset is the realization that the changes will not be occurring at the same rate and on the same scale throughout the coastal zone, but that opportunities will exist to learn from the occurrence of natural hazards and to fashion strategies and institutions to better safeguard the public in the dynamic coastal system.

Options and Future Directions

As the effects of a negative sediment supply are increased by rising sea levels, the present coastline of New Jersey will continue to erode and encroach upon the coastal communities. In the absence of large subsidies from the federal government or the state to rebuild and defend the present shoreline position, it is inevitable that coastal planning will shift toward coastal hazard management rather than coastal stabilization. There is already evidence of this change at the federal level with FEMA's national hazard mitigation policy, where more effort is being directed toward increasing public safety at the shore with the objective of identifying the areas at high risk from natural hazards. Hazard management programs can and should incorporate short-term approaches to the effects of minor storms as well as long-term approaches that may involve relocation from high-risk areas.

For many coastal areas, a short-term approach associated with coastal dune development and limited beach nourishment projects should provide adequate protection against storms of small magnitude. The effects of the slowly developing negative sediment supply and sea-level rise will then be masked by the manipulation of the observable shoreline. Large storms will produce larger displacements that go beyond masking and will require changes in land use or major investments in nourishment to maintain the shoreline's position. The need for major reinvestments will escalate.

As demand for use of the coastal zone grows, better information and more creative management strategies are needed to support continued resource use and stewardship. Further, an integrated, coordinated, management approach has been shown by other coastal states to better address shoreline processes that occur on a regional scale. Partnerships that transcend jurisdictional boundaries are necessary to achieve this goal. A single administrative entity could be developed and charged with the sole responsibility of managing the New Jersey coast, for example. Such an agency could establish well-defined objectives that would be coordinated through a single office, and could function in close cooperation with the public and with county and local planning entities.

There is a need for long-term objectives that strive to increase the public's safety. These long-term mitigation strategies should be developed at the state level and implemented on a regional basis. Long-term approaches, such as targeting high-risk areas for post-storm changes in land use, can achieve the state's coastal management objectives. This is critical, because the post-storm period will be the only opportunity to create land-use changes and alter any coastal development. Of special importance are those low-lying areas severely affected by minor storms and areas subject to repetitive losses. To implement these approaches, new policies need to be initiated, such as enacting zoning ordinances that limit the density in high-hazard areas. The collection of technical data should continue so that local resource managers have access to accurate, current information in their decision making. Additionally, there is a need to foster public awareness of the risks associated with the coast. Alerting the citizens to these risks promotes the concepts of public safety and recognizes the fiscal limitations of attempting to respond to the effects of the very large storms. The *Coastal Hazard Management Report* (Psuty et al. 1996) prepared for the NJDEP points to the development of a public attitude grounded in awareness of coastal hazard issues, stresses safety, and provides disincentives to the occupation of hazardous areas.

The State of New Jersey's Policy and Management Approach

What follows is an assessment of the state of New Jersey's policy and management path with regard to its coast, drawn from past policy actions reviewed in the *NJSPMP* and possible policy actions based on the 1996 reassessment. The philosophical direction and policy actions suggested here parallel national efforts by FEMA. There are many general areas for possible policy actions directed at coastal hazard management. Some of these involve changes in public policies that can be achieved from improved cooperation and coordination of policies at all jurisdictional levels, limiting the subsidization of risk, redirecting disaster relief, and making better use of policy incentives such as tying federal aid to community participation in specific hazard management programs. Similar recommendations were advanced by the Second National Assessment of Natural Hazards in the United States (Mileti 1999). Other major areas, also covered by the national assessment, involve obtaining input from all stakeholders such as public and private groups to promote better private-public partnerships, better monitoring of efforts involving databases, and recognizing the importance of information transfers between researchers, specialists, policy makers, public officials, and the general citizenry.

New Approaches in the 1996 Reassessment

These concerns addressed in the *NJSPMP* and in the 1996 reassessment raise questions about the coastal zone. What is its future? What should the coast look like 50 years in the future (or any future time)? That is, if there were options available to alter land use and redirect the management of resources, what decisions could be made at this time to attain future objectives?

Knowing that management decisions taken at the coast will result in large expenses, it is necessary to establish management objectives and to direct funds toward those alternatives that are part of a consistent and systematic program to achieve long-term goals. A regional approach is critical in this process.

Because enhancing the protection of life and property along the coast ultimately relies on local action by the citizens most directly affected, development of new coastal management and hazard management policies should involve these stakeholders. During the 1996 reassessment process, formal public participation took place through the Citizen Advisory Committees. This structure allowed a two-way flow of information and helped identify and air views and refine local concerns. Success with this process confirmed that a regional approach should be based on local community involvement.

Hazard management or mitigation can be viewed as both a philosophy and a formal approach to coastal management. It is the foundation of recent federal policies designed to reduce the risk associated with natural hazards. In this regard, the development of a National Mitigation Strategy Program by FEMA is especially timely for changes to any state coastal management plan. The incorporation of national objectives to improve public safety and reduce loss within the New Jersey Hazard Mitigation Plan can lead to a more consistent policy and management approach, to a more consistent and more uniform set of incentives, and to an effective and efficient utilization of the coastal zone.

The Coastal Future

The Federal Emergency Management Agency was elevated to Cabinet status in 1996, thereby providing the agency with more federal authority and a direct link to national policy makers. The director of FEMA has announced the creation of a National Mitigation Strategy, whose focus is to remove people from hazard areas, provide support for public safety, reduce the costs of recovery following damage from natural hazards, and reduce payouts from the National Flood Insurance Program by 50% by the year 2010.

The federal Executive Office has not-funded the Army Corps of Engineers for beach stabilization activity on several occasions since 1994 (as recently as 2002). Although Congress has restored some funds for this activity in the near term, it is likely that federal policies will continue to reduce funding for beach stabilization in the long term. Future costs of beach stabilization projects will be increasingly borne at state and local levels.

The private insurance and reinsurance industry is concerned about high-risk coastal areas as sites of unacceptable losses. As a result premium rates (and minimum deductibles) are likely to increase, and flood insurance will become increasingly costly to purchase. It is also possible that some insurers will drop their flood protection policies as did private insurers in the state of Florida.

In light of these effects, coastal planning will become the vehicle to direct and regulate development of the coastal zone. As indicated by federal initiatives and as recommended by leading experts in hazard management, the focus should be directed toward coastal hazard management rather than coastal stabilization. More effort should be generated toward increasing public safety at the shore, identifying high-risk areas associated with coastal natural hazards, adopting development strategies that can reduce the threat of loss and damage, and relocating development out of hazardous areas. Post-storm recovery programs can become integral approaches to reducing future losses in high-risk areas. Identifying and targeting high-risk areas for post-storm changes in land use is important, because the post-storm period is usually the only opportunity to enact land-use changes and alter any coastal development. Heightening citizen awareness of the risks in the coastal zone promotes public safety and recognizes the high cost of attempting to respond to the effects of large-magnitude storms. Any new plan should be based on public input and be grounded in an awareness of coastal hazard issues, stress safety, and provide disincentives for development in hazardous areas. A possible option that may see more widespread use in the future involves acquisition and buyout. Such programs can provide a workable solution that is in the best interests of the community and society at large, as demonstrated by the experience of the 1993 Midwest floods.

An assessment of the major issues includes a recognition that

- Natural processes diminish the geomorphologic and biotic coastal resources
- The rate of loss of coastal resources will increase as the sea level rises with time
- The coastal zone is a well-recognized source of economic activity

- Existing management approaches will be successful with minor storms
- Long-term shoreline management objectives developed by the state are needed to provide leadership in directing the regional management of the coast
- Management strategies should be developed and applied on a regional level
- Post-storm recovery offers the only opportunity to create changes in land use
- Hazard mitigation programs can incorporate both short-term and long-term approaches
- Public attitude and perceptions of shoreline management must be modified to support public safety and risk reduction

New Jersey Coastal Management Organization: State Coordination

New Jersey's citizens remain challenged by increasingly high population density, which continually jeopardizes the natural processes governing stability and change along the coast. One of the traditional approaches for managing coastal systems in developed settings is to implement strategies on a municipality-by-municipality basis. In New Jersey, home rule has fostered a piecemeal approach to coastal management. Although these statutes each seek to achieve the same end—informed management of the coastal zone—no comprehensive administrative framework exists to ensure that the coastal zone is managed on a consistent basis in a fair and reasonable manner.

An administrative framework could be developed under a simple yet effective comprehensive planning statute. A single state entity could be charged with the sole responsibility of managing the New Jersey coast. Consolidation of coastal hazard management efforts within the state government is supported by a 2001 review of the state's coastal management program conducted by the National Oceanic and Atmospheric Administration. Presently, coastal management efforts are divided among several divisions at NJDEP (the Office of Environmental Planning, the Bureau of Coastal Engineering, and the coastal permit group housed in the Office of Land Use Regulation) and the Office of Emergency Management and Preparedness of the New Jersey state police. This diverse organization of coastal management could be reorganized to provide a new regional approach that would function in close cooperation with county and local planning entities. Consolidation would achieve efficiencies in policies and reduce overlapping and redundant efforts that have led to counterproductive, mixed, and opposing policy measures. Reorganization could include the establishment of well-defined objectives coordinated through a single office. The state entity taking on this responsibility could be established in what is now known as the Office of Environmental Planning. This office currently manages federal funds allocated for coastal zone management and possesses a great deal of expertise on coastal issues. In addition to the shift of regulatory and mitigation responsibilities to this office, a liaison must be established with the state Planning Office to ensure that coastal management policy is incorporated into the state's development and redevelopment planning. Consolidation will simplify and enhance coordination of state management efforts.

It would make sense to use the regional Citizen Advisory Committees as a means of gaining local input on coastal issues relevant to their respective regions. The new state "coastal division" could interact with these committees to integrate local and regional coastal policy goals with coastal hazard management strategies. Such a program would ensure a partnership between state and local governments and all stakeholders and create opportunities for consensus regarding the objectives of these diverse groups. Ideally, when agreement on objectives is achieved at several levels, public funds could then be used in support of a regional program; individual local programs would be supported at the local level.

The new coastal division could also develop comprehensive, long-range post-storm response and recovery programs through a process involving all interested groups. One long-range option that follows from FEMA's initiative should be a a coastal lands acquisition program (such as New Jersey's Coastal Blue Acres Program) that will set guidelines for land acquisition and establish a priority list for land acquisition. Such a list is particularly important since many opportunities for acquisition occur quickly, after a storm event. Finally, the coastal division could establish a public information-outreach program pertaining to coastal issues.

Within this general framework, coastal hazard management measures could be implemented effectively at both the state and the local level. The coastal division could implement a set of recommended, uniform guidelines and comprehensive coastal planning requirements, particularly with regard to the beach-dune systems and coastal erosion. High-hazard erosion areas already in mitigation programs at the local and regional level would be placed on the state priority acquisition list. Further, as in the state of Delaware, a program could be implemented that would provide incentives (such as density bonuses and credits) encouraging coastal landowners and developers to refrain from building in high-hazard areas. Simply put, a single state entity could ensure that coastal development and the preservation of coastal resources are pursued in the public interest on the basis of a strong public participation process. For this approach to be effective, however, there must be a clearly defined mechanism enabling citizens and government to solve problems together. The decision-making process should include citizen oversight and monitoring, meetings called jointly by government and citizen groups, and funding to hire technical consultants and to implement projects.

Coastal Management Policy Actions

Long-term objectives need to be developed at the state level to provide leadership in directing regional management of the coastal zone. Such efforts, which can be undertaken by any coastal state, involve:

- Long-term mitigation strategies that support a state's coastal management objectives into the year 2050.[2] What should the coast look like in 2050? Will there be dunes, a beach, similar patterns of land use, the same densities, the same infrastructure?
- Incorporation of sea-level rise and a modified coastal zone in planning for the future when determining objectives; mitigation objectives and strategies should be developed regarding anticipated water levels, rates of erosion, and the degree of protection sought from coastal storms.
- Objectives and strategies created in a manner that recognizes the dynamic nature of coastal systems and thus employs flexible means to interact with a changing system.
- Objectives that incorporate backbay systems and other low-lying coastal areas into regional mitigation efforts aimed at reducing the losses from storms, flooding, and sea-level rise.

Once a state's objectives are developed, coastal management strategies need to be created to achieve those objectives. These efforts can involve the following:

- An emphasis on enhancing public safety and reducing the public's exposure to coastal hazards in keeping with federal and state mitigation efforts.

- A continuous review and assessment of these strategies to ensure that they are fulfilling a state's long-term objectives as the coastal zone changes.
- An incorporation of the various mitigation approaches and techniques described in state documents similar to the *New Jersey Hazard Mitigation Plan,* with recommendations that are periodically updated, into its long-term management strategies.
- Assistance to communities in meeting their mitigation goals and the FEMA National Mitigation Strategy, which not only will reduce public risk but may increase community rating system rankings.

States should apply strategies on a regional basis rather than at the local jurisdictional level. Here policies comprise the following elements:

- Regional planning efforts should be conducted in a manner that recognizes the natural processes operating in a reach and that promotes informed land-use decisions in the coastal zone that are based on hazard management strategies.
- A state management plan should be created that incorporates local input to address the effects of regional coastal processes.
- Regional planning should be used to establish appropriate land uses, land-use densities, and long-term strategies.
- State and federal funding for local programs should be linked to their consistency with state and regional objectives; those programs that agree with a state's objectives of reduction of risk and improvement of public safety should be given funding priority.
- If local communities develop programs that are not consistent with a state's objectives, then the communities should carry these out at their own expense.
- Public funding for projects that support development in high-hazard areas should be eliminated.

High-hazard coastal areas need to become a focus of state and community mitigation efforts. Some suggestions are:

- Characteristics of high-hazard coastal areas should be identified for each region, a process that will help to foster public awareness of the risks at the coast.
- These areas should be prioritized according to the degree of risk to the public, and the evaluation should be periodically updated, particularly following storm events.

- Low-lying areas, including the bayside communities, are of special importance, because they are flooded by even minor storm occurrences.
- State programs similar to New Jersey's Coastal Blue Acres funding could be used to purchase property in identified high-hazard areas.
- The proportion of Blue Acres funding for purchases in high hazard-areas should be a 50/50 cost-share ratio if the purchase meets the local mitigation objectives, but the ratio of the state share should increase to 100% where regional/state mitigation objectives are also achieved.

POLICY APPROACHES

Coastal communities should continue to utilize coastal dunes as a natural barrier to coastal storms and their associated risks. Approaches to dune planning include:

- Development of a coordinated governmental program of dune creation, restoration, awareness, maintenance, and expansion for emergency recovery and long-term buffering.
- Determination of a community's dune preservation objectives, which once established, should guide municipalities' adoption and enforcement of effective dune protection ordinances and maintenance programs to promote dune creation and enhancement.
- Boundaries of dune areas should be scientifically defined and flexible enough to acknowledge the migration of dunes by natural forces. These boundaries need to be periodically reviewed, particularly after a storm event.
- Because setback lines cannot be static, fixed lines if dune buffer areas are to be effective, language should be incorporated into ordinances that requires the periodic revision of setback lines and buffer zones, particularly after a storm event.
- Communities should establish a formal dune buffer area inland of the built dune to permit the inland extension of the dune processes, where possible.
- Other attributes of coastal dunes, such as their sand-storage abilities, habitat, aesthetics, and biodiversity, should be incorporated into municipal dune preservation efforts.

A coastal hazard management plan should be developed that provides a blueprint to guide long-term management of the shoreline and that targets high-hazard areas for postdisaster land-use change. Planning efforts should focus on the reduction of risks to public safety rather than the defense of property. The plan can provide guidelines to develop a more idealized distribution of land uses, avoid development in high-hazard areas, and utilize the environmental, economic, and cultural resources of the area. The elements of the plan should:

- Develop community-based emergency evacuation plans that include analyses of existing evacuation routes, alternative routes or the need for the construction of new routes, population density, and the availability of support facilities.
- Consider options to relocate construction of new infrastructure away from erosion and flood-prone areas. Funding for disaster repairs should be used to relocate infrastructure, such as roads and sanitary sewer lines, away from hazardous areas.
- Enact zoning ordinances to limit the type of development in high-hazard areas, to create buffer areas, and to preserve beach-dune systems.
- Enact building moratoriums as an element of postdisaster plans until an accurate assessment of damages has been completed in the context of long-term mitigation objectives.
- Incorporate postdisaster land-use changes in identified high-hazard areas, which may include prohibiting development, acquiring damaged property, and relocating people in these areas.
- Identify property in high-hazard areas for post-storm acquisition under state programs similar to New Jersey's Coastal Blue Acres funding. The state and coastal municipalities should take advantage of existing funding sources to acquire sites in high-hazard areas and expand the use of this mitigation strategy.
- Develop prescriptive codes to enforce building compliance or introduce enabling legislation to supplement BOCA codes with more stringent flood-proofing controls similar to those used by the state of Florida in its CCCL zone.

Structural solutions may be conditionally employed as a barrier to the effects of coastal storms, but they may cause loss of beach and dunes. They should be used in conjunction with regional approaches to shoreline management. Beach nourishment may be used to reconstruct the protective and recreational features of beaches, but it should be recognized and treated as an expensive short-term strategy that should be part of a regional shoreline management program.

OUTREACH AND EDUCATION POLICY ACTIONS

Efforts in the areas of outreach and education should foster informed planning and community response to coastal hazards. These measures could involve:

- Using and maintaining Citizen Advisory Committees as a vehicle to raise awareness of shoreline-stabilization issues, soliciting local input on the development of regional strategies to implement a state's mitigation objectives, and facilitating the dissemination of information.
- Maintaining the *Coastal Hazard Management Report* bibliography, home page, and background documents on coastal hazard issues at the Rutgers Institute of Marine and Coastal Sciences.
- Establishing and coordinating volunteer mitigation and monitoring efforts.
- Providing local resource managers with easy access to information on management strategies that mitigate the effects of severe storms and coastal erosion.
- Fostering awareness of coastal hazards through formal programs and informal activities.
- Developing programs to raise public awareness on residents' vulnerability to coastal hazards, the potential costs associated with these hazards, and the means to reduce them.

SCIENTIFIC AND TECHNICAL ASSISTANCE

All public expenditure of funds in the coastal zone should be monitored through an evaluation program. Evaluators would:

- Ensure that the technical methods used are credible, replicable, cost-effective, and grounded in scientific research.
- Communicate the results of applied research on coastal system dynamics and associated natural hazards to coastal decision makers.
- Provide evaluations of projects supported by public funds to justify the continuation of successful programs.
- Establish a process to evaluate approaches to sediment management within the regional approach.
- Create regional demonstration projects in support of regional shoreline management strategies.

States should encourage the collection, analysis, and dissemination of data on shoreline change and coastal processes. Elements include:

- Encourage the implementation of regional shoreline management by supporting a fine-scale characterization of sediment distribution and transport processes.
- Conduct regular monitoring to enable the maintenance of a consistent, uniform database for future decision making.
- Monitor of the shoreline through beach-dune profiling to continue to assess the natural dynamics of the coastal zone.
- Periodically update efforts to investigate dune response to varying magnitude storms to reflect any changes.
- Install self-recording instruments to gather data on waves and currents as an essential component to sound decision making.
- Support offshore sidescan sonar mapping of the offshore zone.
- Make efforts to investigate and better understand coastal processes, including the effects of sea-level rise; areas to investigate include sediment dynamics and dune and beach dynamics.
- Investigate the dynamics of estuaries and wetland loss.
- Archive all useful and digitized information in a GIS database that is readily available and usable by regional and local resource managers, planners, and Citizen Advisory Committees.

Programs should to be established to collect appropriate data to measure the economic importance of the coastal economy on a local, regional, and state basis. These efforts include:

- Future economic studies based on accepted research approaches and designs, and the use of appropriate statistical data, to permit an accurate assessment of the relative economic importance of coastal tourism vis-à-vis investment in coastal management.
- Economic techniques, such as cost-benefit analysis and decision models from the field of finance, used in a way that can determine their relative importance and usefulness in policy-oriented studies of coastal management.
- Improved decision models capable of handling tradeoffs involving risk and return, and the effects of risk outside the control of shoreline-stabilization projects (that is, exogenous risk) such as coastal erosion, sea-level rise, and the use of hazard mitigation methods.
- Pertinent databases, created and maintained to manage the appropriate data necessary for specific economic approaches, ultimately allowing a more accurate economic assessment of management approaches and improve regional management strategies.

- Studies conducted on a seasonal basis that would isolate and identify economic activity dependent on the coastal zone and on specific beach nourishment projects. Such studies may require data on economic activity and tourism expenditures that are location specific, in terms of the relative proximity to the shoreline and to beach nourishment projects, in order to incorporate the effects of uncertainty in cost-benefit estimates as well as the element of risk associated with project failure and outcome. Future economic analyses could also incorporate other elements regarded as policy tradeoffs, such as level of development, level of erosion, and storm occurrences.

Model Municipal Dune Protection Ordinance

Introduction

Dune ordinances provide a legal mechanism for communities to establish and maintain certain levels of public safety from the effects of coastal storms. By adopting strong dune ordinances, communities can develop programs that preserve, restore, maintain, and enhance their dunes. The following model ordinance was created from existing dune ordinances to assist communities with the reassessment and development of their own regulations. The municipality that each portion of the ordinance is from is cited, and anything italicized is a suggested change to an original ordinance. Our comments are in squared brackets.

I. Purpose

Although there may be no *economical* long-term defense for fixed oceanfront structures against a constantly rising ocean level *and a decreasing sediment supply, there are methods of* effective protection of the oceanfront and adjacent coastal areas in the *short and middle term against oceanside storm surges* and flooding. *A well-developed dune system provides elevation and breadth to create a level of safety from the effects of the ocean and related hazards.* A well-*developed, coherent coastal foredune will* provide an uninterrupted *storm surge* barrier and a source of sand to mitigate the effect of storm waves for the benefit of the entire *town, inland* as well as oceanfront *properties.* Accordingly, the Borough has a vital interest in the continued maintenance and protection of the ocean beach and dune areas and has the right to cause their restoration in the event of damage or destruction. (Mantoloking 1995)

Ocean and bayfront dunes are *dynamic, valuable* physical features of the natural environment possessing outstanding geological, recreational, scenic, and protective value. Protection and preservation *of the coastal dunes are* vital to this and succeeding generations of the citizens of the City and the State. The dunes are dynamic migrating natural phenomena that help protect lives and property in adjacent landward areas, and *they* buffer barrier islands *and* barrier beach spits from the effect of major natural coastal hazards such as hurricanes, storms, flooding, and erosion. Natural dune systems also provide important habitat for wildlife species. (Brigantine 1986)

Sand dunes are vulnerable to erosion by the natural process of the wind and water as well as by the absence of good husbandry by those responsible for their maintenance and preservation. The best available means of reducing rates of erosion in said dunes is by preventing indiscriminate trespassing, construction, or other acts which might destroy or damage said dunes and by encouraging the use of native plantings, supplemented by sand fencing and other devices designed *to reduce* the free-blowing of sand and to support the maintenance of the surface tensions, root accumulations, normal contours, and other features typical in natural dunes. (Brigantine 1986)

The immediate dune and beach area are not capable of rigid definition or delineation or of completely firm stabilization, so that particular sites, at the time free of dunes, may, as a result of natural forces, become part of the dune area necessary for the continuation of the *buffering* outlined above, and persons purchasing or owning such property shall do so subject to the public interest therein. (Brigantine 1986)

This Ordinance does not attempt to define and regulate all parameters of dune delineation, function, or management and the Borough Council declares its intent to review and update this Ordinance periodically to reflect appropriately new and beneficial knowledge. (Mantoloking 1995)

Because the Bar-Beach-Dune System provides the only viable protection to property, public and private, and persons within the Borough, from the clearly present hazards of erosion and flooding caused by the Atlantic Ocean during periods of storm, and otherwise, all of the provisions of this Ordinance are deemed necessary, material, and substantial; and therefore, they shall not be subject to waiver or variance. (Mantoloking 1995)

II. Definitions

Accretion: *Includes the accumulation of sediment by natural or human-induced means on the beach area.* "Natural accretion" is the buildup of land solely by action of the forces of nature on a beach by deposition of waterborne or airborne material. "Artificial accretion" is a similar buildup of land by reason of *a human act,* such as accretion formed by a breakwater or beachfill deposited by mechanical means. (Atlantic City 1984; Brigantine 1986)

Beach Area: Shall mean gently sloping unvegetated areas of sand or other unconsolidated material that extends landward from the mean high water line to either: the vegetation line; a man-made feature generally parallel to the ocean or bay such as a retaining structure, bulkhead, or road; or the seaward (bayward) foot of dunes, whichever is closest to the ocean or bay waters. (Barnegat Light 1994)

Development Restriction Line: An artificial boundary delineating the extreme limit of allowable development along the shore *inland of* the existing dune fields and beaches. (Brigantine 1986)

Dunes: A dune is a wind *and wave* deposited or man-made formation of sand (mound or ridge) that lies generally parallel to, and landward of, the beach, and between the upland limit of the beach and the foot of the most inland dune slope. "Dune" includes the foredune, secondary, and tertiary dune ridges, as well as man-made dunes, where they exist.

1. Formations of sand *immediately inland from the beaches* which are stabilized by retaining structures, snow fences, planted vegetation, and other measures are considered to be dunes regardless of the degree of modification of the dune by wind or wave action or disturbance by development.

2. A small mound of loose, windblown sand found in a street or on a part of a structure as a result of storm activity is not considered to be a "dune." (This definition is intended to reflect the definitions set forth in CAFRA [Coastal Area Facilities Review Act] regulations *N.J.A.C.* 7:7E-3.16 as it may be amended from time to time.) (Berkeley 1994)

Dune Area: Area between the seaward edge of the dune and the landward edge of the Dune. This shall include all areas within the following districts: Dune Maintenance District, Dune Restoration District, and Dune

Reconstruction District (Brigantine 1986). This Area is considered to have dynamic boundaries which move in response to seasonal winds and storms. Consequently, the boundaries shall be reviewed at the minimum every twelve months and following the winter season and any storm which damages large portions of the Dune Area. The boundary review process will be conducted by the City Council, the Department of Public Works, and representatives of the N.J. Department of Environmental Protection, Division of Coastal Resources. (Atlantic City 1989)

Dune Consultant: Shall mean an expert on dunes and their care retained by the Borough. In any periods which no such expert is regularly retained, it shall mean such other person designated by the Borough Council. (Mantoloking 1995)

Dune Crest: The point or line where the dune's highest elevation is located. (Brigantine 1986)

Dune Inspector(s): Shall mean that person or those persons appointed by the Mayor with the consent of the Borough Council. (Mantoloking 1995)

Dune Maintenance District: A *designated* area delineating dune fields containing *one* or more dune ridges and having a width greater than 300 feet. The dune surface is to be stabilized by natural vegetation. (Brigantine 1986)

Dune Reconstruction District: An artificial area delineating a discontinuous dune field or beach area without dunes, poorly developed, unstable and less than 75 feet in width. (Brigantine 1986)

Dune Restoration District: A *designated* area delineating a dune field containing one or more dune ridges which may be discontinuous between the beach and upland structures and having a width of less than 300 feet. (Brigantine 1986)

Dune and Shoreline Map: A topographic survey of dune fields and beaches delineating the following districts: Dune Maintenance District, Dune Restoration District, and Dune Reconstruction District. The survey will contain, but not be limited to the following information: the dune crestline, dune ridges, and the landward and seaward edge of the dunes. The Map shall be reviewed every twelve months and following any storm which damages large portions of the Dune Area. (modified combination of Brigantine 1986 and Atlantic City 1989)

Dune Vegetation: Shall include all plant species found on beaches and dunes of the northeastern United States, either native or introduced, which can build and stabilize sand dunes. Specifically, it shall include, but not be limited to, such species as American beachgrass *(Ammophila breviligulata),* sea rocket *(Cakile edentula),* seaside spurge *(Euphorbia polygonifolia),* dune cordgrass *(Spartina patens),* seaside goldenrod *(Solidago sempervirens),* dusty miller *(Artemisia stelleriana),* bayberry *(Myrica pennsylvanica),* beach pea *(Lathyrus japonicus),* salt spray rose *(Rosa rugosa),* beach plum *(Prunus maritima),* etc., which normally grow or may be planted on the slopes of the dunes or behind them, with no distinction being made as to how such plants are introduced into their location. (Mantoloking 1995)

Erosion: The wearing away of land by the action of natural forces. On a beach, the carrying away of beach materials by wave action, tidal currents, littoral currents, or by deflation. (Brigantine 1986)

Landward Edge of the Dune: Is the line joining the average Landward Edge of the Dune of the adjoining oceanfront properties, or a line parallel to and 60 feet west of the Seaward Edge of the Dune, whichever is more westerly. (Mantoloking 1995)

Mean High Water Line: Is the line found by the intersection of a plane at the elevation of Mean High Water with the existing slope of the beach. (Mantoloking 1995)

Mean Sea Level: Shall be the average elevation of the sea surface at a site, based on observations of all stages of the tide over a nineteen-year period (tidal epoch).

Pathway: Is an improved, protective access way, at grade across the dune. (Mantoloking 1995)

Person: Natural persons, partnerships, firms, associations, joint-stock companies, syndicates and corporations and any receiver, trustee, conservator, or other officer appointed pursuant to law or by any court, state or federal. "Person" also means the state of New Jersey, counties, municipalities, authorities, other political subdivisions, and all departments and agencies within the aforementioned governmental entities. (Atlantic City 1989)

Sand Fence: Shall include the term "snow fence" *or* a barricade-type *fence* established in a line or a pattern to accumulate sand and aid in the formation of a dune, such as picket type consisting of light wooden fence, 4 feet in height, held together by wire and affixed to wooden posts. Alternate types of "sand fence" may be utilized if approved by the Dune Consultant. (Mantoloking 1995)

Seaward Edge of Dune: Is the intersection line of the foreslope of the dune and the gradient of the Beach Area, or Vegetation Line, or the Upper Driftline, whichever is the more easterly. (Bay Head 1993)

Setback Line: Is the line of the most westward margin of a Landward Edge of the Dune Area. The setback line is delineated on the Dune and Shoreline Map. This Area is considered to have dynamic boundaries which move in response to seasonal winds and storms. Consequently, the boundaries shall be reviewed every five (5) years.

Upper Driftline: Is that line produced by the Winter Spring Tides (highest tides of the year) which contains oceanic debris (flotsam such as seaweed, etc.) and the seeds, rhizomes, or detached plants which can germinate and/or grow to produce a zone of new dune vegetation. (Mantoloking 1995)

Vegetation Line: Is that line connecting the most seaward naturally occurring perennial plants with other such plants. (Mantoloking 1995)

Walkway: A constructed means of crossing the dune area in accordance with approval by the Township. (Brick 1988)

III. Permitted and Prohibited Activities

Construction: Construction *seaward* of the Landward Edge of the Dune and the placement there, except temporarily, of any object that would impede the flow of sand are prohibited, except as provided in this Ordinance and in accordance with any CAFRA regulations, the Land Use Ordinance of the Borough with the following exceptions (Mantoloking 1995):

Dune management programs designed to either maintain, restore, or reconstruct dune fields such as supplemental planting of natural vegetation, placement of sand fences, construction of artificial dunes of berms or any other programs that may be authorized through written approval. (Brigantine 1986)

Dune Platforms: [Although Dune platforms are not recommended because they affect dune stability and sand transport, some communities do permit them. Therefore the following is to serve as a guide to communities that already permit dune platforms, to ensure there is a limited amount of disturbance to the dunes.]

Each oceanfront lot shall be allowed a "dune platform" not to exceed 200 square feet, situated within the dune area and specifically located and delineated by the owner of the premises. The dune platform shall, in all events, be maintained in the same fashion and subject to the same regulations as may govern use of pathways and walkways.

The *specifications for a dune platform* call for neither length nor width greater than 18 feet, an elevation of at least 18 inches above any point of sand surface, supporting post 4 × 4, beams no greater than 6 inches in vertical section, handrails no more than 2 inches in vertical section, and planking no more than 4 inches in width installed with at least a 16% gap area in order to permit dune grass to grow underneath.

The permit application for a Dune Platform is to be accompanied by a sketch, to scale, showing that it is at least 10 feet to landward of the mapped current Dune Crestline. The sketch may be prepared by the Dune Inspector.

Whenever the dune builds to the point that a part of the dune platform surface is less than 6 inches above the sand surface within 5 feet, that platform shall be raised or rebuilt. Raising an existing dune platform does not require a permit. Reconstruction of a dune platform shall be subject to issuance of a permit. (combination of Mantoloking 1995 and Point Pleasant 1994)

Sand Removal: Sand which is transported upon lands by action of wind, tides, storms, or any combination thereof shall not be removed from the lot upon which it is deposited by such action. Surplus sand deposited upon any improved street ends shall be restored into the Beach and Dune area. (Mantoloking 1995)

Trespassing:
   1. No person shall be in the Dune Area unless:
      (a) upon an improved Pathway, Walkway or Dune Platform; or
      (b) in the performance of such activities as may be reasonable and necessarily required to construct or maintain the dune or allowed structures; or

(c) for the purposes of inspection, topographical survey, or enforcement of this Ordinance; entry for these purposes will not be
deemed an actionable trespass. (Mantoloking 1995)

2. No person shall operate a motor vehicle across or upon any Dune
Area except as may be necessitated for allowed construction or for
dune maintenance. (Bay Head 1993)

Tampering with Dune Protection Devices: The removal, cutting, burn-
ing, or destruction of Dune Vegetation, Sand Fences, or such other types
of approved dune protection devices by the Borough Council in the Dune
area is prohibited, except as necessary for construction or maintenance
authorized pursuant to *this ordinance.* (Mantoloking 1995)

Placement of nonliving trees, brush, shrubs, or other debris in the
dune area or beach is prohibited. (Atlantic City 1989)

IV. Dune Systems Creation and Expansion

Beach Access: *During storms and levels of high water, pathways are often weak links in the coastal dune ridge and become sites of overwash and breaches. Allowing a proliferation of man-made structures over the
dunes is also esthetically displeasing and unnecessary. Therefore, the Borough
shall restrict beach access to street ends where feasible. Dune pathways and
steps to permit access across the dunes or berms to the open beach without
damage to the dunes themselves shall be permitted under the following conditions:*

Access to the beaches and dune fields in all districts delineated by the
Dune and Shoreline Map shall be limited to those accessways shown on
the Dune and Shoreline Map. No accessways across a dune field shall be
constructed by a private owner without a permit issued by the Borough
(Brigantine 1986). No walkways or steps, or combination thereof, shall be
approved nor constructed to grant access to the beach on or across any
lot, which lot has a boundary line or lot line adjacent to a public street,
alley, or easement giving access to the open beach. (Long Beach 1994)

One pathway or walkway across the dune area is permitted for each resi-
dence. It shall run, generally, the shortest practical course between the res-
idence and the seaward edge of the dune, and shall not exceed four feet in
width (Bay Head 1993). [If the dune has sufficient width, a path that cuts
through the dunes should incorporate an offset or zig-zag in its planview

to hinder the direct inland transport of sand by wind and waves through the cut.]

In the event that any pathway or walkway shall be or become, in the opinion of the Dune Consultant or Inspector, a substantial detriment to the development and maintenance of the continuous protective dune sought to be achieved by this ordinance, the owner of the premise shall be subject to the provisions of the *Permit section of this ordinance.* (Bay Head 1993)

A walkway is exempt from any provisions requiring a construction permit provided that it: does not extend westward of the Landward Edge of the Dune or to the eastward of the Seaward Edge of the Dune; is at least four inches above the highest point of the Dune over which it passes; is not wider than four feet; provides at least 16% of the walkway surface as gap space between the walkway surface boards in order to permit dune grass to grow underneath; has a walkway surface including lateral supporting members with a vertical cross-section of not more than five inches. (Bay Head 1993)

Where an elevated walkway is constructed and sand has accreted to a point where the walkway is on the surface of the sand at the dune crest and is below the adjacent crest, the natural accretion of the dune is impeded, unless the dune height exceeds the *acceptable dune height.* In such case, the Dune Inspector shall serve written notice, certified by mail return receipt requested, upon the record owner westward of the dune at his last known address, directing that the walkway be raised in compliance with the standards of this ordinance and, if the walkway is not raised within a period of six months from the date of such notice, the Borough may raise the walkway at the expense of the owner. In such case, the cost of construction shall become a lien on the real property situated immediately westward of the Dune. (Bay Head 1993)

If an elevated walkway is not used to access the ocean, the pathway shall be protected by placing suitable material on the sand surface, to be removed when the premises are not occupied. The depth of a crestline gap is the vertical distance between the bottom of the pathway through the crest of the dune and a line connecting the highest points of the dune with 20 feet on either side of the pathway (Mantoloking 1995). *The construction of elevated walkways over the crest of the dune is encouraged to preserve the Dune Area* (Bay Head 1993). If the crestline gap depth is two feet or more at any time, the pathway shall be replaced by an elevated walk-way (Mantoloking 1995).

[Dune Dimensions: One of the purposes of this ordinance is to achieve the maintenance of sand dunes at the highest level of protection from erosion and the effects of coastal storms. To this end, no dune shall be directly or indirectly reduced in height or width by the action or inaction of any owner or his/her agent. However, if any dune shall be or become lower or narrower than the dimensions deemed materially significant by the Dune Consultant, applying recognized criteria, with due regard to the intent of this ordinance and reasonable use of the premise, the owner thereof shall be obliged to install such Sand Fence and plantings in accordance with the specifications set forth in this ordinance at his/her expense. The owner shall have an obligation to maintain and replace, if necessary, such fences and plantings. In such case, the cost of construction shall become a lien on the real property situate immediately landward of the Dune. The dimensions of the dune area shall be as follows:]

> Dunes should be located on the order of 100 feet landward of the Mean High Tide (MHT) line.
> *The slope of a dune should be maintained not steeper than a 1:5 (vertical horizontal) slope.*
> The height/width ratio should be maintained no less than 1:10.
> *Dunes will be maintained at a minimum elevation of fourteen feet (14') above mean sea level at the bulkhead line and an elevation of sixteen feet (16') above mean sea level at the oceanfront building line.*

Dune Vegetation and Planting: In order to provide for effective protection and/or restoration of the Dune areas, each owner shall plant or cause to be planted in the Dune Area adjoining his property suitable vegetation and erect, or cause to be erected, in the Dune Area, suitable sand fencing all *in accordance with the specifications set forth in this ordinance and in conformance with the current standards of the U.S. Natural Resources Conservation Service.* (Brick 1988)

For initial planting, or replanting sparse areas, "Cape" American beachgrass *(Ammophila breviligulata)* should be used. The entire Dune Area shall be planted.

Planting may take place any time between October 15th and April 1st, if the ground is not frozen. Spring planting should be accompanied by frequent watering. Initial and subsequent fertilization is recommended at the rate of about 2 pounds of slow-release 10-10-10 per 1000 square feet.

Only fresh planting stock cut back to 16–18 inches long shall be utilized. Spacing shall be no greater than 18 inches, two stems to a hole, at least 7 inches deep. If not planted with a water flooding method, the sand shall be compacted by rains before planting *is commenced, to eliminate any air pockets.*

After beachgrass has been established, other appropriate vegetation may be added. (Mantoloking 1995)

Sand Fencing: Fencing shall be standard 4-foot wood sand (snow) fence in good condition, secured to wooden posts of a minimum cross-section of 4 square inches and a minimum length of 61/2 feet, with maximum span between posts of 12 feet. Alternate fencing, as approved by the Dune Inspector with advice of the Dune Consultant, prior to installation, may be utilized.

There shall be at least two lines of fencing the length of the Dune Area of each property. At least one line of fencing should be in a zig-zag pattern with alternate posts offset by at least 5 feet. Half-height fencing may be used on the dune back.

A straight (or zig-zag) line of fencing may be erected adjacent to the seaward toe of the dune to prevent incursion into the dune area, but if it is more than 3 feet to seaward, a permit is required and such fencing must be removed during the winter months. (Mantoloking 1995)

The construction of fencing along the western limits of the backshore and dunes areas and the provision of suitable markings to identify the *dune area is permitted to prevent damage to the dunes from indiscriminate passage.* (Beach Haven 1994)

V. Repairing and Maintaining Coastal Dunes

Vegetation Maintenance: *Fertilization should be applied each spring after regrowth begins for established dune vegetation. Yearly fertilization for maintenance should not exceed 50 lb. of nitrogen per acre per year. Where vegetation has been destroyed, American beachgrass* (Ammophila breviligulata) *will be replanted following the specifications set forth by the U.S. Natural Resources Conservation Service.*

Signage: Because the population of a summer resort is transient, it is necessary to remind visitors that the dunes are fragile and that it is illegal to walk on them. Signs must be placed to notify and educate beach users; consideration should be given to placing some signs right on the sand fencing. (Ocean City 1994)

Fixing Blowouts: *Blow-outs in the dune system will be repaired by placing a sand fence between the existing dune parts. One or more fences may be required. It is essential to tie the ends of the fence into the existing dune to keep the wind from whipping around the ends.*

Dune Replenishment: The municipality shall not undertake any mechanical manipulation, including but not limited to bulldozing, grading and scraping, of the beach and dune area unless written authorization is received from the New Jersey Department of Environmental Protection, Division of Coastal Resources. (Brick 1988)

Permits:

1. A permit shall not be required for the planting of dune grass or other appropriate vegetation, or for the erection of sand fencing or the placement of temporary walkway protection in the Dune area in compliance with approved standards set forth in this Ordinance. All other construction, modification, alterations, or like activity in the Dune area, unless specifically exempted in this Ordinance, shall require that the owner or his/*her* agent obtain a Dune Area Permit. Activities requiring a Permit include but are not limited to elevated walkways, dune platforms, and the placement of sand fencing more than 3 feet seaward of the Seaward Edge of the Dune. All permits are subject to revocation, suspension, or modification in the event of changed site conditions, as determined by the Dune Inspector with advise of the Dune Consultant. The permittee or any agent shall promptly, upon request, allow any Borough official to examine the permit or a certified true copy thereof at any time.

2. The Dune Inspector shall make periodic inspections and shall provide written advice to owners. These writings shall not be deemed as notice of violations of this Ordinance, but shall be maintained as part of the record for the subject property and may be considered by the Court in the imposition of penalty upon conviction under any subsequent complaint for violation of this Ordinance. Further, the Dune Inspector shall coordinate his or her efforts with those of the Dune Consultant(s) to the

end that the purposes of this ordinance may be achieved. (Mantoloking 1995)

Conditions for Issuance of Permit *for Removal of Sand:* No such permit shall be issued without a determination by the City Engineer, based upon an inspection of the area involved, that such removal *of sand* will not create or increase a danger or hazard to life or property. No permit will be granted if the proposed moving or displacement will:

Adversely affect the littoral drift in the districts delineated by the Dune and Shoreline Map *or other municipalities within this reach.*

Result in a reduction of dune protection as provided for in the Dune and Shoreline Map.

Interfere with the general configuration of the districts as delineated by the Dune and Shoreline Map.

Otherwise substantially impair or interfere with the intent and purpose *and objectives* of this ordinance. (Brigantine 1986)

Enforcement: The Borough Dune Inspector or, in his/*her* absence, the Chief of Police and in all events the Borough Council shall enforce the affirmative duty of each oceanfront owner, as set forth in this Ordinance. Owners have 30 days after receiving a written notice to begin to comply with the ordinance or they will be subject to pay the cost with interest at the highest legal rate via a lien on the property. (Mantoloking 1995)

Violations and Penalties: For each and every violation of this Ordinance, *the regulations or standards set forth in the Ordinance, or the terms and conditions of any permit issued hereunder,* the owner of lands abutting the beach or Dune Area where such violation has been committed, or the trespasser *and any contractor or agent of the owner* shall for each and every violation be subject to a fine of not more than one thousand dollars ($1000) or imprisonment for a term not to exceed ninety (90) days, or both, at the discretion of the Court having jurisdiction in this matter. Each and every day that a violation continues shall be considered a separate offense. (Bay Head 1993)

Estimated Expenditures on Travel and Tourism in New Jersey, 1987–1994

Year	Item	Cost (millions of dollars)	Cost92 (millions of 1992 dollars)	Est. visitors (thousands)
1987	Seasonal rent	1.7	2.10	
1987	Daily rent	1.8	2.22	
1987	Beach tag fees	58.0	71.63	
1987	Parking fees	52.0	64.22	
1987	Daytime entertainment	482.0	595.29	
1987	Groceries	625.0	771.90	
1987	Food and restaurant	612.0	755.84	
1987	Nighttime entertainment	653.0	806.48	
1987	Entertainment	1135.0	1401.76	
1987	Gifts	224.0	276.65	
1987	Total			6.0
1988	Seasonal rent	1.2	1.42	
1988	Daily rent	1.7	2.02	
1988	Beach tag fees	43.0	51.00	
1988	Parking fees	56.0	66.41	
1988	Daytime entertainment	453.0	537.24	
1988	Groceries	657.0	779.18	
1988	Food and restaurant	466.0	552.66	
1988	Nighttime entertainment	666.0	789.85	
1988	Entertainment	1119.0	1327.10	
1988	Gifts	191.0	226.52	
1988	Total			4.8
1989	Seasonal rent	2.0	2.26	
1989	Daily rent	2.7	3.05	
1989	Beach tag fees	33.0	37.34	
1989	Parking fees	57.0	64.49	
1989	Daytime entertainment	658.0	744.50	
1989	Groceries	495.0	560.07	
1989	Food and restaurant	477.0	539.70	

Year	Item	Cost (millions of dollars)	Cost92 (millions of 1992 dollars)	Est. visitors (thousands)
1989	Nighttime entertainment	643.0	727.52	
1989	Entertainment	1301.0	1472.02	
1989	Gifts	279.0	315.68	
1989	Total			5.7

Year	County	Item	Cost (millions of dollars)	Cost92 (millions of 1992 dollars)	Est. visitors (thousands)
Type: County level					
1990	Atlantic	Lodging	502.23	517.35	4115.3
1990	Atlantic	Food and restaurant	1186.94	1222.67	4115.3
1990	Atlantic	Entertainment	286.70	295.33	4115.3
1990	Atlantic	Automobile	584.83	602.44	4115.3
1990	Atlantic	Local transportation	36.19	37.28	4115.3
1990	Atlantic	Retail	1025.34	1056.21	4115.3
1990	Atlantic	Gambling	2598.03	2676.24	4115.3
1990	Cape May	Lodging	228.14	235.01	2583.9
1990	Cape May	Food and restaurant	502.51	517.64	2583.9
1990	Cape May	Entertainment	116.61	120.21	2583.9
1990	Cape May	Automobile	232.99	240.00	2583.9
1990	Cape May	Local transportation	14.48	14.92	2583.9
1990	Cape May	Retail	364.20	375.16	2583.9
1990	Cape May	Gambling	0.00	0.00	2583.9
1990	Monmouth	Lodging	78.66	81.03	1990.3
1990	Monmouth	Food and restaurant	277.74	286.10	1990.3
1990	Monmouth	Entertainment	54.93	56.58	1990.3
1990	Monmouth	Automobile	117.53	121.07	1990.3
1990	Monmouth	Local transportation	7.81	8.05	1990.3
1990	Monmouth	Retail	221.89	228.57	1990.3
1990	Monmouth	Gambling	0.00	0.00	1990.3
1990	Ocean	Lodging	69.83	71.93	1990.3
1990	Ocean	Food and restaurant	251.99	259.58	1990.3
1990	Ocean	Entertainment	51.60	53.15	1990.3
1990	Ocean	Automobile	105.09	108.25	1990.3
1990	Ocean	Local transportation	6.56	6.76	1990.3
1990	Ocean	Retail	192.69	198.49	1990.3
1990	Ocean	Gambling	0.00	0.00	1990.3
1991	Atlantic	Lodging	473.82	488.08	4115.3
1991	Atlantic	Food and restaurant	1121.24	1154.99	4115.3
1991	Atlantic	Entertainment	270.65	278.80	4115.3
1991	Atlantic	Automobile	552.30	568.93	4115.3
1991	Atlantic	Local transportation	34.19	35.22	4115.3
1991	Atlantic	Retail	968.95	998.12	4115.3
1991	Atlantic	Gambling	2489.79	2564.74	4115.3

Year	County	Item	Cost (millions of dollars)	Cost92 (millions of 1992 dollars)	Est. visitors (thousands)
1991	Cape May	Lodging	236.84	243.97	2583.9
1991	Cape May	Food and restaurant	518.82	534.44	2583.9
1991	Cape May	Entertainment	120.32	123.94	2583.9
1991	Cape May	Automobile	241.36	248.63	2583.9
1991	Cape May	Local transportation	15.08	15.53	2583.9
1991	Cape May	Retail	378.29	389.68	2583.9
1991	Cape May	Gambling	0.00	0.00	2583.9
1991	Monmouth	Lodging	78.74	81.11	1990.3
1991	Monmouth	Food and restaurant	275.43	283.72	1990.3
1991	Monmouth	Entertainment	54.63	56.27	1990.3
1991	Monmouth	Automobile	116.83	120.35	1990.3
1991	Monmouth	Local transportation	7.76	7.99	1990.3
1991	Monmouth	Retail	220.17	226.80	1990.3
1991	Monmouth	Gambling	0.00	0.00	1990.3
1991	Ocean	Lodging	72.99	75.19	1990.3
1991	Ocean	Food and restaurant	256.58	264.30	1990.3
1991	Ocean	Entertainment	52.90	54.49	1990.3
1991	Ocean	Automobile	107.57	110.81	1990.3
1991	Ocean	Local transportation	6.71	6.91	1990.3
1991	Ocean	Retail	195.91	201.81	1990.3
1991	Ocean	Gambling	0.00	0.00	1990.3
1992	Atlantic	Lodging	657.72	638.60	.
1992	Atlantic	Food and restaurant	1198.31	1163.48	.
1992	Atlantic	Entertainment	247.67	240.47	.
1992	Atlantic	Automobile	567.37	550.88	.
1992	Atlantic	Local transportation	42.56	41.32	.
1992	Atlantic	Retail	861.12	836.09	.
1992	Atlantic	Gambling	3129.76	3038.79	.
1992	Cape May	Lodging	237.20	230.31	.
1992	Cape May	Food and restaurant	385.26	374.06	.
1992	Cape May	Entertainment	71.07	69.00	.
1992	Cape May	Automobile	170.57	165.61	.
1992	Cape May	Local transportation	13.98	13.57	.
1992	Cape May	Retail	215.92	209.64	.
1992	Cape May	Gambling	0.00	0.00	.
1992	Monmouth	Lodging	87.42	84.88	.
1992	Monmouth	Food and restaurant	432.78	420.20	.
1992	Monmouth	Entertainment	107.97	104.83	.
1992	Monmouth	Automobile	202.58	196.69	.
1992	Monmouth	Local transportation	10.29	9.99	.
1992	Monmouth	Retail	421.63	409.38	.
1992	Monmouth	Gambling	0.00	0.00	.
1992	Ocean	Lodging	58.30	56.61	.
1992	Ocean	Food and restaurant	317.35	308.13	.

Year	County	Item	Cost (millions of dollars)	Cost92 (millions of 1992 dollars)	Est. visitors (thousands)
1992	Ocean	Entertainment	77.51	75.26	.
1992	Ocean	Automobile	142.02	137.89	.
1992	Ocean	Local transportation	6.96	6.76	.
1992	Ocean	Retail	290.39	281.95	.
1992	Ocean	Gambling	0.00	0.00	.
1993	Atlantic	Lodging	662.23	642.98	4317.20
1993	Atlantic	Food and restaurant	1112.83	1080.48	4317.20
1993	Atlantic	Entertainment	214.59	208.35	4317.20
1993	Atlantic	Automobile	517.68	502.63	4317.20
1993	Atlantic	Local transportation	41.62	40.41	4317.20
1993	Atlantic	Retail	736.80	715.38	4317.20
1993	Atlantic	Gambling	3167.74	3075.67	4317.20
1993	Cape May	Lodging	243.79	236.70	2394.80
1993	Cape May	Food and restaurant	476.75	462.89	2394.80
1993	Cape May	Entertainment	102.50	99.52	2394.80
1993	Cape May	Automobile	220.59	214.18	2394.80
1993	Cape May	Local transportation	15.40	14.95	2394.80
1993	Cape May	Retail	331.20	321.57	2394.80
1993	Cape May	Gambling	0.00	0.00	2394.80
1993	Monmouth	Lodging	90.18	87.56	1392.50
1993	Monmouth	Food and restaurant	372.69	361.86	1392.50
1993	Monmouth	Entertainment	84.00	81.56	1392.50
1993	Monmouth	Automobile	166.83	161.98	1392.50
1993	Monmouth	Local transportation	9.65	9.37	1392.50
1993	Monmouth	Retail	332.58	322.91	1392.50
1993	Monmouth	Gambling	0.00	0.00	1392.50
1993	Ocean	Lodging	59.94	58.20	1392.50
1993	Ocean	Food and restaurant	281.80	273.61	1392.50
1993	Ocean	Entertainment	63.02	61.19	1392.50
1993	Ocean	Automobile	120.58	117.08	1392.50
1993	Ocean	Local transportation	6.60	6.41	1392.50
1993	Ocean	Retail	237.08	230.19	1392.50
1993	Ocean	Gambling	0.00	0.00	1392.50
1994	Atlantic	Lodging	645.75	611.33	6365.90
1994	Atlantic	Food and restaurant	1216.34	1151.50	6365.90
1994	Atlantic	Entertainment	263.74	249.68	6365.90
1994	Atlantic	Automobile	578.98	548.12	6365.90
1994	Atlantic	Local transportation	41.38	39.17	6365.90
1994	Atlantic	Retail	916.95	868.07	6365.90
1994	Atlantic	Gambling	3202.84	3032.11	6365.90
1994	Cape May	Lodging	649.49	614.87	8984.10
1994	Cape May	Food and restaurant	722.93	684.39	8984.10
1994	Cape May	Entertainment	179.68	170.10	8984.10

Year	County	Item	Cost (millions of dollars)	Cost92 (millions of 1992 dollars)	Est. visitors (thousands)
1994	Cape May	Automobile	331.99	314.29	8984.10
1994	Cape May	Local transportation	18.14	17.17	8984.10
1994	Cape May	Retail	625.28	591.95	8984.10
1994	Cape May	Gambling	0.00	0.00	8984.10
1994	Monmouth	Lodging	451.39	427.33	1344.26
1994	Monmouth	Food and restaurant	451.39	427.33	1344.26
1994	Monmouth	Entertainment	113.77	107.71	1344.26
1994	Monmouth	Automobile	210.15	198.95	1344.26
1994	Monmouth	Local transportation	10.33	9.78	1344.26
1994	Monmouth	Retail	446.18	422.40	1344.26
1994	Monmouth	Gambling	0.00	0.00	1344.26
1994	Ocean	Lodging	404.50	382.94	1279.52
1994	Ocean	Food and restaurant	404.50	382.94	1279.52
1994	Ocean	Entertainment	101.57	96.16	1279.52
1994	Ocean	Automobile	180.26	170.65	1279.52
1994	Ocean	Local transportation	8.42	7.97	1279.52
1994	Ocean	Retail	385.59	365.04	1279.52
1994	Ocean	Gambling	0.00	0.00	1279.52

Type: Barrier islands, by county

Year	County	Item	Cost (millions of dollars)	Cost92 (millions of 1992 dollars)	Est. visitors (thousands)
1992	Atlantic	Lodging	18.310	17.778	117.8
1992	Atlantic	Food and restaurant	4.756	4.618	117.8
1992	Atlantic	Entertainment	2.648	2.571	117.8
1992	Atlantic	Automobile	0.603	0.586	117.8
1992	Atlantic	Local transportation	–	–	117.8
1992	Atlantic	Retail	3.481	3.380	117.8
1992	Atlantic	Gambling	0.951	0.923	117.8
1992	Cape May	Lodging	363.275	352.716	2488.7
1992	Cape May	Food and restaurant	62.265	60.456	2488.7
1992	Cape May	Entertainment	13.732	13.333	2488.7
1992	Cape May	Automobile	11.734	11.393	2488.7
1992	Cape May	Local transportation	–	–	2488.7
1992	Cape May	Retail	58.233	56.540	2488.7
1992	Cape May	Gambling	0.00	0.00	2488.7
1992	Monmouth	Lodging	20.619	20.020	90.9
1992	Monmouth	Food and restaurant	3.534	3.431	90.9
1992	Monmouth	Entertainment	0.779	0.757	90.9
1992	Monmouth	Automobile	0.666	0.647	90.9
1992	Monmouth	Local transportation	–	–	90.9
1992	Monmouth	Retail	3.305	3.209	90.9
1992	Monmouth	Gambling	0.00	0.00	90.9
1992	Ocean	Lodging	138.233	134.215	579.9
1992	Ocean	Food and restaurant	23.693	23.004	579.9
1992	Ocean	Entertainment	5.225	5.073	579.9
1992	Ocean	Automobile	4.465	4.335	579.9

Year	County	Item	Cost (millions of dollars)	Cost92 (millions of 1992 dollars)	Est. visitors (thousands)
1992	Ocean	Local transportation	–	–	579.9
1992	Ocean	Retail	22.159	21.515	579.9
1992	Ocean	Gambling	0.00	0.00	579.9
1993	Atlantic	Lodging	19.460	18.894	110.7
1993	Atlantic	Food and restaurant	5.032	4.886	110.7
1993	Atlantic	Entertainment	2.802	2.721	110.7
1993	Atlantic	Automobile	0.638	0.620	110.7
1993	Atlantic	Local transportation	–	–	110.7
1993	Atlantic	Retail	3.684	3.577	110.7
1993	Atlantic	Gambling	1.006	0.977	110.7
1993	Cape May	Lodging	431.595	419.050	2954.1
1993	Cape May	Food and restaurant	73.975	71.825	2954.1
1993	Cape May	Entertainment	16.314	15.840	2954.1
1993	Cape May	Automobile	13.941	13.535	2954.1
1993	Cape May	Local transportation	–	–	2954.1
1993	Cape May	Retail	69.185	67.174	2954.1
1993	Cape May	Gambling	0.00	0.00	2954.1
1993	Monmouth	Lodging	21.303	20.684	94.0
1993	Monmouth	Food and restaurant	3.651	3.545	94.0
1993	Monmouth	Entertainment	0.805	0.782	94.0
1993	Monmouth	Automobile	0.688	0.668	94.0
1993	Monmouth	Local transportation	–	–	94.0
1993	Monmouth	Retail	3.415	3.316	94.0
1993	Monmouth	Gambling	0.00	0.00	94.0
1993	Ocean	Lodging	147.972	143.671	621.4
1993	Ocean	Food and restaurant	25.362	24.625	621.4
1993	Ocean	Entertainment	5.593	5.431	621.4
1993	Ocean	Automobile	4.780	4.641	621.4
1993	Ocean	Local transportation	–	–	621.4
1993	Ocean	Retail	23.720	23.030	621.4
1993	Ocean	Gambling	0.00	0.00	621.4
1994	Atlantic	Lodging	21.297	20.161	117.5
1994	Atlantic	Food and restaurant	5.461	5.170	117.5
1994	Atlantic	Entertainment	3.087	2.922	117.5
1994	Atlantic	Automobile	0.698	0.661	117.5
1994	Atlantic	Local transportation	–	–	117.5
1994	Atlantic	Retail	4.018	3.804	117.5
1994	Atlantic	Gambling	10.216	9.671	117.5
1994	Cape May	Lodging	395.086	374.025	2881.0
1994	Cape May	Food and restaurant	68.118	64.487	2881.0
1994	Cape May	Entertainment	14.909	14.114	2881.0
1994	Cape May	Automobile	12.745	12.065	2881.0
1994	Cape May	Local transportation	–	–	2881.0

Year	County	Item	Cost (millions of dollars)	Cost92 (millions of 1992 dollars)	Est. visitors (thousands)
1994	Cape May	Retail	63.724	60.327	2881.0
1994	Cape May	Gambling	0.00	0.00	2881.0
1994	Monmouth	Lodging	27.886	26.400	140.2
1994	Monmouth	Food and restaurant	4.808	4.552	140.2
1994	Monmouth	Entertainment	1.052	0.996	140.2
1994	Monmouth	Automobile	0.900	0.852	140.2
1994	Monmouth	Local transportation	–	–	140.2
1994	Monmouth	Retail	4.498	4.258	140.2
1994	Monmouth	Gambling	0.00	0.00	140.2
1994	Ocean	Lodging	127.366	120.576	–
1994	Ocean	Food and restaurant	21.960	20.789	–
1994	Ocean	Entertainment	4.806	4.550	–
1994	Ocean	Automobile	4.109	3.890	–
1994	Ocean	Local transportation	–	–	–
1994	Ocean	Retail	20.543	19.448	–
1994	Ocean	Gambling	0.00	0.00	–

SOURCES: 1987–89, Opinion Research Corporation 1989; 1990–91, Longwoods International (1992); 1992–93, Longwoods International (1994a); 1994, Longwoods International (1995). See also references in chapter 8.

NOTE: Cost is in millions of current dollars associated with the year of the study and refers to the projected expenditures on travel and tourism; Cost92 is in millions of 1992 dollars adjusted by the relevant consumer price index; Est. visitors refers to the estimated number of visitors in thousands. For the 1992 tourism expenditure estimates from the Longwoods study, the 1992 estimates under the Cost column are in 1993 dollars and had to be deflated to express them in 1992 dollars under the Cost92 column. For the number of visitors, estimates were available only on a regional basis, and hence Ocean and Monmouth counties are considered as the Shore region (see Longwoods studies for details).

∽ Notes

Chapter 2. *New Jersey's Shoreline*

1. All original data are in English units and are rounded to a near large whole number. These numbers are retained without conversion.

Chapter 4. *Coastal Storms*

1. The information obtained from these sources can vary because of differences in the collection and processing of these data. Location of gauges, reference plane (such as Mean Low Level Water, Mean Sea Level, National Geodetic Vertical Datum of 1929), measuring units (feet or meters), and duration and time of measurements may also vary. Thus there may be slight inconsistencies in the measurement data for each storm. Adjustments have been made to present this information in similar terms and to reconcile differences where possible. However, some small differences remain.

2. The spread of dates in the heading represents the duration of the elevation of the water level under the influence of the storm. The peak storm effects usually have a much smaller duration.

Chapter 6. *Hard and Soft Approaches to Coastal Stabilization*

1. All of the data on sand volumes are originally in English units and are usually rounded to a near large whole number. Rather than produce spurious accuracy by converting these values to metric units that do not retain the concept of general rounding, all the sand volumes are reported as they were initially measured or projected in English units.

Chapter 11. *Coastal and Hazard Management*

1. An estimate of total losses (insured and uninsured losses) versus insured losses from catastrophic events for the 1995–1998 period of $96.2 billion-total versus $32.6 billion-insured in the United States yields a ratio of 1:0.34, or 34% of the total (Munich Reinsurance Group, various years). Using this percentage as an estimate of the insured to uninsured losses, one can project that the level of insured losses from table 11.2 for the 1989–1999 period of $97.48 billion represents estimated total losses of $286 billion.

2. The year 2050 is selected because of its similarity to the U.S. ACOE 50-year maintenance agreement. Further, all of the recommendations are achievable within a 50-year time frame.

~ References

Asbury Park Press. 1992. Storm '92: The Nor'easter That Changed the Face of the New Jersey Shore. *Asbury Park Press,* December 20: special section.

Atlantic City. 1989. *Article VII: Beach Protection Code (Ordinance # 22-1989).* Atlantic City, N.J.

Auerbach, A. J. 1985. The Theory of Excess Burden and Optimal Taxation. In *Handbook of Public Economics,* ed. A. J. Auerbach and M. Feldstein. Vol. 1, 61–127. New York: North-Holland.

Barth, M. C., and J. G. Titus, eds. 1984. *Greenhouse Effect and Sea Level Rise.* New York: Van Nostrand Reinhold Co.

Bates, T., and K. Moore. 1996. Erosion of Some Beaches Severe. *Asbury Park Press,* January 9: 8.

Bay Head, City of. 1993. *Ordinance 1993–8: Dune and Beach Regulation, Protection, and Preservation.* Bay Head, N.J.

Beatley, T. 1998. The Vision of Sustainable Communities. In *Cooperating with Nature: Confronting Natural Hazards with Land-Use Planning for Sustainable Communities,* ed. R. J. Burby, 233–262. Washington, D.C.: Joseph Henry Press.

Beatley, T., D. Brower, and A. Schwab. 1994. *An Introduction to Coastal Zone Management.* Covelo, Calif.: Island Press.

Belcher, C. 1986. *Fertilization of American Beachgrass on Sand Dunes.* Somerset, N.J.: Northeast Plant Materials Center.

Bell, F. W., and V. R. Leeworthy. 1985. An Economic Analysis of Saltwater Recreational Beaches in Florida, 1984. *Shore and Beach* 52 (April): 16–21.

———. 1986. *An Economic Analysis of the Importance of Saltwater Beaches in Florida.* SGR-82. Gainesville, Fla.: Florida Sea Grant College.

———. 1990. Recreational Demand by Tourists for Saltwater Beach Days. *Journal of Environmental Economics and Management* 18: 189–205.

Benston, G. J., and G. G. Kaufman. 1997. FDICIA after Five Years. *Journal of Economic Perspectives* 11 (3): 139–158.

Berkeley Township. 1994. *Ordinance #94-40-OAB: Beach Protection Amendment.* Berkeley, N.J.

Beukema, J. J., W. J. Wolff, and J.J.M.N. Browns, eds. 1990. *Expected Effects of Climatic Change on Marine Coastal Ecosystems.* Amsterdam: Kluwer Academic Publishers.

BOCA. *See* Building Officials Code Administrators.

Bockstael, N. E. 1995. Economic Concepts and Issues: Social Costs and Benefits of Beach Nourishment Projects. Appendix E. In National Research Council, *Beach Nourishment and Protection.* Washington, D.C.: National Academy Press.

Brealey, R. A., and S. C. Myers. 1991. *Principles of Corporate Finance.* New York: McGraw-Hill.

Brick, Township of. 1988. *Chapter 134: Dune Preservation Ordinance.* Brick, N.J.

Brigantine, City of. 1986. *Ordinance #7 of 1986: An Ordinance Modifying Ordinance #28 of 1981.* Brigantine, N.J.

Brigham, E. F., and J. F. Houston. 1998. *Fundamentals of Financial Management*. New York: Dryden Press.

Broadway, R. W., and N. Bruce. 1984. *Welfare Economics*. Oxford: Basil Blackwell Publishers.

Bruno, M. S., T. O. Harrington, and K. L. Rankin. 1996. The Use of Artificial Reefs in Erosion Control: Results of the New Jersey Pilot Reef Project. *Proceedings*. National Conference on Beach Preservation Technology.

Building Officials Code Administrators. 1993. *The BOCA National Building Code–1993*. Washington, D.C.: BOCA.

Burby, R. J., ed. 1998a. *Cooperating with Nature: Confronting Natural Hazards with Land-Use Planning for Sustainable Communities*. Washington, D.C.: Joseph Henry Press.

————. 1998b. Introduction. In *Cooperating with Nature: Confronting Natural Hazards with Land-Use Planning for Sustainable Communities*, ed. R. J. Burby, 1–26. Washington, D.C.: Joseph Henry Press.

Carter, R.W.G. 1989. *Coastal Environments*. New York: Academic Press.

Carter, R.W.G., T.G.F. Curtis, and M. J. Sheehy-Skeffington, eds. 1992. *Coastal Dunes: Geomorphology, Ecology, and Management for Conservation*. Rotterdam: A. A. Balkema.

Clayton, G. 1987. An Acquisition Program for Storm-Damaged Properties on Coastal Barriers: The State Role. In *Cities on the Beach: Management Issues of Developed Coastal Barriers*, ed. R. Platt, S. Pelczarski, and B. Burbank, 281–288. Chicago: University of Chicago Press.

Cordes, J. J., and A. M. Yezer. 1995. *Shore Protection and Beach Erosion Control Study: Economic Effects of Induced Development in Corps-Protected Beachfront Communities*. IWR Report 95-PS-1. U.S. ACOE, Institute for Water Resources, Alexandria, Va.

Culliton, T., M. A. Warren, T. Goodspeed, D. Remer, C. Blackwell, and J. McDonough. 1990. *The Second Report of a Coastal Trends Series: 50 Years of Population Change along the Nation's Coasts, 1960–2010*. Rockville, Md.: National Oceanic and Atmospheric Administration.

Curtis, T. D., and E. W. Shows. 1982. *Economic and Social Benefits of Artificial Beach Nourishment Civil Works at Delray Beach*. Prepared for Florida Department of Natural Resources, Division of Beaches and Shores, Tallahassee, Fla.

————. 1984. *A Comparative Study of Social Economic Benefits of Artificial Beach Nourishment—Civil Works in Northeast Florida*. Prepared for Florida Department of Natural Resources, Divison of Beaches and Shores, Tallahassee, Fla.

Cutter, S. L., ed. 2001. *American Hazardscapes: The Regionalization of Hazards and Disasters*. Washington, D.C.: Joseph Henry Press.

Dinwiddy, C., and F. Teal. 1996. *Principles of Cost-Benefit Analysis for Developing Countries*. New York: Cambridge University Press.

Doehring, F., I. W. Duedall, and J. M. Williams. 1994. Introduction. *Florida Hurricanes and Tropical Storms, 1871–1993: An Historical Survey*. Technical Paper 71. Division of Marine and Environmental Systems, Florida Institute of Technology, Melbourne, Fla.

Dolan, R., and R. Davis. 1992. An Intensity Scale for Atlantic Northeast Storms. *Journal of Coastal Research* 8 (4): 840–853.

————. 1994. Coastal Hazards: Perception, Susceptibility, and Mitigation. *Journal of Coastal Research*, Special Issue no. 12: 103–114.

Eckstein, O. 1958. *Water Resource Development: The Economics of Project Evaluation*. Cambridge: Harvard University Press.

Ekelund, R. B., and R. D. Tollison. 1991. *Economics*. New York: Harper Collins.

Falk, J. M., A. R. Graefe, and M. E. Suddleson. 1994. *Recreational Benefits of Delaware's Public Beaches: Attitudes and Perceptions of Beach Users and Residents of the Mid-Atlantic Region*. DEL-SG-05-94. Newark, Del.: University of Delaware, Sea Grant College Program.

Farrell, S. C., P. Venanzi, D. Inglin, S. Hafner, and S. Tatar. 1988. *Final Report for 1987 on New Jersey Beach Profiles Network: A Series of FEMA Monitoring Survey Stations*. Pomona, N.J.: Coastal Research Center, Richard Stockton College of New Jersey.

Farrell, S. C., and S. Leatherman. 1989. *Computer-based Coastal Erosion Rate Maps for the State of New Jersey and Its Inlets*. Trenton: Division of Coastal Resources, New Jersey Department of Environmental Protection.

Farrell, S. C., B. Sullivan, S. Hafner, and T. Lepp. 1994. *New Jersey Beach Profile Network Analysis of the Shoreline Changes in New Jersey, Coastal Reaches One through Fifteen, Raritan Bay to Delaware Bay.* Pomona, N.J.: Coastal Research Center, Richard Stockton College of New Jersey.

Farrell, S. C., B. Sullivan, S. Hafner, T. Lepp, and K. Cadmus. 1995. *New Jersey Beach Profile Network Analysis of the Shoreline Changes in New Jersey, Coastal Reaches One through Fifteen, Raritan Bay to Delaware Bay.* Pomona, N.J.: Coastal Research Center, Richard Stockton College of New Jersey.

Farrell, S. C., B. Speer, S. Hafner, and T. D. Lepp. 1997. *New Jersey Beach Profile Network Analysis of the Shoreline Changes in New Jersey, Coastal Reaches One through Fifteen, Raritan Bay to Delaware Bay.* Pomona, N.J.: Coastal Research Center, Richard Stockton College of New Jersey.

Farrell, S. C., S. Hafner, B. Speer, T. D. Lepp, S. E. Ebersold, and C. Constantino. 1999. *New Jersey Beach Profile Network Annual Report on Monitoring New Jersey Beaches Fall of 1997 through Spring of 1998.* Pomona, N.J.: Coastal Research Center, Richard Stockton College of New Jersey.

Federal Emergency Management Agency. 1986a. *Coastal Construction Manual.* FEMA-55. Washington, D.C.: FEMA.

———. 1986b. *Retrofitting Flood-prone Residential Structures.* Washington, D.C.: FEMA.

———. 1990. *Post-Disaster Hazard Mitigation Planning Guidance for State and Local Governments.* Washington, D.C.: FEMA.

———. 1991a. *Disaster Mitigation: Reducing Losses of Life and Property through Model Codes.* FEMA-209. Washington, D.C.: FEMA.

———. 1991b. *Answers to Questions about Substantially Damaged Buildings.* Washington, D.C.: FEMA.

———. 1993a. *FEMA's Disaster Management Program: A Performance Audit after Hurricane Andrew.* Washington, D.C.: FEMA.

———. 1993b. *Hazard Mitigation Grant Program.* Washington, D.C.: FEMA.

———. 1994a. *Flood Insurance Reform Act of 1994.* Washington, D.C.: FEMA.

———. 1994b. *Small Business Administration (SBA).* Washington, D.C.: FEMA.

———. 1995a. *National Mitigation Strategy: Partnerships for Building Safer Communities.* Washington, D.C.: FEMA.

———. 1995b. *An Integrated Approach to Natural Hazard Risk Mitigation.* FEMA-261. Washington, D.C.: FEMA.

———. 1995c. *Federal Emergency Management Agency: 44CFR Part 65 Review of Determinants for Required Purchase of Flood Insurance.* Washington, D.C.: FEMA.

———. 1995d. *Letter to National Flood Insurance Program Policy Issuance 8-95—Subject: 30-day Waiting Period.* Washington, D.C.: FEMA.

———. 1995e. *Disaster Assistance: A Guide to Recovery Programs.* FEMA-229(4). Washington, D.C.: FEMA.

———. 1995f. *Out of Harm's Way: The Missouri Buyout Program.* Mitigation Directorate. Washington, D.C.: FEMA.

———. 1996. *Hurricane Opal in Florida: A Building Performance Assessment.* Mitigation Directorate. Washington, D.C.: FEMA.

———. 1997. *Report on Costs and Benefits of Natural Hazard Mitigation.* Mitigation Directorate. Washington, D.C.: FEMA.

FEMA. *See* Federal Emergency Management Agency.

Fisher, J. J. 1967. Origin of Barrier Island Chain Shorelines: Middle Atlantic States. *Geological Society of America Special Paper* 115: 66–67.

Fisher, R., and S. Brown. 1989. *Getting Together: Building Relationships as We Negotiate.* New York: Penguin Books.

Fisher, R., W. Ury, and B. Patton. 1991. *Getting to Yes: Negotiating Agreement without Giving In.* New York: Penguin Books.

Freeman, A. M. 1979. *The Benefits of Environmental Improvement.* Baltimore: Johns Hopkins University Press.

———. 1993. *The Measurement of Environmental and Resource Values: Theory and Methods.* Washington, D.C.: Resources for the Future.

Gares, P., K. Nordstrom, and N. P. Psuty. 1982. *Coastal Dunes: Their Function, Delineation, and Management.* Trenton, N.J.: NJDEP.

Garofalo, J. 1992. *Coastal Storm, January 4, 1992.* Trenton, N.J.: NJDEP.

Gittinger, J. P. 1972. *Economic Analysis of Agricultural Projects.* Baltimore: Johns Hopkins University Press.

Godschalk, D. R., T. Beatley, P. Berke, D. J. Brower, and E. J. Kaiser. 1999. *Natural Hazard Mitigation: Recasting Disaster Policy and Planning.* Covelo, Calif.: Island Press.

Hamer, D., B. Cluster, and C. Miller. 1992. *Restoration of Sand Dunes along the Mid-Atlantic Coast.* Somerset, N.J.: U.S. Department of Agriculture, Soil Conservation Service.

H. John Heinz III Center for Science, Economics and the Environment. 2000a. *Evaluation of Erosion Hazards.* Prepared for the Federal Emergency Management Agency, Washington, D.C.

———. 2000b. *The Hidden Costs of Coastal Hazards: Implications for Risk Assessment and Mitigation.* Covelo, Calif.: Island Press.

Herfindahl, O. C., and A. V. Kneese. 1974. *Economic Theory of Natural Resources.* Columbus, Ohio: Charles E. Merrill Publishing Co.

Hillyer, T. M. 1996. *Shore Protection and Beach Erosion Control Study: Final Report: An Analysis of the U.S. Army Corps of Engineers Shore Protection Program.* IWR Report 96-PS-1. U.S. ACOE, Institute for Water Resources, Alexandria, Va.

Hotta, S., N. Kraus, and M. A. Horiwaka. 1987. Function of Sand Fences in Controlling Wind-Blown Sand. *Coastal Sediments '87* 1: 772–787.

Houghton, J. T., G. J. Jenkins, and J. J. Ephraums, eds. 1991. *Climate Change: The IPCC Scientific Assessment.* Cambridge: Cambridge University Press.

Houston, J. R. 1995a. Beach Nourishment. *Shore and Beach* 63 (January): 21–24.

———. 1995b. The Economic Value of Beaches. *The CERCular* CERC-95-4: 1–4.

Howe, C. W. 1971. *Cost-Benefit Analysis for Water System Planning.* Washington, D.C.: American Geophysical Union.

Hull, C.H.J., and J. G. Titus, eds. 1986. *Greenhouse Effect, Sea Level Rise, and Salinity in the Delaware Estuary.* Washington, D.C.: U.S. Environmental Protection Agency.

ICF, Incorporated. 1989. *Developing Policies to Improve the Effectiveness of Coastal Flood Plain Management: Executive Summary to New England/New York Coastal Zone Task Force,* Fairfax, Va.: ICF, Inc.

Insurance Research Council. 1995. *Coastal Exposure and Community Protection: Hurricane Andrew's Legacy.* Malvern, Pa.: IRC.

Insurance Services Office, Inc. 1999. Insured Catastrophic Loss Data. Property Claims Division, Jersey City, N.J.

IRC. *See* Insurance Research Council.

ISO. *See* Insurance Services Office, Inc.

Janin, L. F. 1987. Simulation of Sand Accumulation around Fences. *Coastal Sediments '87* 1: 202–212.

Jensen, R. E. 1983. *Atlantic Coast Hindcast, Shallow-Water Significant Wave Information.* Wave Information Studies Report 9. Vicksburg, Miss.: U.S. Army Engineer Waterways Experiment Station.

Johansson, P. O. 1987. *The Economic Theory and Measurement of Environmental Benefits.* New York: Cambridge University Press.

———. 1991. *An Introduction to Modern Welfare Economics.* New York: Cambridge University Press.

———. 1993. *Cost-Benefit Analysis of Environmental Changes.* New York: Cambridge University Press.

Johnston, J. 1984. *Econometric Methods.* New York: McGraw-Hill.

Jones, G. V., and R. Davis. 1995. Climatology of Nor'easters and the 30 kPa Jet. *Journal of Coastal Research* 11 (4): 1210–1220.

Judge, G. G., R. C. Hill, W. E. Griffiths, H. Lutkepohl, and T. C. Lee. 1988. *Introduction to the Theory and Practice of Econometrics*. New York: John Wiley & Sons.

Just, R. E., D. L. Hueth, and A. Schmitz. 1982. *Applied Welfare Economics and Public Policy*. Englewood Cliffs, N.J.: Prentice Hall.

Kohli, K. N. 1993. *Economic Analysis of Investment Projects: A Practical Application*. New York: Oxford University Press.

Koppel, R. 1994. *Report on Five Studies for the United States Army Corps of Engineers: Absecon Island and Seven Mile Island, New Jersey; Stone Harbor, Avalon, Atlantic City, Longport, Margate, Ventnor—Surveys of Beach Users, Businesses, and Homeowners*. Camden, N.J.: Forum for Policy Research and Public Service, Rutgers University.

Kriebel, D. L. 1995. *Users Manual for Dune Erosion Model: EDUNE*. Research Report no. CACR-95-05. Newark, Del.: Ocean Engineering Laboratory, University of Delaware.

Kriebel, D. L., and R. G. Dean. 1985. Numerical Simulation of Time-Dependent Beach and Dune Erosion. *Coastal Engineering* 9: 221–245.

Kucharski, S. 1995. Beach Policy: New Jersey's Great Shore Debate. M.S. thesis, Rutgers University, Camden, N.J.

Kunreuther, H. 1998. Introduction. In *Paying the Price: The Status and Role of Insurance against Natural Disasters in the United States*, ed. H. Kunreuther and R. J. Roth, Sr., 1–16. Washington, D.C.: Joseph Henry Press.

Kunreuther, H., and R. J. Roth, Sr., eds. 1998. *Paying the Price: The Status and Role of Insurance against Natural Disasters in the United States*. Washington, D.C.: Joseph Henry Press.

Lecomte, E., and K. Gahagan. 1998. Hurricane Insurance Protection in Florida. In *Paying the Price: The Status and Role of Insurance against Natural Disasters in the United States*, ed. H. Kunreuther and R. J. Roth, Sr., 97–124. Washington, D.C.: Joseph Henry Press.

Lelis, L. 1995. Low Tide Blunts Storm's Blow. *Asbury Park Press*, November 15: 1, 5.

———. 1996. Scarred Beaches Prompt Concern for Future. *Asbury Park Press*, January 10: 1, 3.

Leontieff, W. 1986. *Input-Output Economics*. 2d ed. New York: Oxford University Press.

Levy, M., and M. Sarnat. 1994. *Capital Investment and Financial Decisions*. Hertfordshire: Prentice Hall, International.

Lewis, P. D. 1993. *Governor's Disaster Planning and Response Review Committee*. Governor's Disaster Planning and Response Review Committee, Tallahassee, Fla.

Lindsay, B. E., and H. C. Tupper. 1989. Demand for Beach Protection and Use in Maine and New Hampshire: A Contingent Valuation Approach. *Coastal Zone '89*, 79–87.

Lloyd, J. B. 1994. *Eighteen Miles of History of Long Beach Island*. Harvey Cedars, N.J.: Down the Shore Publishing.

Long Beach Township. 1994. *Ordinance #94-16C*. Long Beach, N.J.

Longwoods International. 1992. *The Economic Impact, Performance, and Profile of the New Jersey Travel and Tourism Industry, 1990–91*. Prepared for N.J. Division of Travel and Tourism, Trenton, N.J.

———. 1994a. *The Economic Impact, Performance, and Profile of the New Jersey Travel and Tourism Industry, 1992–93*. Prepared for N.J. Division of Travel and Tourism, Trenton, N.J.

———. 1994b. *The New Jersey 1993 Travel Research Program*. Prepared for N.J. Division of Travel and Tourism, Trenton, N.J.

———. 1995. *The Economic Impact, Performance, and Profile of the New Jersey Travel and Tourism Industry, 1993–94*. Prepared for N.J. Division of Travel and Tourism, Trenton, N.J.

Ludlum, D. 1983. *New Jersey Weather Book*. New Brunswick, N.J.: Rutgers University Press.

———. 1991. Weatherwatch. *Weatherwise*, December: 48–49.

Lyles, D., L. E. Hickman, and H. A. Debaugh, Jr. 1988. *Sea Level Variations for the United States*. Washington, D.C.: U.S. Department of Commerce, Office of Oceanography and Marine Assessment.

Manheim, T., and T. J. Tyrrell. 1986a. *Social and Economic Impacts from Tourism in Block Island, Rhode Island*. R.I. Sea Grant Report. Narragansett, R.I.

———. 1986b. *Social and Economic Impacts from Tourism in Newport, Rhode Island*. R.I. Sea Grant Report. Narragansett, R.I.

Mantoloking, City of. 1995. *Ordinance #348: Beach and Dune Protection.* Mantoloking, N.J.

Markowitz, H. M. 1959. *Portfolio Selection.* New York: John Wiley.

———. 1989. *Mean Variance Analysis in Portfolio Choice and Capital Markets.* Cambridge: Basil Blackwell.

Meltz, R., D. H. Merriam, and R. M. Frank. 1998. *The Takings Issue: Constitutional Limits on Land Use Control and Environmental Regulation.* Covelo, Calif.: Island Press.

Methot, J. 1988. *Up and Down the Beach.* Navesink, N.J.: Whip Publishers.

Meyerson, A. L. 1972. Pollen and Paleosalinity Analysis from a Holocene Tidal Marsh Sequence, Cape May County, New Jersey. *Marine Geology* 12: 335–357.

Mileti, D. S. 1999. *Disasters by Design: A Reassessment of Natural Hazards in the United States.* Washington, D.C.: Joseph Henry Press.

Mishan, E. J. 1976. *Cost-Benefit Analysis.* New York: Praeger Publishers.

Mitsch, W. J., and J. G. Gosselink. 1994. *Wetlands.* New York: John Wiley and Sons.

Moore, K. 1996. Home Quality May Help Insured. *Asbury Park Press,* April 6 (A): 5.

Munich Reinsurance Group. 1995. *Topics: Annual Review of Natural Catastrophes, 1995.* Washington, D.C.: Division of American Reinsurance Group.

———. 1996. *Topics: Annual Review of Natural Catastrophes, 1996.* Washington, D.C.: Division of American Reinsurance Group.

———. 1997. *Topics: Annual Review of Natural Catastrophes, 1997.* Washington, D.C.: Division of American Reinsurance Group.

———. 1998. *Topics: Annual Review of Natural Catastrophes, 1998.* Washington, D.C.: Division of American Reinsurance Group.

Murray, J. D. 1993. *Mathematical Biology.* New York: Springer-Verlag.

Najjar, R. G., H. A. Walker, P. J. Anderson, E. J. Barron, R. Bord, J. Gibson, V. S. Kennedy, C. G. Knight, P. Megonigal, R. O'Connor, C. D. Polsky, N. P. Psuty, B. Richards, L. G. Sorenson, E. Steele, and R. S. Swanson. 2000. The Potential Impacts of Climate Change on the Mid-Atlantic Coastal Region. *Climate Research* 14: 219–233.

National Oceanic and Atmospheric Administration. 1987–1994 (various years). *Yearly Mean Sea Levels and Monthly Tidal Summary Reports for: Atlantic City, NJ; Battery, NY; Lewes, DE; Philadelphia, PA; and Sandy Hook, NJ.* Rockville, Md.: U.S. Department of Commerce, National Ocean Service.

———. 1992. *Natural Disaster Report: The Halloween Nor'easter of 1991, East Coast of the United States—Maine to Florida and Puerto Rico.* Beltsville, Md.: NOAA, National Weather Service.

———. 1993. *Comparison of Historical Highest Water Levels at NOS Tide Stations along North Eastern Coast of the U.S. and Highest Recorded as a Result of the Dec. 1992 Northeaster.* Washington, D.C.: NOAA.

———. National Ocean Service. 1996. *Tide Tables: High and Low Water Predications for the East Coast of North and South America including Greenland, 1985–1996.* Washington, D.C.: NOAA.

———. National Buoy Data Center. 1998. Delaware Bay Data Buoy Station 44009. Website: *www.ndbc.noaa.gov.*

National Research Council. 1987. *Responding to Changes in Sea Level.* Engineering and Technical Systems Commission. Washington, D.C.: National Academy Press.

———. 1990. *Managing Coastal Erosion.* Washington, D.C.: National Academy Press.

———. 1995. *Beach Nourishment and Protection.* Washington, D.C.: National Academy Press.

———. 1999a. *The Impacts of Natural Disasters: A Framework for Loss Estimation.* Washington, D.C.: National Academy Press.

———. 1999b. *Our Common Journey: A Transition Towards Sustainability.* Washington, D.C.: National Academy Press.

New Jersey Beach Erosion Commission. 1950. *Report on the Protection and Preservation of the New Jersey Beaches and Shorefront.* Trenton, N.J.: NJBEC.

New Jersey Board of Commerce and Navigation. 1922. *The Erosion and Protection of New Jersey Beaches.* Trenton, N.J.: Engineering Advisory Board, NJBCN.

———. 1930. *Report on the Erosion and Protection of the New Jersey Beaches.* Trenton, N.J.: NJBCN.

New Jersey Department of Environmental Protection. 1977. *Statewide Comprehensive Outdoor Recreation Plan (SCORP).* Trenton, N.J.: Office of Green Acres, NJDEP.

———. 1980. *Proposed New Jersey Coastal Management Program.* Trenton, N.J.: Division of Coastal Resources, NJDEP.

———. 1981. *New Jersey Shore Protection Master Plan.* Prepared by Dames and Moore for Division of Coastal Resources, NJDEP, Trenton, N.J.

———. 1984a. *Assessment of Dune and Shore Protection Ordinances in New Jersey.* Trenton, N.J.: NJDEP.

———. 1984b. *Coastal Storm Hazard Mitigation: Atlantic County Barrier Islands and Ocean City, NJ.* Trenton, N.J.: NJ DEP.

———. 1985a. *Coastal Storm Hazard Mitigation: A Handbook on Coastal Planning and Legal Issues.* Trenton, N.J.: NJDEP.

———. 1985b. *Guidelines and Recommendations for Coastal Dune Restoration and Creation Projects.* Trenton, N.J.: NJDEP.

———. 1986. *Hazard Mitigation Plan: Section 406.* Trenton, N.J.: NJDEP.

———. 1995. *Digitized Historical Shoreline Profiles.* CD-ROM. Trenton, N.J.: NJDEP.

———. 1997. *GIS Tools for Decision Making: Mapping the Present to Protect New Jersey's Future.* Series 2, volume 1. Trenton, N.J.: NJDEP.

New Jersey Office of Emergency Management. 1993. *Coastal Storm 1992 (DR-936-NJ) Interagency Hazard Mitigation Team Report.* Trenton, N.J.: NJOEM.

———. 1994. *Hazard Mitigation Plan for New Jersey.* DR-973-NJ. Trenton, N.J.: New Jersey State Police.

New Jersey Senate and General Assembly. 1995. *Green Acres, Farmland and Historic Preservation, and Blue Acres Bond Act of 1995 (P.L. 1995, c.204).* Trenton, N.J.: Legislative Services.

New Jersey State Legislator. 1993. *P.L. 1993 c. 190, CAFRA.* Trenton, N.J.: N.J. Office of Administrative Law.

NJBCN. *See* New Jersey Board of Commerce and Navigation.

NJBEC. *See* New Jersey Beach Erosion Commission.

NJDEP. *See* New Jersey Department of Environmental Protection.

NJOEM. *See* New Jersey Office of Emergency Management.

NJSGA. *See* New Jersey Senate and General Assembly.

NOAA. *See* National Oceanic and Atmospheric Administration.

Nordstrom, K. F. 1994. Beaches and Dunes of Human Altered Coasts. *Progress in Physical Geography* 18: 497–516.

Nordstrom, K. F., S. Fisher, M. Burr, E. Frankel, T. Buckalew, and G. Kucma. 1977. *Coastal Geomorphology of New Jersey.* Technical Report 77-1. 2 vols. New Brunswick, N.J.: Center for Coastal and Environmental Studies, Rutgers University.

Nordstrom, K. F., P. A. Gares, N. P. Psuty, O. Pilkey, Jr., W. Neal, and O. H. Pilkey, Sr. 1986. *Living with the New Jersey Shore.* Durham, N.C.: Duke University Press.

Nordstrom, K. F., N. P. Psuty, and B. Carter, eds. 1990. *Coastal Dunes: Form and Process.* Chichester: John Wiley & Sons.

NRC. *See* National Research Council.

Oakes, T. A. 1994. The Role of Regional Coastal Groups in Planning Coastal Defense. *Littoral* 1: 63–76.

Ocean City, City of. 1994. *Beachfront Construction and Maintenance Plan.* Ocean City, N.J.

Ofiara, D. D., and B. Brown. 1999. Assessment of Economic Losses to Recreational Activities from 1988 Marine Pollution Events and Assessment of Economic Losses from Long-term Contamination of Fish within the New York Bight to New Jersey. *Marine Pollution Bulletin* 38 (11): 990–1004.

Ofiara, D. D., and N. P. Psuty. 2001. Suitability of Decision-Theoretic Models to Public Policy Issues Concerning the Provision of Shore Stabilization and Hazard Management. *Coastal Management* 29: 271–294.

Ofiara, D. D., and J. J. Seneca. 2000. *Economic Losses from Marine Pollution: A Handbook for Assessment.* Covelo, Calif.: Island Press.

Olshansky, R. B., and J. D. Kartez. 1998. Managing Land Use to Build Resilience. In *Cooperating with Nature: Confronting Natural Hazards with Land-Use Planning for Sustainable Communities*, ed. R. J. Burby, 167–201. Washington, D.C.: Joseph Henry Press.

Opinion Research Corporation. 1989. *The Economic Impact of Visitors to the New Jersey Shore: The Summer of 1989*. Prepared for N.J. Division of Travel and Tourism, Trenton, N.J.

Ostle, B., and R. W. Mensing. 1975. *Statistics in Research*. Ames: Iowa State University Press.

Palm, R. 1998. Demand for Disaster Insurance: Residential Coverage. In *Paying the Price: The Status and Role of Insurance against Natural Disasters in the United States*, ed. H. Kunreuther and R. J. Roth, Sr., 51–66. Washington, D.C.: Joseph Henry Press.

Patton, A. 1993. *From Harm's Way: Flood-Hazard Mitigation in Tulsa, Oklahoma*. Tulsa: City of Tulsa Public Works Department.

Philadelphia Inquirer. 1987. Storm Belts Area with Snow, Rain: New Jersey Shore Girds for Tides and Flooding. January 2 (B): 7, 9.

Platt, R. H., H. C. Miller, T. Beatly, J. Melville, and B. G. Mathenia. 1992. *Coastal Erosion: Has Retreat Sounded?* Boulder: Institute of Behavioral Science, University of Colorado.

Podufaly, E. T. 1962. Operation Five-High. *Shore and Beach* 30 (2): 9–17.

Prichard, M. 1995. Felix Likely Another Near Miss. *Asbury Park Press*, August 17: 1, 3.

Property Claim Services, Insurance Services Office, Inc. 2000. PCS Catastrophe History Database. Kearney, N.J.

Psuty, N. P. 1986. Holocene Sea-Level in New Jersey. *Physical Geography* 7: 154–167.

———. 1992. Estuaries: Challenges for Coastal Management. In *Ocean Management in Global Change*, ed. P. Fabbri. New York: Elsevier Applied Science.

———, ed. 1988. Dune/Beach Interaction. *Journal of Coastal Research*, Special Issue no. 3. Lawrence, Kansas: Allen Press.

———, panel chair. 1991. *The Effects of an Accelerated Rise in Sea Level on the Coastal Zone of New Jersey, U.S.A.* Contribution 91-55. New Brunswick, N.J.: Institute of Marine and Coastal Sciences, Rutgers University.

Psuty, N. P., D. Collins, M. DeLuca, M. Grace, W. Keppe, G. Klein, H. Mattioni, G. Martinelli, J. McDonnell, D. D. Ofiara, M. Padula, S. Pata, M. Siegel, and E. Spence. 1996. *Coastal Hazard Management Report*. Final report to NJDEP. New Brunswick, N.J.: Institute of Marine and Coastal Sciences, Rutgers University.

Psuty, N. P., Q. Guo, and N. Suk. 1993. *Sediments and Sedimentation in the Proposed ICWW Channels, Great Egg Harbor Bay, NJ*. New Brunswick, N.J.: Institute of Marine and Coastal Sciences, Rutgers University.

Psuty, N. P., and T. A. Piccola. 1991. *Foredune Profile Changes, Fire Island, New York*. New Brunswick, N.J.: Institute of Marine and Coastal Sciences, Rutgers University.

Psuty, N. P., and C. Tsai. 1997. *Coastal Storms and Higher Frequency Storm Events: Dimensional Responses to 10-, 20-, and 30-Year Recurrence Interval Storms*. Report to New Jersey Office of Emergency Management, Trenton, N.J.

Pye, K. 1993. *The Dynamics and Environmental Context of Aeolian Sedimentary Systems*. London: Geological Society.

R L Associates. 1987. *Economic Impact of Tourism to the New Jersey Shore in 1987*. Prepared for N.J. Division of Travel and Tourism, Trenton, N.J.

———. 1988. *The Economic Impact of Visitors to the New Jersey Shore: The Summer of 1988*. Prepared for N.J. Division of Travel and Tourism, Trenton, N.J.

Ross, N. 1995. Houses That Stand Up to Hurricanes. *Washington Home*, October 12 (A): 8–9.

Roth, R. J., Sr. 1998. Earthquake Insurance Protection in California. In *Paying the Price: The Status and Role of Insurance against Natural Disasters in the United States*, ed. H. Kunreuther and R. J. Roth, Sr., 67–96. Washington, D.C.: Joseph Henry Press.

Savadove, L., and M. T. Buchholz. 1993. *Great Storms of the Jersey Shore*. Harvey Cedars, N.J.: Down the Shore Publishing.

Seymour, R. J. 1996. An Introduction to the Marine Board Study on Beach Nourishment and Protection. *Shore and Beach* 64: 3–4.

Shelby, K. 1992. Nor'easter Pounds Region Flooding Worst in 30 Years. *Asbury Park Press*, January 12: 1, 5.

Silberman, J., D. A. Gerlowski, and N. A. Williams. 1992. Estimating Existence Value for Users and Nonusers of New Jersey Beaches. *Land Economics* 68 (2): 225–236.

Silberman, J., and M. Klock. 1988. The Recreational Benefits of Beach Renourishment. *Ocean and Shore Management* 11: 73–90.

Smith, R. 1991. *Final Tabulations of Significant Tidal Events in Relation to Mean Lower Low Water and Mean Low Water at the Steel Pier and Ventnor Pier.* Washington, D.C.: National Ocean Service.

State Hazard Mitigation Team. 1993. A Comparison of Historic NGVD/MLLW Tidal Storm Surge Levels at Atlantic City. Unpublished report. New Jersey Department of Environmental Protection, Trenton, N.J.

Stronge, W. B. 1994. Beaches, Tourism, and Economic Development. *Shore and Beach* 62 (2): 6–8.

———. 1995. The Economics of Government Funding for Beach Nourishment Projects: The Florida Case. *Shore and Beach* 63: 4–6.

Sugden, R., and A. Williams. 1990. *The Principles of Practical Cost-Benefit Analysis.* Oxford: Oxford University Press.

Teal, J. M., and M. Teal. 1969. *Life and Death of the Saltmarsh.* Boston: Little, Brown.

Thomas, P. 1990. Fraud Is Called Small Factor in S&L Cost. *Wall Street Journal,* July 20.

Tierney, K. J., M. K. Lindell, and R. W. Perry. 2001. *Facing the Unexpected: Disaster Preparedness and Response in the United States.* Washington, D.C.: Joseph Henry Press.

Titus, J. G., ed. 1988. *Greenhouse Effect, Sea Level Rise, and Coastal Wetlands.* Washington, D.C.: U.S. Environmental Protection Agency.

Titus, J. G., S. P. Leatherman, C. H. Everts, D. L. Kreibel, and R. G. Dean. 1985. *Potential Impacts of Sea Level Rise on the Beach at Ocean City, Maryland.* Washington, D.C.: U.S. Environmental Protection Agency.

Titus, J. G., and V. K. Narayanan. 1995. *The Probability of Sea Level Rise.* Washington, D.C.: U.S. Environmental Protection Agency.

U.S. ACOE. *See* U.S. Army Corps of Engineers.

U.S. Army Corps of Engineers. 1964. A Pictorial History of Selected Structures along the New Jersey Coast. Miscellaneous paper no. 54. Coastal Engineering Research Center, U.S. ACOE, Fort Belvoir, Va.

———. 1971. *National Shoreline Study: Regional Inventory Report for North Atlantic Region.* Vols. 1–2. New York: North Atlantic Division.

———. 1972. *Tide Flood Plain Information, Sandy Hook Bay, and Raritan Bay Shore Areas, Monmouth County, New Jersey.* New York: U.S. Army Corps of Engineers.

———. 1985. *Post-Storm Report, Coastal Storm 1984, Delaware and New Jersey.* Philadelphia: U.S. Army Corps of Engineers.

———. 1986. *Engineering and Design: Storm Surge Analysis.* EM-1110-2-1412. Washington, D.C.: U.S. Army Corps of Engineers.

———. 1989a. *Atlantic Coast of New Jersey, Sandy Hook to Barnegat Inlet Beach Erosion Control Project Section I—Sea Bright to Ocean Township, New Jersey. General Design Memorandum, Main Report with Environmental Impact Statement.* Vol. 1. New York District, New York.

———. 1989b. *Atlantic Coast of New Jersey, Sandy Hook to Barnegat Inlet Beach Erosion Control Project Section I—Sea Bright to Ocean Township, New Jersey. General Design Memorandum, Technical Appendices.* Vol. 2. New York District, New York.

———. 1990. *New Jersey Shore Protection Study: Report of Limited Reconnaissance Study.* Philadelphia District, Philadelphia.

———. 1991a. *Delaware Bay Coastline, New Jersey, and Delaware: Reconnaissance Report.* Philadelphia District, Philadelphia.

———. 1991b. *Delaware Bay Coastline, New Jersey, and Delaware: Reconnaissance Report—Technical Appendices.* Philadelphia District, Philadelphia.

———. 1991c. *National Economic Development Procedures Manual—Coastal Storm Damage and Erosion.* IWR Report 91-R-6. Fort Belvoir, Va.: Institute for Water Resources.

———. 1992. *New Jersey Shore Protection Study: Barnegat Inlet to Little Egg Inlet—Reconnaissance Report,* Philadelphia District, Philadelphia.

————. 1993a. *Coastal Storm 1992, Delaware and New Jersey.* Philadelphia: U.S. Army Corps of Engineers.

————. 1993b. *Post-Storm Report, Coastal Storm of December 1992, Delaware and New Jersey.* Region II, Philadelphia.

————. 1993c. *Raritan Bay and Sandy Hook Bay, New Jersey Combined Flood Control and Shore Protection Reconnaissance Study.* New York District, New York.

————. 1994a. *Atlantic Coast of New Jersey, Sandy Hook to Barnegat Inlet Beach Erosion Control Project Section II—Asbury Park to Manasquan, New Jersey: General Design Memorandum, Main Report and Environmental Impact Statement.* Vol. 1. New York District, New York.

————. 1994b. *Atlantic Coast of New Jersey, Sandy Hook to Barnegat Inlet Beach Erosion Control Project Section II—Asbury Park to Manasquan, New Jersey: General Design Memorandum, Technical Appendices.* Vol. 2. New York District, New York.

————. 1994c. *Comparative Storm Data since 1991: Delaware and New Jersey.* Philadelphia: U.S. Army Corps of Engineers.

————. 1994d. *New Jersey Protection Study: Lower Cape May Meadows—Cape May Point, Reconnaissance Study.* Philadelphia District, Philadelphia.

————. 1994e. *Shoreline Protection and Beach Erosion Control Study—Phase I: Cost Comparison of Shoreline Protection Projects of the U.S. Army Corps of Engineers.* IWR Report 94-PS-1. Alexandria, Va.: Water Resources Support Center, Institute of Water Resources.

————. 1995. *New Jersey Shore Protection Study: Barnegat Inlet to Little Egg Inlet—Reconnaissance Study,* Philadelphia District, Philadelphia.

————. 1996. *New Jersey Shore Protection Study Update.* Philadelphia District, Philadelphia.

U.S. Department of Housing and Urban Development. 1996. *Guide to HUD's Community Planning and Development Programs.* Washington, D.C.: HUD.

U.S. General Accounting Office. 1995. Disaster Assistance: Information on Expenditures and Proposals to Improve Effectiveness and Reduce Future Costs. GAO/T-RCED-95-140. Washington, D.C.: GAO.

U.S. Geological Survey. 1981. *Guidelines for Determining Flood Flow Frequency.* Bulletin no. 17b. Washington, D.C.: U.S. Geological Survey.

U.S. HUD. *See* U.S. Department of Housing and Urban Development.

U.S. Natural Resource Conservation Service. 1992. *Restoration of Sand Dunes along the Mid-Atlantic Coast.* Somerset, N.J.: U.S. Department of Agriculture.

————. 1995. *Plant Sources from Cape May Plant Materials Center.* Cape May, N.J.: U.S. Department of Agriculture.

————. 1996. *Initial Evaluation of Sea Oats (Uniola paniculata).* Somerset, N.J.: U.S. Department of Agriculture.

University of North Carolina. 1991. *Evaluation of the National Coastal Zone Management Program.* Chapel Hill: Center for Urban and Regional Studies, University of North Carolina.

Uptegrove, J., L. Mullikin, J. Waldner, G. Ashely, R. Sheridan, D. Hall, F. Gilroy, and S. Farrell. 1995. *Characterization of Offshore Sediments in Federal Waters as Potential Sources of Beach Replenishment Sand—Phase I.* New Jersey Geological Survey Open-File Report, OFR 95-1. Trenton, N.J.: New Jersey Department of Environmental Protection.

Warrick, R. A., E. M. Barrow, and T. M. L. Wigely, eds. 1993. *Climate and Sea Level Change: Observations, Projections, and Implications.* Cambridge: Cambridge University Press.

Weimer, D. L., and A. R. Vining. 1991. *Policy Analysis: Concepts and Practice.* Englewood Cliffs, N.J.: Prentice Hall.

White, L. J. 1991. *The S&L Debacle: Public Policy Lessons for Bank and Thrift Regulation.* New York: Oxford University Press.

Williams, J. M., and I. W. Duedall. 1997. *Florida Hurricanes and Tropical Storms.* Gainesville, Fla.: University of Florida Press and Florida Sea Grant College Program.

Woodhouse, W. W., Jr. 1978. *Dune Building and Stabilization with Vegetation.* Special Report no. 3. Fort Belvoir, Va.: U.S. Army Corps of Engineers, Coastal Engineering Research Center.

～ Index

~ About the Authors

Norbert P. Psuty is a professor of geography, marine and coastal sciences, and geological sciences at Rutgers University and associate director of the Institute of Marine and Coastal Sciences at Rutgers. He is the coeditor of *Coastal Dunes: Form and Process* and coauthor of *Living with the New Jersey Shore*. He received his Ph.D. in geography from Louisiana State University.

Douglas D. Ofiara is an assistant professor of public policy and management at the Edmund S. Muskie School of Public Service at the University of Southern Maine, and a visiting scholar in the Institute of Marine and Coastal Sciences, Rutgers University. He is coauthor of *Economic Losses from Marine Pollution: A Handbook for Assessment.* He received his Ph.D. in economics from the City University of New York.